Ecology of arable land – organisms, carbon and nitrogen cycling

Ecological Bulletins No. 40

Ecology of arable land – organisms, carbon and nitrogen cycling

Edited by
Olof Andrén, Torbjörn Lindberg, Keith Paustian
and Thomas Rosswall

Ecological Bulletins

ECOLOGICAL BULLETINS are published in cooperation with the ecological journals Holarctic Ecology and Oikos. Ecological Bulletins consist of monographs, reports and symposia proceedings on topics of international interest, published on a non-profit making basis, in many cases in cooperation with agencies such as Unesco, UNEP, and SCOPE. Orders for the volumes should be placed with the publisher. Discounts are available for standing orders.

Editor-in-Chief and Editorial Office:
Pehr H. Enckell
Ecology Building
University of Lund
S-22362 Lund
Sweden

Published and distributed by:
Munksgaard International Booksellers
and Publishers,
P. O. Box 2148, DK-1016 Copenhagen K,
Denmark

Under the auspices of the
Swedish Natural Science Research Council
and the
Swedish Council for Planning and Coordination of Research

Suggested citation:
Author's name. 1989. Number and title of chapter. In: Andrén, O., Lindberg, T., Paustian, K. and Rosswall, T. (eds). Ecology of arable land – organisms, carbon and nitrogen cycling. – Ecol. Bull. (Copenhagen) 40: 000–000.

Cover: Ornament found on a helmet in a boatgrave at Vendel, close to the experimental field. See Chapter 2 for more information.

© 1990, ECOLOGICAL BULLETINS

ISBN 90-16-10227-4

Preface

This volume presents the results of the project "Ecology of Arable Land. The Role of Organisms in Nitrogen Cycling". The project was carried out during 1979–1988 with field studies concentrated to 1980–1985. The main project objective was to investigate and synthesize the contributions of the soil organisms to nitrogen and carbon circulation in four contrasting cropping systems. More than 25 scientists, from disciplines including soil hydrology, crop production ecology, soil microbiology, soil zoology, soil chemistry and systems analysis, were involved in the project.

In view of the integrated nature of the studies of carbon and nitrogen cycling carried out within the project and the large number of separate project publications in the scientific literature, it was considered important to summarize the results in one volume. This volume may be considered as a final report of project findings, although additional scientific papers from the project will be published in the years to come.

We hope that the volume will be enjoyable, or at least informative, for as wide an audience as possible, but our main target groups are ecologists, agronomists and biologists in general. We feel that the book, as a whole or selected parts of it, could be of use in courses in subjects related to agronomy and ecology. The specialized expert in any of the many areas the book covers may be disappointed by a lack of detail, but the cited original papers may then be of interest.

As part of the writing process, a combined book review workshop/international conference was held. A number of leading scientists were invited to comment on the book manuscript and discuss improvements with us. The book manuscript was sent to the reviewers in April 1987, and the review workshop in Gysinge near Uppsala was held 4–7 June. Following the workshop, we used the opportunity to arrange an international conference "Ecology of Arable Land – The Role of Organisms in Nitrogen Cycling – Perspectives and Challenges", which took place 9–12 June 1987 at the Swedish University of Agricultural Sciences in Uppsala. The conference was not primarily aimed at presenting the project results, but rather to look ahead to the future challenges of agroecosystem research (Clarholm and Bergström 1989).

The synthesis volume manuscript was revised in light of the reviewers' comments and further integration and restructuring of the material resulted in the present volume. The senior author of each chapter was responsible for synthesizing the contributions from individual co-authors, and in all instances the senior authors have written large parts. The editors' responsibilities were to make a coherent text out of the chapter manuscripts, to ascertain that there were no major overlaps or gaps and to be responsible for all details concerning language, layout, printing, etc.

Since the synthesis volume was not included in the original research proposal, no ear-marked funding was available for its preparation. However, through reallocation of project funding and some parasitizing on other projects and university funds most authors and editors received some financial help during the production of this volume. Much of the work was made possible only by the authors' willingness to sacrifice evenings, weekends and vacations. The restricted time and funding also resulted in the postponement of several scientific papers, which partly are synthesized in this volume before being published. The only way to totally avoid this would have been to wait several years before writing the synthesis volume, which would have resulted in no volume at all.

The project received financial support from the Swedish Council for Planning and Coordination of Research (FRN), the Swedish Council for Forestry and Agricultural Research (SJFR), the Swedish Natural Science Research Council (NFR), the Swedish Environment Protection Board (SNV) and the Swedish University of Agricultural Sciences (SLU). The publication of this book was made possible by grants from FRN, SJFR and NFR. In general, it is very difficult to arrange financial support from several funding agencies, and we are grateful to the chairman of the Project Committee, established by the funding agencies, Professor Bert Bolin, for making this task possible. During the early stages of the planning, the late Dr Bengt Lundholm, then at FRN, also played a major role in piecing the funding together with his eternal optimism.

The late Professor Börje Norén was, with Professor Eliel Steen, the stimulus behind the original concepts and planning. We would have liked Börje to have seen the developments during the active experimental period as well as the final results. We miss him both as a scientist and friend.

The Scientific Advisory Committee, Professors James P. Curry, Klaus H. Domsch, Martin J. Frissel, Peter Newbould, James M. Tiedje and Robert G. Woodmansee made several trips to Sweden for review of plans and results. We are greatly indebted to them for their constructive criticism and encouragement over the years.

We are indebted to the late Countess Metta von Rosen. Not only did she show a great interest in the work on her land but also let us use parts of the castle at Örbyhus for lunches in connection with important meetings. The managers at Örbyhus Estate, Viggo Madsen, Tage Wigren and Sven Bäärnhielm, helped us with field history and management. Gert and Göran Enander skillfully performed the field cultivation under the guidance of the site manager Ruben Johansson.

The project could not have been carried out without the committed help of technical personnel; they did a great job. Our project secretaries, Karin Åberg, Helena Wiklund and Ann-Cristine Lundquist, served long hours without complaining. Patricia Sweanor improved the English, and contributed to the logic as well. The drawings were made, often under severe time constraints, by Britt-Marie Björkman, Torgny Fremböck and Kajsa Göransson.

Last, but not least, we thank the participants at the book review workshop for encouragements and criticism, and their thorough work in spite of the extremely rough manuscripts they had received. We are thus indebted to: C. V. Cole, D. C. Coleman, J. P. Curry, M. J. Frissel, D. O. Hall, O. W. Heal, H. W. Hunt, P. Lavelle, S. Linder, S. Long, J. M. Melillo, P. Newbould, E. I. Newman, D. R. Nielsen, N. E. Nielsen, W. J. Parton, T. Persson, H. Petersen, J. Pettersson, P. S. C. Rao, P. G. Risser, L. Ryszkowski, M. J. Swift, J. W. B. Stewart, B. Söderström, J. M. Tiedje, J. A. van Veen, P. M. Vitousek, F. Warembourg, L. Wasilewska, J. W. Woldendorp and R. G. Woodmansee.

Uppsala
February 1989

Olof Andrén, Torbjörn Lindberg, Keith Paustian and Thomas Rosswall.

Contents

Preface 5
1. **Introduction.** O. Andrén, T. Lindberg, K. Paustian and T. Rosswall 9
    Background to the project "Ecology of arable land" 11
    Project characteristics and experimental design 11
    The synthesis volume 13

2. **The experimental field.** E. Steen, L. Bergström, P.-E. Jansson, R. Johansson, H. Johnsson and J. Persson 15
    Introduction 17
        Historical perspective 17
        Early land use 18
    Climate 18
        Regional climate 18
        Local climate 20
    Soil 20
        Soil physical properties 21
        Soil chemical properties 22
    Design of the field experiment 24
    Effects of the cropping systems on soil physical and chemical properties 27
    Field installations 27
        Lysimeters 27
        Meteorological and soil abiotic measurements 29
    Management 30

3. **Heat and water processes.** P.-E. Jansson, H. Johnsson and G. Alvenäs 31
    Introduction 33
    Radiation and heat balance 33
        Radiation balance 33
        Heat balance 34
        Soil temperature 34
    Water conditions 35
        Adaptation of the model 36
        Annual water balance 37
        Soil water-flow paths 38
        Soil moisture conditions 38
    Concluding remarks 40

4. **Structure of the agroecosystem.** A.-C. Hansson, O. Andrén, S. Boström, U. Boström, M. Clarholm, J. Lagerlöf, T. Lindberg, K. Paustian, R. Pettersson and B. Sohlenius 41
    Introduction 43
    Methods 43
    Components 47
        The crop 47
        Herbivores 51
        Soil microorganisms, including protozoa .. 51
        Nematodes 55
        Microarthropods and enchytraeids 57
        Soil and soil surface macroarthropods .. 60
        Earthworms 61
    Seasonal dynamics of the components 61
        The crop 62
        Herbivores 63
        Litter 63
        Decomposer organisms 64
    Long-term dynamics of the components 74
        Litter 76
        Herbivores 76
        Decomposer organisms 76
    Summary 83

5. **Organic carbon and nitrogen flows.** O. Andrén, T. Lindberg, U. Boström, M. Clarholm, A.-C. Hansson, G. Johansson, J. Lagerlöf, K. Paustian, J. Persson, R. Pettersson, J. Schnürer, B. Sohlenius and M. Wivstad 85
    Introduction 87
    Primary production 87
        Photosynthesis 88
        Plant biomass production 89
        Isotope carbon budget 91
    Plant uptake of nitrogen 94
        Symbiotic nitrogen fixation 94
        Uptake of fertilizer and soil nitrogen .. 97
    Decomposition and nitrogen mineralization .. 98
        Mass loss and abiotic control 99
        Heterogeneous resource 100
        A decomposition model including resynthesis 102
        Nitrogen mineralization and immobilization .. 103
        The decomposer organisms 107
        A synthesis of organism activities and decomposition processes 118
    Summary 125

6. **Inorganic nitrogen cycling processes and flows.** T. Rosswall, P. Berg, L. Bergström, M. Ferm, C. Johansson, L. Klemedtsson and B. H. Svensson 127
    Introduction 129
    The organisms 130
        Introduction 130

Nitrifiers 130
Denitrifiers 134
Effects of roots on nitrification and denitrification 134
  Nitrification 134
  Denitrification 135
Water as a regulator of transformation rates . 136
  Introduction 136
  Nitrogen mineralization 137
  Nitrification 137
  Nitrous oxide production 138
  Nitric oxide production 139
  Denitrification 140
Water as carrier of substances 143
  Introduction 143
  Drainage conditions 144
  Nitrate in drainage water 144
Nitrogen losses from four cropping systems.. 145
  Introduction 145
  Leaching 145
  Denitrification 146
  Nitric oxide losses 148
  Ammonia losses 150
  Total losses 151
Management impacts on nitrogen losses 151

7. **Ecosystem dynamics.** K. Paustian, L. Bergström, P.-E. Jansson and H. Johnsson 153
Introduction 155
Carbon and nitrogen budgets 155
  Data base and calculations 155
  Input/output balances 157
  Internal fluxes 159
  Comparisons with other agroecosystems... 163
Simulation of nitrogen dynamics 165
  Model description 166
  Barley 168
  Grass ley 171
  Three crop rotations in southern Sweden.. 174
  Long-term trends in nitrate leaching from an agricultural watershed 176
Summary 178

8. **Agricultural outlook.** E. Steen 181
Introduction 183
  Transitions in Swedish agriculture 183
Overview of the Kjettslinge field experiment in relation to practical agriculture 184
  General 184
  Components of the soil 184
  Production and harvest 185
  The balance of organic matter in arable soil 185
  The role of nitrogen 186
Swedish agriculture – problems and challenges 189
  Regional differences in Sweden 189
  Swedish agriculture – the present situation 189
  Future changes in Swedish agriculture 189

9. **Epilogue.** T. Rosswall, O. Andrén, T. Lindberg and K. Paustian 193
Project history 195
Project administration 197
Post-graduate education 199

References 201
Appendix 1: Management calendar 216
Appendix 2: Scientific and technical/administrative personnel 217
Appendix 3: Project publication list 218

1. Introduction

Olof Andrén, Torbjörn Lindberg, Keith Paustian and Thomas Rosswall

1. **Introduction.** O. Andrén, T. Lindberg, K. Paustian and T. Rosswall
Background to the project "Ecology of arable land" 11
Project characteristics and experimental design 11
The synthesis volume 13

Background to the project "Ecology of Arable Land"

Beginning with the pioneering investigations of the early naturalists, ecologists have traditionally concentrated their efforts on understanding relationships in natural environments. While man's influences on natural ecosystems were recognized early on, environments that were largely man-created, such as agricultural lands, were generally not considered to be suitable objects for study. Such systems were, in short, "unnatural". Disturbances of natural ecosystems due to human activities, such as agriculture, were studied at the "receiving end", i.e., in the native ecosystem, rather than at the source of the disturbance.

More recently, these traditions have changed and the attentions of both basic and applied ecologists have been drawn to the study of agroecosystems. The impetus for this development has largely been due to increasing concerns about negative environmental consequences of modern industrial agriculture. Older environmental issues included the use of mercury compounds as seed dressing, and DDT accumulation in the food chain. More current environmental issues include the increased overall use of various pesticides, the addition to soil of heavy metals and other toxic substances from sewage sludge and fertilizers, and the pollution of non-agricultural environments from nutrients lost from agricultural land. In the latter case, nitrogen losses from agricultural soils have been of particular concern, due to 1) nitrogen leaching, causing eutrophication of surface waters and pollution of wells and municipal water supplies, 2) emissions of gaseous nitrogen compounds which function as greenhouse gases and are active in ozone depletion, and 3) ammonia emissions, which can contribute to soil and water acidification. A more efficient use of the nitrogen applied to agricultural land and a reduction of nitrogen losses necessitates a better understanding of what regulates the biogeochemical cycles in agroecosystems.

The multidisciplinary research project "Ecology of Arable Land – The Role of Organisms in Nitrogen Cycling" was initially conceived by Professor Eliel Steen and the late Professor Börje Norén through a series of discussions during 1975–1979. Towards the end of that period Professor Thomas Rosswall, together with many colleagues, joined the discussions, and the project started on a small scale in the fiscal year of 1979/80. At the time the project was conceived, there were few research projects studying agricultural cropping systems in an ecosystem context. Although the International Biological Programme (IBP) had initiated ecosystem research in many countries, the main focus was on native ecosystem types or biomes, such as grassland, forest and tundra. The need for an ecosystem-oriented study of agricultural production systems was thus a basic reason for initiating the project.

Project characteristics and experimental design

The main project objective was formulated as follows:
– to investigate the functions of soil microorganisms and soil fauna, with particular attention to their importance in circulation of nitrogen and carbon in four cropping systems with differing nitrogen input, above- and below-ground primary production, and organic matter incorporation into the soil.

Three subordinate objectives were added:
– to quantify and analyze nitrogen budgets and dynamic processes regulating nitrogen circulation in the four cropping systems
– to quantify the influence of the cropping systems on the abundance and biomass of soil microorganisms and soil fauna
– to assess the impact of this influence, induced by the different cropping systems, on decomposition and nutrient cycling.

Because the objective of the project was to do a detailed study of carbon and nitrogen cycling processes, it was necessary to have a limited number of treatments in the field experiment. For the same reason, field studies were concentrated at one experimental site, Kjettslinge (Chapter 2). Four cropping systems were selected to provide contrasts in primary production and nitrogen inputs:

1) B0; barley (*Hordeum distichum* L.) with no nitrogen fertilizer for six years (1980–1985). Straw was removed from the field.

2) B120; barley with 120 kg ha^{-1} yr^{-1} of nitrogen applied as calcium nitrate for six years (1980–1985). Straw was removed from the field.

3) GL (grass ley); meadow fescue (*Festuca pratensis* Huds.) was undersown in barley with 60 kg ha^{-1} yr^{-1} of nitrogen applied in 1980 and 200 (120+80) kg ha^{-1} yr^{-1} of nitrogen applied during each of the following four years (1981–84). The ley was ploughed after four years (August 1984).

4) LL (lucerne ley); lucerne, (*Medicago sativa* L.) was undersown in barley with 30 kg ha^{-1} yr^{-1} of nitrogen applied in 1980 and no nitrogen fertilizer thereafter (1981–1984). The ley was ploughed after four years (August 1984).

Barley was chosen as the annual crop because it is spring sown, eliminating the risk of out-wintering, it has a relatively short growth period, and it is the second most important cereal in Swedish agriculture. Grain and straw were removed from the barley plots to minimize carbon input to the soil. Meadow fescue, representing a perennial crop receiving nitrogen fertilizer, was grown as a monoculture to facilitate the production studies. Lucerne was selected for its symbiotic fixation of atmospheric nitrogen, and its longevity, i.e., there was a good probability it would last through the four year ley period. Clover (usually in mixtures with grass species) is a more common ley crop in the region, but has a poorer longevity compared with lucerne. It was desirable to have a long ley period to allow greater differentiation of the treatments. The experiment was performed in a randomized block design with four blocks. A full description of the field, climate, and experimental installations is given in Chapter 2.

The contrasts underlying the experimental design can be summarized as: (1) low vs high nitrogen inputs (B0 vs the others), (2) fertilizer-N vs biological nitrogen fixation (B120 and GL vs LL), (3) low vs high carbon input to the soil (B0 and B120 vs GL and LL), and (4) annual vs perennial crops (B0 and B120 vs GL and LL). These contrasts relate to two assumptions concerning the influence of carbon and nitrogen inputs to fertile agricultural soils, namely, that the soil heterotrophs in general are carbon-limited while plants are mainly nitrogen-limited. Thus, the main hypothesis expressed in the field experiment was that high N additions would stimulate primary production and C delivery to the soil, resulting in an increased and more active microbial and faunal biomass and more rapid flows of N through the system.

Subordinate to this hypothesis were several more specific questions and hypotheses. The leys were expected to represent more "closed" systems with respect to N losses via leaching, due to the continuous plant cover taking up water and nutrients. It was also hypothesized that the fertilized systems (B120 and GL) would be more susceptible to nitrogen losses, particularly immediately after fertilization, when nitrate concentrations in the soil would be high. However, it was also stated that nitrogen losses from lucerne could be high in spring and autumn, due to rapid mineralization of the nitrogen-rich litter. Nitrogen loss was expected to be particularly great when the lucerne ley was ploughed.

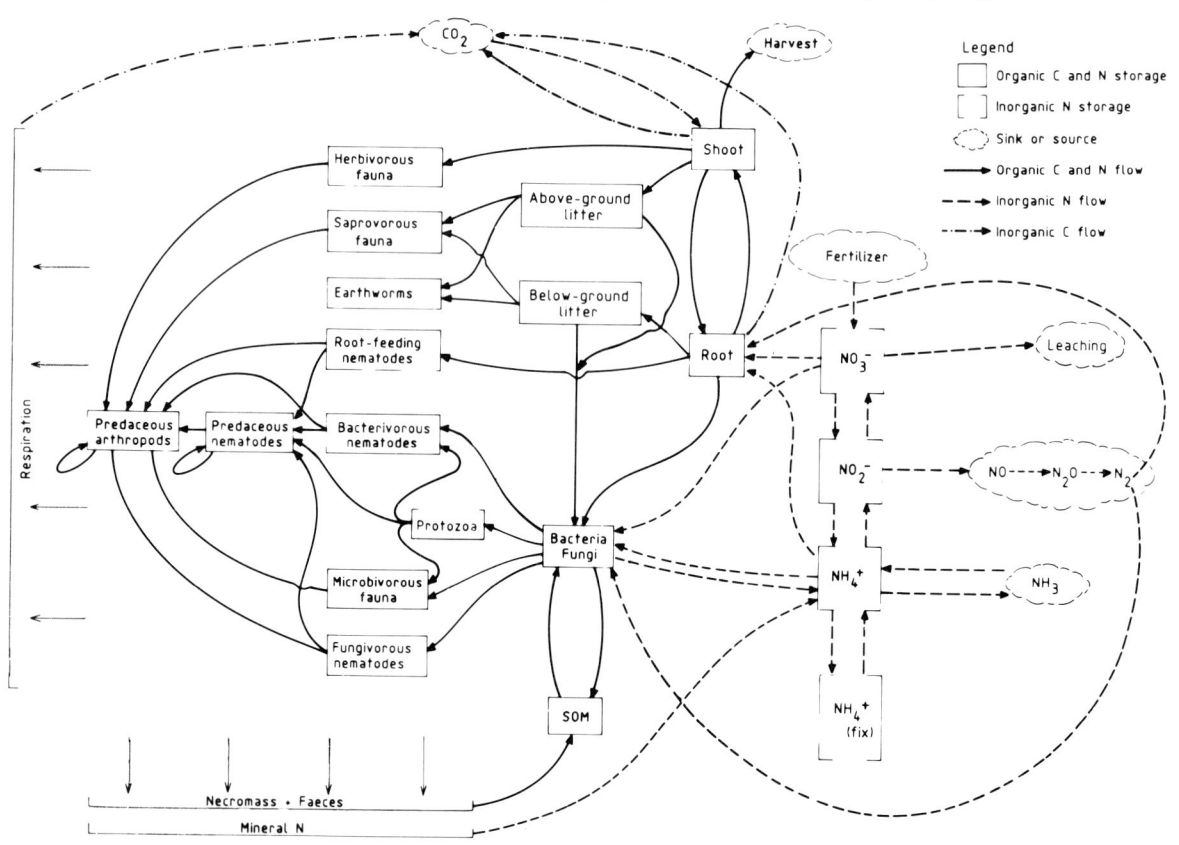

Fig. 1.1. Conceptual model of an agroecosystem, serving as a guideline for the planning and realization of the project.

The lower degree of disturbance of the soil in the leys was expected to significantly affect the faunal community. It was stated that the leys would offer a wider range of niches, supporting a larger and more diverse faunal community in comparison with the barley. Ploughing and tillage would, however, increase decomposition rates by mixing and distributing crop residues through the depth of the plough layer and by increasing soil porosity. Animals with smaller body sizes, less susceptible to mechanical damage, would be favoured in the ploughed systems.

The project investigations can be summarized in a carbon and nitrogen flow diagram (Fig. 1.1). Beginning on the left side of the diagram, C and N flows through different trophic levels of the detrital community are shown. This part of the project included the most detailed population studies, both in the field and laboratory. While a few experiments focused on individual species (e.g. in the earthworm subproject), most of the information on soil fauna was organized at higher functional and taxonomic levels. Microorganisms were mainly treated as aggregate populations of fungi and bacteria. In addition to organism-directed studies, process-oriented approaches to quantify decomposition and N mineralization/immobilization were given major attention.

Primary production (upper center of Fig. 1.1) studies were largely concerned with quantifying net production, nitrogen uptake and the input of organic C and N to the soil. For determining the organic matter input to the soil, studies of above- and below-ground production were closely linked to decomposition experiments. Isotope experiments, using ^{14}C and ^{15}N were used to further clarify production and turnover processes belowground.

On the right hand side of Fig. 1.1, transformations and fluxes involving inorganic nitrogen forms are shown. Investigations of these processes were devoted towards quantifying flux rates in the field and determining the influence of various environmental factors on process rates. Abiotic measurements and hydrological modelling were most closely linked to the leaching studies, but also provided important background information to other subprojects.

The project inherited many concepts from the International Biological Programme (IBP) (see, e.g., Rosswall and Heal 1975, Breymeyer and Van Dyne 1980), and benefitted from people, ideas and experiences from the Swedish Tundra Biome Project (Sonesson 1980) and the Swedish Coniferous Forest Project (Persson 1980b).

As in these earlier projects, our principal aim was to gain an understanding of ecosystem function and thus there was a strong commitment to conducting coordinated experiments and collecting data for the purpose of ecosystem-level synthesis. However, a high priority was also placed on more specialized research within various disciplines, even if these would not contribute directly to ecosystem-level results. The perceived role of systems analysis and modelling was also different from that in some of the IBP projects in that experiments and data collection were not expressly dictated by the requirements of a single overall ecosystem simulation model. Modelling was nevertheless an essential part of the project and was utilized at several different levels of resolution (see Chapters 3–7).

The most important aspects of the project can be summarized as integration, completeness and parallelity. Close integration between subdisciplines was an objective which gradually grew stronger as the project progressed. Initially, the project was divided into several sub-projects grouped under the headings: primary production, faunal investigations, decomposition and humification, mineralization, nitrogen losses and overall carbon and nitrogen budgets. These delineations were necessary during the early stages, but over time the communication between sub-projects increased and they merged into larger working units. In our efforts to thoroughly investigate the major processes affecting carbon and nitrogen cycling, we used several parallel methods (see Chapter 4). Similar results from the different methods strengthened the results, and conflicting results yielded new insights or revision of concepts and hypotheses. As examples of parallel methods used, litter mass loss measurements were made using both ^{14}C-labelled litter and litter-bags, leaching was measured by the use of lysimeters and tile-drained plots of different sizes and nitrogen fixation measurements utilized three different methods.

The project's goal of including all major flows in carbon and nitrogen budgets (Fig. 1.1) resulted in "perhaps the most detailed and complete whole-ecosystem study so far" (Long and Hall 1987), but the wide scope also resulted in restricted attention to details. Data collection and analyses to be used for the overall system budgets have been a major task for the project. Over 100 man-years have been spent on the project and yet detailed investigations of, e.g., the factors governing the population biology of a single species were not possible within the project's framework.

The synthesis volume

Although there were no plans at the start of the project to publish a comprehensive volume of research results, the need for a synthesis volume became more and more apparent, as the work progressed. In particular, some of our colleagues from outside the project (including the Scientific Advisory Committee, see Chapter 9) should be credited for promoting the idea of a synthesis book. It was our intention to present as complete a report as possible of project results, including both scientific highlights as well as more descriptive data. The emphasis on our own results means that extensive reviews of the international literature are not included, although frequent references to results from other projects and investigations are made.

It was decided to go beyond the traditional ecosystem-related division into abiotic factors, primary producers, consumers etc., when deciding on the structure of the book. Instead a structure was chosen that encouraged synthesis to a greater extent, with chapters treating the experimental field (2), heat and water processes (3), ecosystem structure (4), organic C and N flows (5), inorganic C and N flows (6), ecosystem dynamics (7), and an agricultural outlook (8).

Chapters 2–4 give an extensive description of the four cropping systems. The field site is presented in detail in Chapter 2, including a historical perspective, and background information on climate, soil physics and soil chemistry. In Chapter 2 the experimental design, field installations and the effect of the treatments on the physical properties of the soil are also discussed.

Abiotic processes, involving heat and water fluxes, are discussed in Chapter 3. This chapter focuses on the influence of biotic components on the abiotic conditions, especially radiation, heat balance and water conditions. Integration and analysis of the results was aided by a water and heat simulation model.

Chapter 4 is divided into three main sections. The first section gives a basic description of the organisms in the agroecosystem, i.e., their morphology, physiology, habitat requirements etc. It also contains comparisons between Kjettslinge and other temperate ecosystems in the overall levels of organism abundance and biomass present. In the two following sections the dynamic nature of the organism community is emphasized. First, the seasonal, i.e., within-year abundance dynamics are reported, discussed and compared between the organism groups. Finally the long-term dynamics, induced by the different cropping systems, are presented. Each section is divided into sub-sections following the path from producer to consumer to decomposer. Due to the diversity of the decomposer community both from taxonomic and functional viewpoint (Fig. 1.1), different groups within this community are presented separately.

Chapter 5 presents the flows of organic carbon and nitrogen through the ecosystem and is structured similarly to Chapter 4, according to the carbon pathway from atmosphere to plant, from plant to decomposer organisms, and back to the atmosphere. The nitrogen transfers are, when possible, discussed in the same order. Primary production studies, including photosynthesis measurements, field estimates of net primary production, and assimilation chamber experiments employing ^{14}C-labelling are reported in the first section. Root nitrogen uptake, either from soil or through symbiotic nitrogen fixation is included in the section on primary production. Decomposition studies, using litter-bags and labelled plant material, and nitrogen mineralization/immobilization are discussed in the next section. Soil organisms are presented and discussed, with emphasis on their respective contributions to the C and N flows of the cropping systems. Finally, the interactions of different soil organisms in litter decomposition and nitrogen mineralization are discussed.

Nitrogen mineralization and immobilization are the processes that connect Chapters 5 and 6, where the latter chapter deals in detail with inorganic N transformations and flows. Chapter 6 discusses the factors regulating inorganic N flows with special reference to nitrification, denitrification and leaching. Results from measurements of NO_x and NH_3 losses from the cropping systems are also presented. In general, the studies were designed to determine process rates, but results on population dynamics of autotrophic nitrifiers are also included. Special attention is given to the role of water in regulating inorganic N flows, since soil moisture is a primary determinant of the oxygen status of the soil affecting the aerobic nitrification and anaerobic denitrification processes. The role of water in transporting nitrogen from the rooting zone is also discussed.

The highest level of synthesis in the volume, an integrated presentation of carbon and nitrogen amounts and flows in the four cropping systems, is presented in Chapter 7. First, the agroecosystems are presented as annual budgets based on data from 1982–1983. Then the nitrogen dynamics of the systems at Kjettslinge are examined and discussed, aided by a simulation model. Finally, the model is used to analyze mineral N dynamics and nitrate leaching for some experimental sites elsewhere in Sweden.

The main efforts of the project were concentrated on thorough ecosystem analyses at one site. In chapter 8, the results from Kjettslinge are discussed in the context of practical agriculture in Sweden, and in agriculturally similar regions of Europe and North America. We also take some steps along the path from the project's results to recommendations for a future agricultural policy in Sweden. This path is long and winding, and thus Chapter 8 is divided into steps – each step representing a scenario further away from the present and, consequently, more speculative.

Finally, there are other project data than those found in the scientific publications. In Chapter 9, project history, administration and educational activities are presented. A management calendar, reporting the agricultural measures performed, is enclosed as Appendix 1. Appendix 2 lists the project personnel, and Appendix 3 is a list of project publications.

2. The experimental field

Eliel Steen, Lars Bergström, Per-Erik Jansson, Ruben Johansson, Holger Johnsson and Jan Persson

2. **The experimental field.** E. Steen,
L. Bergström, P.-E- Jansson, R. Johansson,
H. Johnsson and J. Persson

Introduction	17
Historical perspective	17
Early land use	18
Climate	18
Regional climate	18
Local climate	20
Soil	20
Soil physical properties	21
Soil chemical properties	22
Design of the field experiment	24
Effects of the cropping systems on soil physical and chemical properties	27
Field installations	27
Lysimeters	27
Meteorological and soil abiotic measurements	29
Management	30

Introduction

The project field was established at the Kjettslinge farm on Örbyhus Estate (60°10′N, 17°38′E) in south central Sweden, 40 km north of Uppsala (Fig. 2.1). Soil characteristics were an important consideration in the selection of the field. The loamy top soil facilitated soil sampling and root washing. A less permeable clay below the top soil and an intermediate sand layer were advantageous for the nitrogen leaching measurements within the large, tile-drained plots. Moreover, the field's proximity to the university (45 km) made daily round trips, for field data collection, possible.

Due to the intensive nature of the research effort, it was decided to restrict the six years of field samplings to just the Kjettslinge site. Many agricultural field experiments are repeated for several years and in several locations. Indeed, cultivar tests and fertilizer experiments where only crop yield is estimated are, in accordance with Swedish standards, repeated annually for at least five years in ten locations, i.e., 50 field trials. However, the experiments at Kjettslinge were designed not only to estimate crop yields; a multitude of studies were carried out throughout the year for six years to provide the basis for a detailed description of processes and functions in the crop – arable soil system.

Historical perspective

The region has a rich and varied history of human settlement and agricultural land use, as documented by archaeological findings. There are several burial sites from the Stone Age (3000 BC), the subsequent Bronze Age (2000 BC) and especially from the Iron Age, which comprised the Vendel culture, AD 600–800, and the Viking period AD 800–1000.

Essential to the Iron Age culture were the waterways leading southwards, i.e., the Vendel and Fyris rivers. Near Uppsala, these rivers drain into Mälaren, which is now a lake (Fig. 2.1) but at that time was part of the Baltic Sea. These waterways were fundamental transit routes for the local Vikings who used 15–20 m long boats, with a sail and 10–20 pairs of oars, for voyages to the Baltic's southern and eastern coasts, to Ladoga (near present-day Leningrad) and further on to the Russian rivers of Dnjepr and Volga and as far as Istambul.

The Vendel settlement was probably one of the most important, especially during the Vendel period. The boat graves found outside the church (Fig. 2.2), 2 km from the field site, in 1881 are of great archaeological interest and show striking similarities to the boat graves at Sutton Hoo, England (Bruce-Mitford 1979). The 14 graves contained weapons, tools, pieces of jewelry in bronze and gold, and bones of humans, horses, cattle, sheep, pigs, dogs and poultry. The helmets, swords and shields that were found in these graves were elaborately adorned with bronze and gilded plates (Fig. 2.3). The grave boats, which were 8–10 m long and 2 m wide, were probably used for river transport to Uppsala (named Aros at that time). The early Vendel settlers were probably farmers and fur hunters. Fishing was an important supplement. Later (two graves were dated to AD 1000–1100, i.e., late Viking period) Viking contacts were extended across the Baltic Sea. The people in these areas also became merchants and warriors (Davidson 1976). Aros was their local seaport, since the Vendel area by that time had risen about 15 m above sea level.

There are several archaeological finds on the Kjettslinge farm. Most of them are from the late Iron Age, i.e., AD 800–1000. The apparent prosperity of the settlement at Vendel AD 500–1000 was probably due to iron. This part of the province of Uppland is very rich in iron, with ores and lake ores, "nuggets", in the lakes and mires. The people of that time knew how to process the iron ore and make it malleable. The energy source and reduction agent used for this process was charcoal. The southern fringe of the taiga extended, as it still does, through this area, and the raw material for charcoal production was inexhaustible at that time. Malleable iron was the most important commodity produced by the Vendel settlement, even more important than furs, hides, and farm products. The iron was exported with the Viking ships and was the primary article of exchange. The iron trade can best explain the rich treasures of gold, silver, glass, pearls and coins found in this area. Recent studies have shown that the Vikings of this area were primarily tradesmen and suggest that their reputation as plunderers has been exaggerated (Ambrosiani 1983).

Approximately three hundred years after the Viking period, farmers had settled in villages. In AD 1312 Kjettslinge is mentioned in land taxation records as a village with 15 farms. Later the village became a *säteri* – a large farm, owned by a nobleman, that was passed on by inheritance. In AD 1538 Kjettslinge was incorporated into the Örby estate, later called Örbyhus, which was one of the properties of King Gustav Vasa. A notable event at Örbyhus was the death (allegedly from eating arsenic-laced pea soup) in AD 1577 of Erik XIV, son of Gustaf Vasa, who was imprisoned there by his brother Johan III.

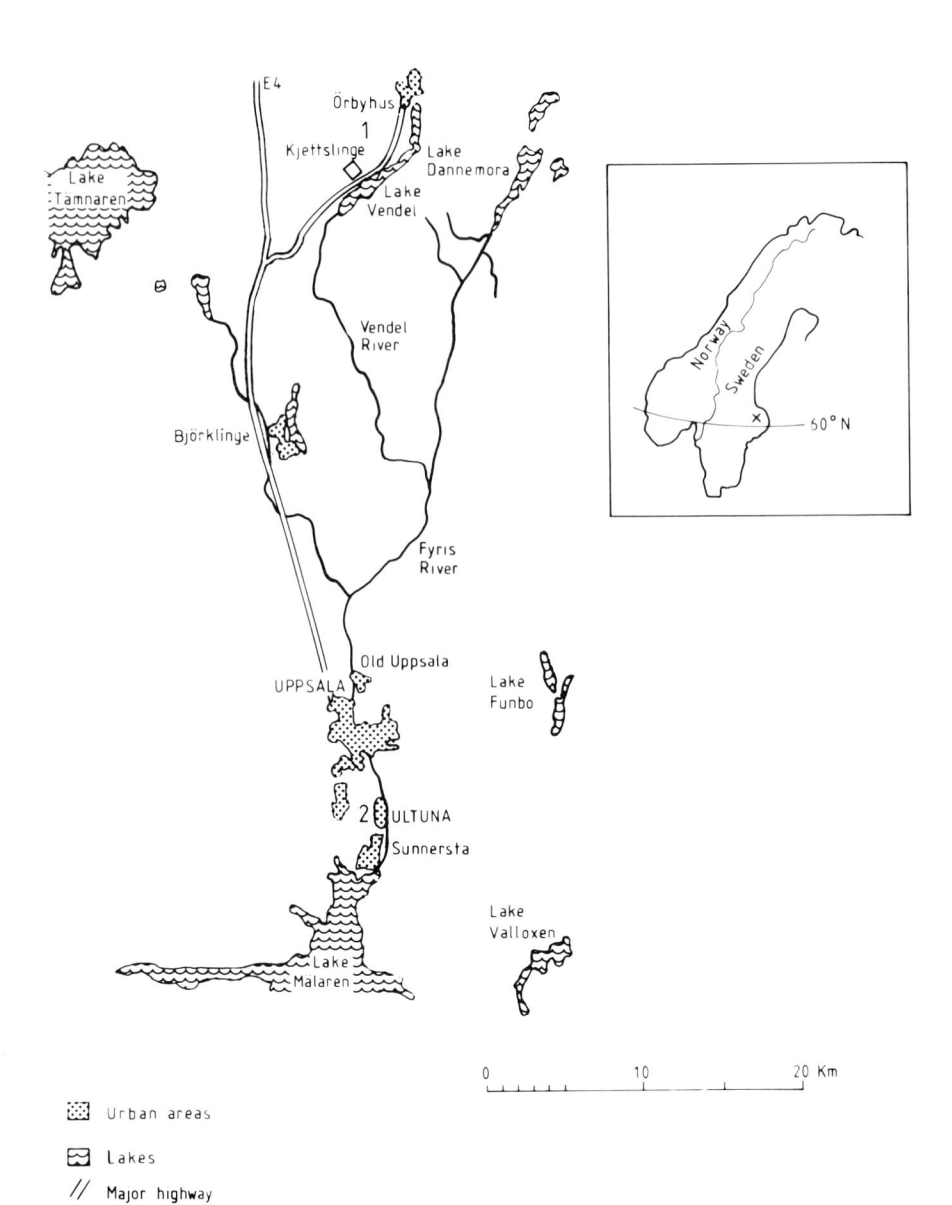

Fig. 2.1. Location of the experimental field at Kjettslinge (1) and the Swedish University of Agricultural Sciences, Ultuna, Uppsala (2).

Early land use

The field has been cultivated for about 130 years, which is a comparatively short time in this region of Sweden. During the 50 years prior to the start of the project, the field was managed according to seven-year crop rotations with spring-sown cereals (oats, barley), three-year leys (red clover, timothy, meadow fescue), winter rye, spring-sown rapeseed and turnips or potatoes. Farmyard manure, 40×10^3 kg ha^{-1} (fresh weight) was applied once in each rotation. The regrowth of the leys was normally grazed by dairy cows. Previously, the field had been used for at least 1 500 years as a hay meadow or pasture. The alteration in land use initiated a substantial change in the humus balance. The amount of carbon probably started to decline, after a long period of stability, as it did in other fields in the Uppsala area when they were cultivated (Persson 1974). On a map, dated 1809, the field was characterized as a wet meadow with occasional, scattered birches (*Betula pubescens* L.). About 1840 there was an attempt to lower the water level of nearby Lake Vendel to improve drainage, but afterwards spring flooding still occurred frequently. When the experiment was started in 1979 there was still a need for drainage. The field was also infested with couch grass (*Agropyron repens* (L.) Beauv.).

Climate

Regional climate

The climate of the region is temperate and humid. The normal mean annual air temperature is about 5°C, and

Fig. 2.2. The Vendel church and the commemoratory obelisk at the site of the boat graves. Photo: Jan Norrman, Riksantikvarieämbetet, Stockholm.

the coldest and warmest months are January (–5°C) and July (17°C), respectively (Taesler 1972; Fig. 2.4). The growing season (defined as days with a mean air temperature above 5°C) is around 200 days (Odin et al. 1983).

Day length varies greatly, from about 6 h in December to 18 h in June. Correspondingly, normal mean values of total radiation range between about 30 MJ m^{-2} month^{-1} in December to 540 in June (Stockholm; Ångström et al. 1974). In winter the ground is normally frozen, to a maximum depth of 80 cm, from the beginning of December to the end of March, and snow covers the ground for ca. 110 d (normal value; Ångström et al. 1974). Prevailing winds are south to southwest and northwest to north (Fig. 2.5).

Normal annual precipitation has been estimated at about 600 mm when corrected for wind losses (1951–80;

Fig. 2.3. Ornament found on a helmet in a boat grave at Vendel. From H. Stolpe and T.J. Arne (1912), Gravfältet vid Vendel. Kungl. Vitterhets historie och antikvitetsakademien, Monografier 3 (Stockholm).

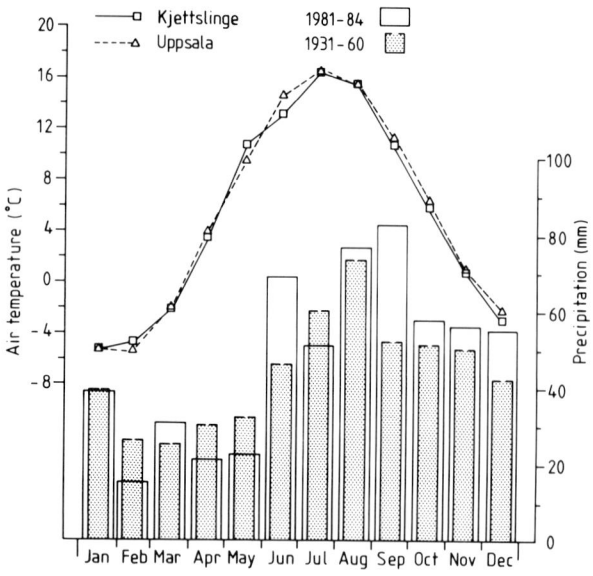

Fig. 2.4. Monthly mean air temperature (top) and precipitation (bottom) in Kjettslinge, 1981-1984, and Uppsala (40 km S), 1931-1960.

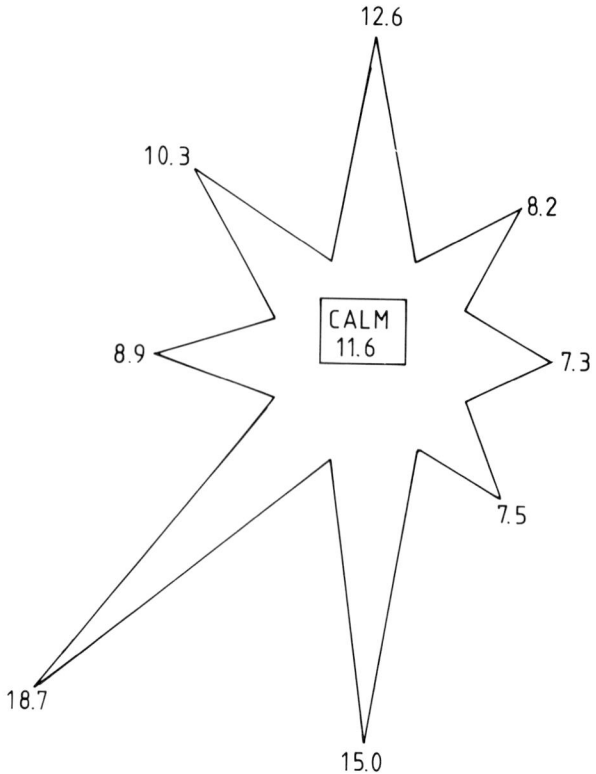

Fig. 2.5. Frequency (%) of wind directions at Uppsala, 40 km S of Kjettslinge. Mean of 1931-1960 (Taesler 1972).

Eriksson 1983). Autumn is generally wet with higher amounts of precipitation compared with spring, which is normally dry (Fig. 2.4). Annual potential evapotranspiration is around 550 mm (Eriksson 1981) while annual runoff is ca. 200 mm (Tryselius 1971), leaving 350 mm for actual evapotranspiration. Peak runoff occurs during the spring-snowmelt and in autumn. Summer runoff is low.

Local climate

The mean monthly air temperatures during the experimental period 1981–84 were close to normal for the region (Fig. 2.4). Air temperatures during 1981 were considerably lower than normal, and the growing season was unusually short, while 1983 was comparatively warm (Fig. 2.6).

During the experimental period, the springs were somewhat drier than normal and autumns wetter than normal (Fig. 2.4). The dry springs of 1980, 1981 and 1984 were followed by wet summers and autumns. The year 1982 was drier than normal (Tab. 2.1). The growing seasons of 1982 and 1983 were mostly sunny, while 1984 was comparatively cloudy with low quantities of incoming global radiation (Tab. 2.2).

The comprehensive abiotic program at Kjettslinge resulted in large amounts of data about the local field climate from 1981–1985. Alvenäs et al. (1986) report these data in detail and also give instructions about how to access the computerized data base. Results concerning soil heat and water are discussed in Chapter 3.

Soil

The area is a denuded peneplain of archaic rocks, sediments and moraines, and the land is flat with patches of moraine hills and eskers. The last glaciation ended about 10 000 yr ago. As late as 2 000 BC the area was a shallow bay of the Littorina Sea. The land is currently rising 6 mm yr^{-1}. The flatland is about 30 m a.s.l. The bottom sediment in the experimental field is glacial clay, 2–5 m thick. Above it is a layer of post-glacial clay, 1–3 m thick. On top of this there are lighter deposits of resedimented material from surrounding moraines and eskers.

The soil profile, described by Öborn and Johnsson (1990), is characterized by distinct textural horizons, loamy top soil (plough layer: 0–27 cm) followed by sand layers overlaying clay loam and clay (Fig. 2.7). The top soil is brownish black. Sand and clay horizons are lighter in colour but have brown mottles (iron-oxides). The sand horizons have no structure, while the top soil and clay horizons have moderate to strong structure. Below the clay loam layer the clay content steadily increases, and at 1 m the texture class is clay.

The soil is classified as a *Typic Haplaquoll*, sandy over clayey, mixed, frigid, according to the USDA classifica-

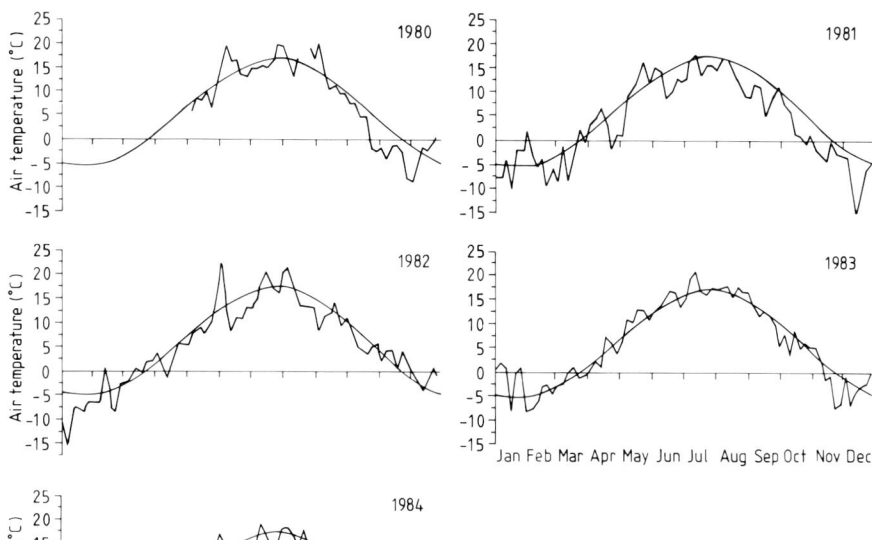

Fig. 2.6. Five-day means of air temperature at 1.5 m above ground at Kjettslinge (1980–1984), and monthly means at Ultuna (45 km S, smooth curve) 1931–1960 (Rodskjer and Tuvesson 1975).

tion system (Soil Survey Staff 1975) and as a *Mollic Gleysol* according to the FAO classification system (FAO 1974).

Soil physical properties

Mapping of textural horizons and soil sampling for physical and chemical soil properties were carried out in 1978 and 1979 in areas undisturbed by the tile drainage and the permanently installed equipment.

The thickness of textural horizons in the soil profile (0–100 cm) was determined at 261 locations along perpendicular transects. At each point a 1-m-deep core of the soil profile was taken using a small diameter auger tube. The boundaries between the different textural horizons were determined visually.

Undisturbed soil cores (7 cm diam.) were taken from 12 soil profiles, covering blocks A–D (Fig. 2.12), for laboratory analyses of soil-water characteristics, hydraulic conductivity, texture and structure. Samples were identified according to the textural horizons from which they were taken.

There are five horizons with distinguishable texture (Fig. 2.8). The thickness of and the degree of cover by the different horizons varies irregularly across the field except for the top soil (Fig. 2.9., Tab. 2.3), which has

Tab. 2.1. Monthly precipitation (mm) at Kjettslinge (1980–1984). Normal values are given for Uppsala (1931–1960; Taesler 1972). ND = not determined.

Month	Uppsala Normal value	\multicolumn{6}{c}{Kjettslinge}
		1980	1981	1982	1983	1984	1981–84
Jan	39	ND	12	25	64	53	38
Feb	26	ND	11	24	7	19	15
Mar	25	ND	30	27	52	13	30
Apr	30	ND	9	36	26	13	21
May	32	11	12	32	32	14	22
Jun	46	51	114	36	68	65	71
Jul	60	62	41	44	16	95	49
Aug	73	116	139	77	22	73	78
Sep	52	27	14	28	136	146	81
Oct	51	101	92	33	37	80	60
Nov	50	44	88	64	31	34	54
Dec	42	52	103	28	54	34	55
Jan–Dec	526	ND	666	453	546	639	576

Tab. 2.2. Monthly sums of global radiation (MJ m^{-2}) at Kjettslinge, 1981–1984.

Month	1981	1982	1983	1984
Jan	54	55	45	33
Feb	98	111	132	75
Mar	231	217	190	220
Apr	423	384	287	370
May	619	539	464	516
Jun	469	554	566	512
Jul	554	649	598	485
Aug	434	452	495	439
Sep	271	278	250	211
Oct	140	125	154	107
Nov	61	71	74	38
Dec	35	39	37	18
Jan–Dec	3338	3475	3292	3023

A_p 0–27 cm Brownish black (10 YR 2.5/2) moist and grayish yellow brown (10 YR 4/2) dry, loam; moderate fine to medium granular; slightly sticky, slightly plastic, very friable, hard; medium discontinuous pores; very frequent very fine and fine roots; abrupt smooth boundary

2C_g 27–43 cm Grayish yellow brown (10 YR 6/2) moist and light gray (10 YR 7/1) dry, common coarse faint diffuse yellowish brown (10 YR 5/6) mottles, loamy fine sand; single grain; nonsticky nonplastic, loose, soft; few very fine roots; abrupt smooth boundary.

3B_{wg1} 43–75 cm Brownish gray (10 YR 4/1) moist and light gray (10 YR 7/1) dry, many medium distinct clear brown (7.5 YR 4/4) mottles, clay loam; strong to medium subangular blocky; sticky, plastic, very firm, very hard; sesquioxides both on ped surfaces and in peds; common very fine continuous vertical inped tubular pores; common very fine roots; diffuse, smooth boundary.

3B_{wg2} 75–100 cm Brownish gray (10 YR 4/1) moist and light gray (10 YR 7/1) dry, many medium distinct clear brown (7.5 YR 4/4) mottles, clay; strong medium and coarse subangular blocky; sticky, plastic, very firm, very hard; sesquioxides both on ped surfaces and in peds; common very fine continuous vertical inped tubular pores; common very fine roots.

Fig. 2.7. The soil profile with horizon designations according to FAO (1977). Photo: H. Johnsson.

been homogenized by ploughing and harrowing during the last century. The thickness of the loamy fine sand horizon varies substantially. This horizon covers 94% of blocks A–D, with a mean thickness of 15 cm and a maximum of 53 cm. The thickest loamy fine sand horizon, with a mean of 24 cm, occurs in block A and is approximately twice as thick as that in the other three blocks. Both the coarse sand and the loam horizons are fairly thin, 3–8 cm. The loamy coarse sand is rather homogeneous, but there is a gradual decrease in thickness from block A towards block D (Fig. 2.9, Tab. 2.3).

Data on soil water characteristics were fitted to the equation proposed by Brooks and Corey (1964):

$$S_e = (\Psi/\Psi_e)^{-\lambda} \quad (1)$$

where Ψ = soil water tension, Ψ_e = air entry pressure and λ = pore size distribution index. The effective saturation, S_e, is given by

$$S_e = (\Theta - \Theta_r)/(\Theta_s - \Theta_r) \quad (2)$$

where Θ = volumetric water content, Θ_r = residual water content and Θ_s = water content at pore saturation.

The soil water characteristics as well as bulk density varied between horizons (Fig. 2.10, Tab. 2.4). The variation within each horizon was generally small.

Soil chemical properties

Samples for chemical analyses were taken at the start of the experiment (1979) and five years later (1984) in the top soil, the sand layer, and the post-glacial clay layer. The soil chemical status in the top soil is summarized in Tab. 2.5. Initially the top soil had a homogeneous lime (pH), phosphorus, potassium and magnesium content, although block A had a somewhat lower P, and block D a slightly lower K and Mg content than other blocks

Tab. 2.3. Thickness and degree of cover of each textural horizon. Mean of four replicates (blocks) and SE in parentheses.

Treatment	Loamy top soil Lower boundary (cm)	Loamy fine sand Degree of cover	Loamy fine sand Thickness (cm)	Loamy coarse sand Degree of cover	Loamy coarse sand Thickness (cm)	Loam Degree of cover	Loam Thickness (cm)	Clay loam Upper boundary (cm)
B0	27 (2)	0.99	19 (7)	0.34	4 (2)	0.21	8 (2)	49 (7)
B120	27 (2)	0.90	10 (3)	0.25	3 (2)	0.32	8 (3)	41 (5)
GL	27 (2)	0.89	15 (7)	0.30	3 (2)	0.24	7 (3)	43 (6)
LL	28 (2)	1.00	15 (5)	0.22	4 (1)	0.43	9 (2)	48 (6)
Average	27 (2)	0.94	15 (6)	0.28	3 (1)	0.30	8 (3)	45 (6)

Fig. 2.8. Soil texture in the five layers. Mean and SE.

Fig. 2.9. The thickness of the loamy fine sand layer in the main experimental field. Note the exaggerated vertical scale.

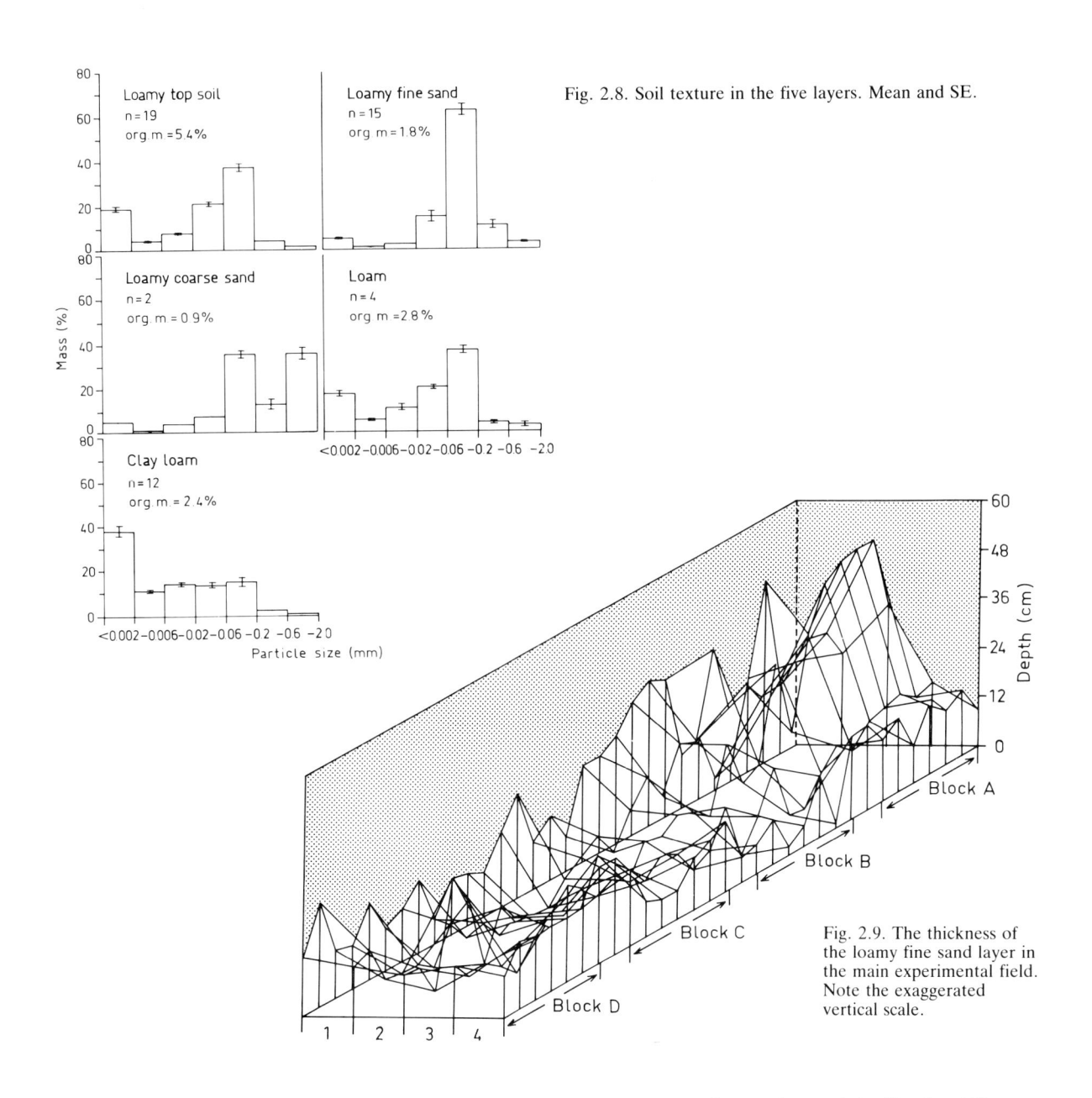

Tab. 2.4. Areal means of hydraulic saturated conductivity, k_s and parameters of soil water characteristics (Eq. 1 and 2), water content at wilting point (150 m water), and dry bulk density, for blocks A–D. SD within parentheses. Sampling date: 8 August, 1979.

Layer	Hydraulic conductivity k_s (cm hr^{-1})	Water content at saturation porosity Θ_s (vol%)	Water content at wilting point Θ_w (vol%)	Residual water content Θ_r (vol%)	Pore size distribution index λ	Air entry pressure Ψ_e (cm water)	Dry bulk density ϱ_b (g cm^{-3})	No. of samples n
Loamy top soil	0.59 (0.82)	45.3 (3.4)	14.9 (2.2)	0.15 (0.02)	0.15 (0.04)	33.1 (24.4)	1.45 (0.09)	23
Loamy fine sand	1.04 (0.50)	39.7 (1.5)	4.1 (2.6)	3.9 (2.4)	0.50 (0.11)	28.6 (10.8)	1.60 (0.04)	8
Clay loam	0.48 (0.93)	52.9 (4.1)	27.8 (2.2)	0.28 (0.02)	0.10 (0.02)	48.4 (29.2)	1.25 (0.11)	9

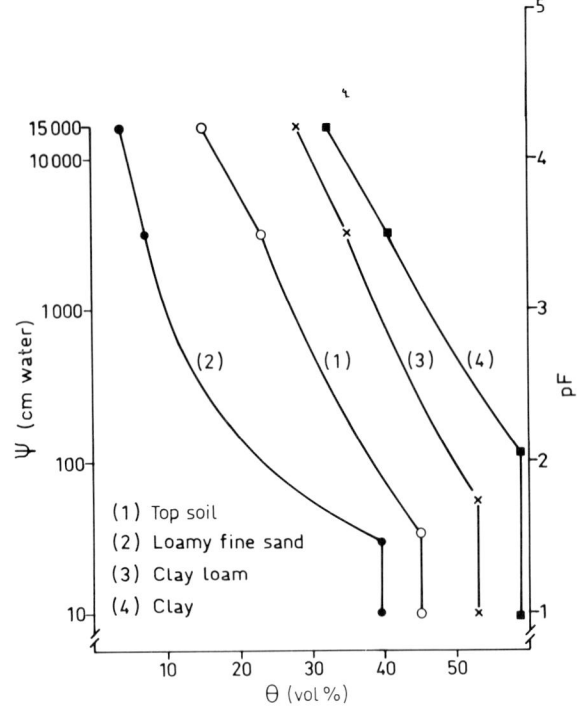

Fig. 2.10. Soil water characteristics of the soil layers in the main experimental field.

(Steen et al. 1984). Block differences in carbon and nitrogen were also small. On average there was 2.2% C and 0.23% N in the top soil. However, the four plots later grown with unfertilized barley had a lower initial mean content of C and N, 1.8 and 0.19%, respectively.

Soil chemical data for the soil profile in Fig. 2.7 are given in Tab. 2.6. Calcium was the dominating base-cation in the exchange complex. The sand layer had lower cation exchange capacity and base saturation than the other three layers. There was very little carbon and nitrogen below the top soil. The pH increased somewhat with depth. The two horizons in the subsoil differed in K and Mg which both had higher concentrations in the clay than in the sand (Steen et al. 1984).

The textural composition (15–20% clay) and the organic matter content (about 4.5%) in the Kjettslinge field is normal for arable soils in central Sweden. The crops grown at Kjettslinge produced similar, or higher, yields than the average for central Sweden (Tab. 2.7).

Design of the field experiment

Barley was uniformly sown across the entire field in spring 1978 and fertilized (NPK 20–6–6, 500 kg ha^{-1}) to obtain as homogeneous nutrient conditions as possible

Tab. 2.5. Soil chemical data for the top soil. The analyses were performed as described in Steen et al. (1984). P, K and Mg were extracted with ammonium lactate (pH 3.75). Seven soil cores were collected and pooled in each block A–D. Mean values of blocks A–D (n = 4) and standard errors in parentheses.

Treatment	Year	pH (aq.)	P (mg 100 g^{-1})	K (mg 100 g^{-1})	Mg (mg 100 g^{-1})	C (%)	N (%)	C/N
B0	1979	6.3 (0.02)	4.5 (0.7)	6.4 (0.4)	9.7 (0.7)	1.8 (0.1)	0.19 (0.01)	9.5
	1984	6.1 (0.03)	5.7 (0.8)	5.6 (0.6)	10.3 (0.5)	1.8 (0.2)	0.18 (0.02)	10.0
B120	1979	6.4 (0.1)	5.5 (0.1)	6.7 (0.6)	10.4 (1.1)	2.6 (0.2)	0.26 (0.02)	10.0
	1984	6.4 (0.1)	6.7 (0.5)	6.2 (0.3)	10.8 (0.6)	2.6 (0.2)	0.25 (0.02)	10.4
GL	1979	6.3 (0.1)	4.8 (0.5)	7.1 (0.6)	10.2 (0.9)	2.3 (0.1)	0.24 (0.02)	9.6
	1984	6.4 (0.1)	6.4 (0.4)	5.2 (0.3)	10.2 (0.4)	2.6 (0.2)	0.24 (0.02)	10.8
LL	1979	6.1 (0.1)	5.5 (0.4)	7.1 (0.7)	10.0 (0.9)	2.2 (0.1)	0.22 (0.02)	10.0
	1984	6.1 (0.1)	6.0 (0.4)	6.0 (0.4)	10.8 (0.8)	2.2 (0.2)	0.23 (0.02)	9.6
Average	1979	6.3	5.1	6.8	10.8	2.2	0.23	9.6
	1984	6.3	6.2	5.8	10.5	2.3	0.23	9.6

Tab. 2.6. Soil chemical data for the soil profile in Fig. 2.7, sampled on 1 Sep 1987.

	Depth (cm)	Cation exchange capacity (meq. 100 g^{-1})	Base saturation (%)	C (%)	N (%)	C/N	pH (aq.)
Loamy top soil	0–27	34	76	2.2[a]	0.23[a]	9.6[a]	6.3[a]
Loamy fine sand	27–43	11	62	0.3	0.04	7.5	6.6[a]
Clay loam	43–75	30	77	0.3	0.04	7.5	7.0[a]
Clay	75–100	33	78	0.3	0.06	5.0	7.0[a]

[a]Mean values for the whole field (Tab. 2.5 and Steen et al. 1984).

Tab. 2.7. Harvested yield of grain (15% water) and forage (dry matter) at Kjettslinge compared with means from lowland areas in central Sweden (kg ha^{-1} yr^{-1}).

Area	Barley	Grass ley	Lucerne
Kjettslinge	4470[a]	8720[c]	8580[c]
Central Sweden	4630[b]	8320[d]	9020[d]

[a]1981–85, B120 (Pettersson 1989, Hansson et al. 1987). – [b]1976–85 (Bengtsson 1985). – [c]1982–83 (Pettersson et al. 1986, Pettersson and Hansson, submitted, Hansson and Pettersson, 1989). – [d]1976–85 (Jönsson and Nilsdotter-Linde 1987).

before establishing the various cropping systems. The field was ploughed in October after harvest and fallowed during 1979. However, patches of couch grass remained, and the field was treated in August with glyphosate (4 l ha^{-1}). This treatment was successful, and no further couch grass growth was observed. Owing to the risk of a high water table during the experimental period, the field was tile-drained in May 1980. The drainage pipes were placed 1 m deep at 14 m intervals.

The intensive field measurement program made it necessary to limit the number of experimental treatments. Four cropping systems were selected to represent high and low levels of nitrogen input as well as annual and perennial crops. Barley was chosen as the annual crop because it is spring sown, eliminating the risk of outwintering, and it has a relatively short growth period. This crop is grown on 24% of the Swedish agricultural lands and is the dominant cereal. Meadow fescue, representing a perennial crop receiving nitrogen fertilizer, was sown as a monoculture to make the stand simple and well defined. A mixture of meadow fescue with timothy would have been an alternative that is often chosen by farmers. Lucerne (alfalfa) was chosen to represent crops with symbiotic fixation of atmospheric nitrogen. Red clover is an obvious alternative for Swedish conditions, but lucerne leys last longer, and the soil (pH 6.3, fairly good drainage) was suitable for lucerne. It was essential that the leys could be kept for four years to create a perennial system that differed as much as possible from the annual one. In the six-year sequence of barley, grain and straw were removed to minimize carbon input to the soil. The four-year grass and lucerne leys were intended to have high inputs of carbon. With these prerequisites, the following cropping systems were established:

1) B0 (unfertilized barley); barley (*Hordeum distichum* L.) without N-fertilizer for six years, 1980–1985.
2) B120 (fertilized barley); barley with 120 kg N ha^{-1} yr^{-1} applied as calcium nitrate for six years, 1980–1985.
3) GL (grass ley); meadow fescue (*Festuca pratensis* Huds.) was undersown in barley with 60 kg N ha^{-1} yr^{-1} applied in 1980 and 200 (120+80) kg N ha^{-1} yr^{-1} applied in early May and after the first harvest in June during each of the following four years (1981–84).
4) LL (lucerne ley); lucerne (*Medicago sativa* L.) was undersown in barley with 30 kg N ha^{-1} yr^{-1} applied in 1980 and no nitrogen fertilizer thereafter (1981–1984).

In spring 1981 the leys (GL- and LL-plots) were patchy owing to winter kill. They were therefore resown after disc harrowing. The lucerne (LL) plots in blocks C-G were ploughed in July 1982, due to poor growth. The remaining ley plots, except for the controls, were ploughed after four years (autumn 1984), and barley with 120 kg ha of N-fertilizer was grown in 1985 (Fig. 2.11).

The experiment was performed in a randomized block design with four blocks, A-D (Fig. 2.12). This design was chosen to enable analyses of variance (ANOVA, see Sokal and Rohlf 1969 for details) of the results as follows:

1) Single sampling occasions in all four cropping systems were analyzed with a two-way ANOVA (Factors: Cropping system, Block; Error term: Cropping system * Block). If overall significant (p<0.05) differences were found between cropping systems, the LSD test was used to separate cropping system differences.
2) Repeated samplings in the crops were analyzed using a split-plot ANOVA (Main factors: Cropping system, Block; "Split-plot": Time; Interactions: Cropping system * Block, Cropping system * Time). The main factors were also in this case tested using Cropping system * Block. Time and the interaction Cropping system * Time were tested against the remaining error term. Thus a sensitive analysis was obtained of Cropping system * Time, showing the different dynamics in the four cropping systems, starting from a common initial treatment.

Each plot measured 40×14 m (Fig. 2.13). All individual plots were subdivided into annual subplots (1980–1985). Sampling was thus restricted to 1/6 of the main plot each year. Since the subplots were rather small, while the number of destructive samplings of various kinds was large, it was necessary to allot a fixed part of each subplot to each subproject (Fig. 2.14).

The experimental field contained additional plots for large-scale leaching measurements (F), the nitrogen fixation subproject (I), ^{15}N and ^{14}C studies of decomposition (J), humification and primary production studies, earthworm studies and litter-bag experiments (Fig. 2.12).

Fig. 2.11. View of the experimental field towards southeast during rotary cultivation, 11 August 1984 (Photo J. Lagerlöf).

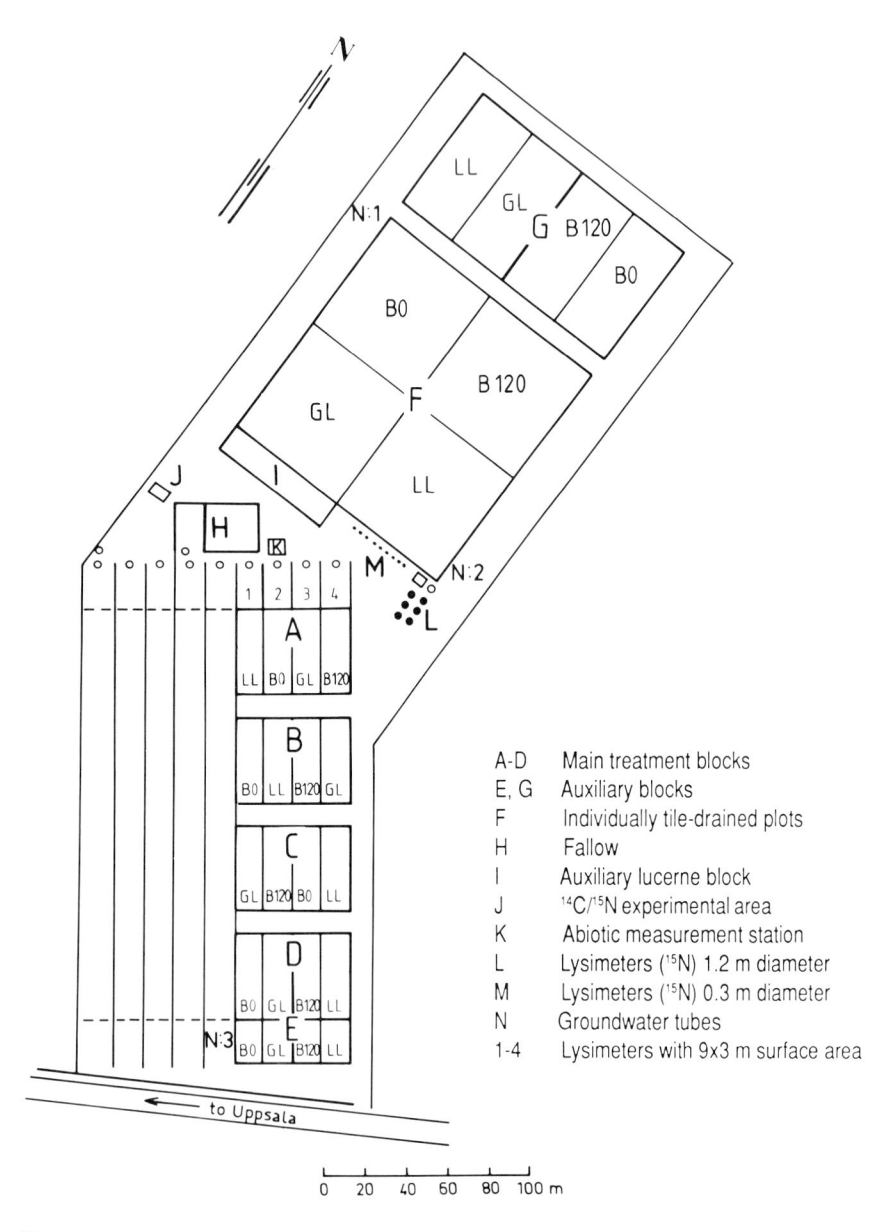

A–D	Main treatment blocks
E, G	Auxiliary blocks
F	Individually tile-drained plots
H	Fallow
I	Auxiliary lucerne block
J	$^{14}C/^{15}N$ experimental area
K	Abiotic measurement station
L	Lysimeters (^{15}N) 1.2 m diameter
M	Lysimeters (^{15}N) 0.3 m diameter
N	Groundwater tubes
1–4	Lysimeters with 9×3 m surface area

Fig. 2.12. The experimental field. The five long plots to the left of blocks A-D belong to an associated project not discussed here.

Effects of the cropping systems on soil physical and chemical properties

To check if any major differences in dry bulk density had been created by the cropping systems, the top soil of the fertilized barley (B120) and the grass ley (GL) was sampled in September 1984, before ploughing the leys. In 1984, the dry bulk density increased with depth under both cropping systems (Fig. 2.15). The grass ley had lower dry bulk density than the barley. There were no significant differences compared with the start of the experiment (Tab. 2.4). However, there were tendencies for decreased dry bulk density near the surface under the grass ley, and for increased dry bulk density under barley, compared with 1979. Little or no change in chemical composition had occurred in the top soil five years after the establishment of the cropping systems (Tab. 2.5). There was a small overall increase of P and a decline of K. The carbon content increased under the grass ley.

Field installations

The permanently installed equipment in the experimental field consisted of three types of lysimeters, separate tile-drained plots with measurement units, and the equipment for the abiotic measurement programme.

Lysimeters

The main purposes of the lysimeters was for studying nitrogen leaching (Fig. 2.16) and for ^{15}N-studies. Large-scale lysimeters (9×3×1 m) were placed in the four plots in block A. N-leaching was further studied with the ^{15}N-technique in two smaller types of lysimeters: concrete cylinders (1.2 m diam.) filled with soil and plastic cylinders (30 cm diam.) with undisturbed soil cores, 1.2 m deep. In addition to these lysimeters, four large tile-drained plots, 60×60 m, were used (cf. Patni 1978).

The use of large lysimeters, allowing a full-scale simulation of the field situation, is an alternative to using large plots with tile-drains. Large lysimeters have, for instance, been used by Uhlen (1978) and by Bolton et al. (1970). However, in principle, the methods are very different. While all percolating water is measured in a lysimeter, only part of the inflow water is usually captured by tile-drains in the field. Whether or not a measuring method yields representative results can only be evaluated by comparing the results obtained with different methods (Chapter 6). Several methods used in parallel may provide additional information on leaching and drainage patterns in the soil.

The use of small-scale lysimeters was due to the high cost of ^{15}N-labelled fertilizer. These lysimeters only contained the top soil and the sand layer owing to problems involved in removing an undisturbed clay column and because most of the N-cycling processes, which were

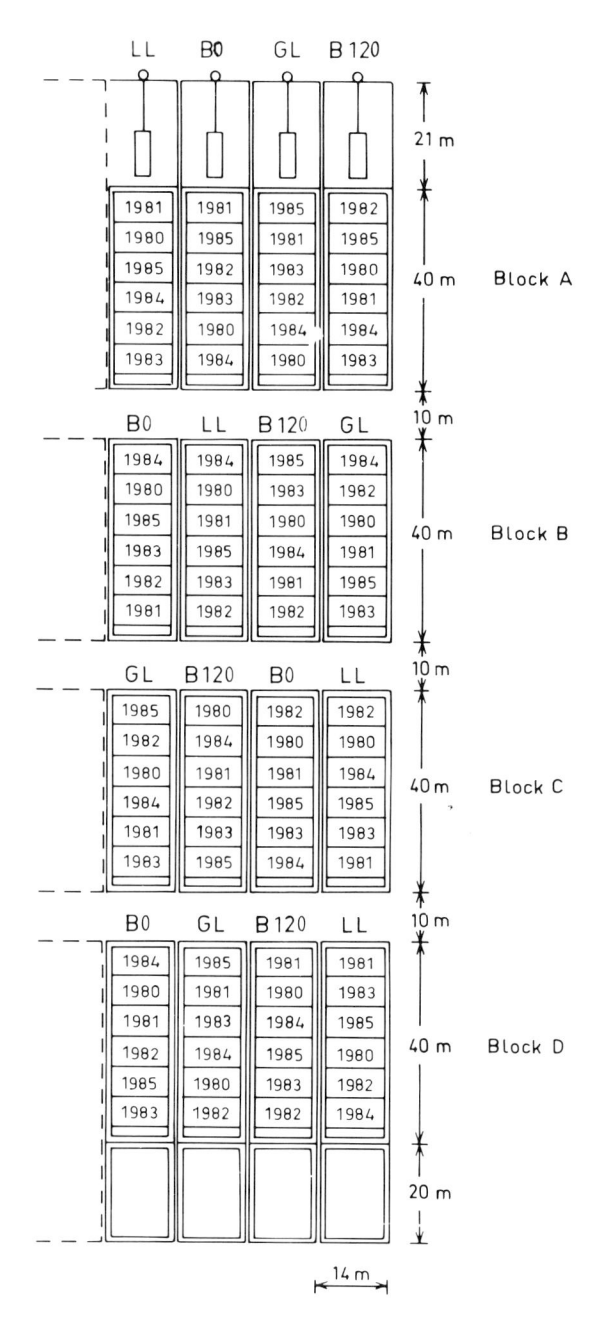

Fig. 2.13. The main experimental field divided into blocks, cropping systems and annual subplots.

studied in the ^{15}N experiments, occurred in these two layers.

The large lysimeters were sealed vessels made of rubber sheeting and were placed in block A (Figs 2.12, 2.13, and 2.16a). They were constructed in the field in 100 cm deep pits 9×3 m in size. Drainage pipes were placed in the bottom, and the soil was refilled with the clay on the bottom, the sand layer in the middle, and then the top soil. The upper edge of the wall was placed

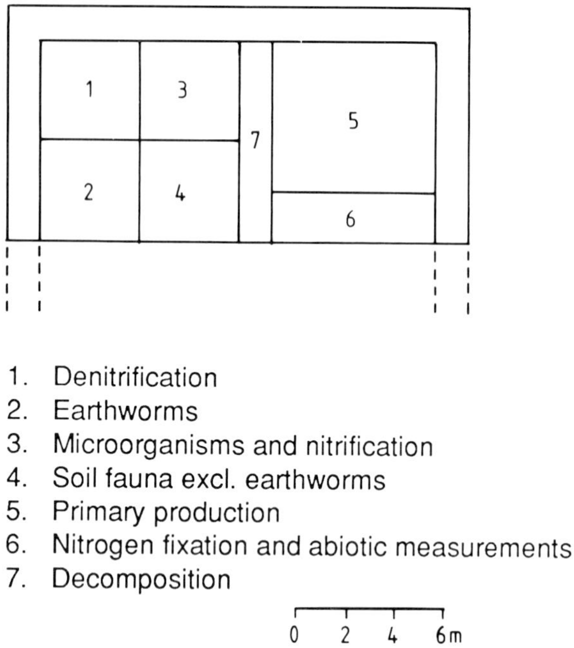

1. Denitrification
2. Earthworms
3. Microorganisms and nitrification
4. Soil fauna excl. earthworms
5. Primary production
6. Nitrogen fixation and abiotic measurements
7. Decomposition

Fig. 2.14. Division of an annual subplot into areas for different samplings.

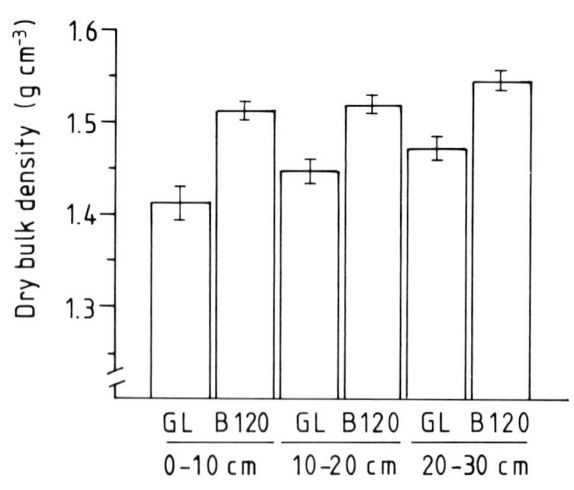

Fig. 2.15. Bulk densities in B120 and GL. Sampling date: 10 September, 1984. Mean values (n=24) and SE of two soil cores from three soil profiles in each of block A-D.

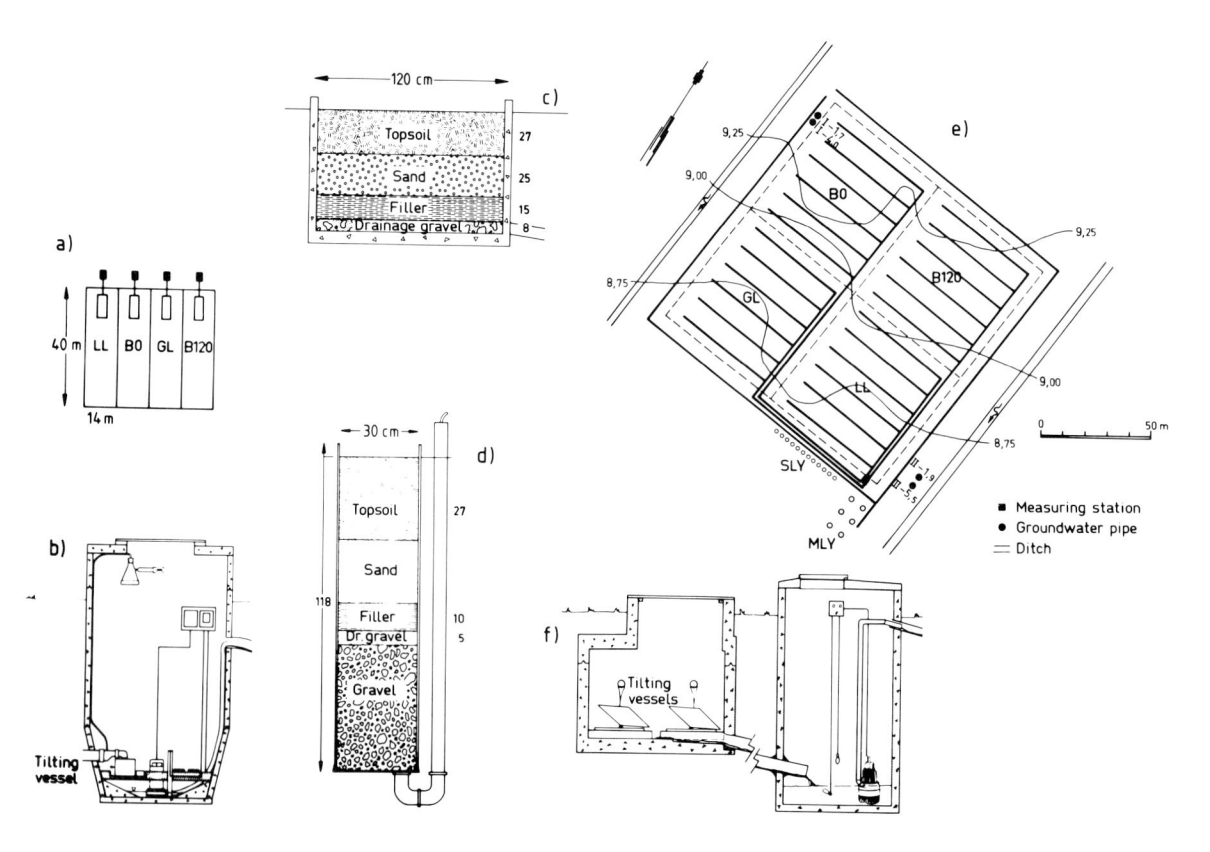

Fig. 2.16. The lysimeter installations: a) rubber lysimeters, 9×3 m b) measuring station for the rubber lysimeters c) cement ring lysimeter 1.2 m diam. d) plastic pipe lysimeter 0.30 m diam. e) large plots (60×60 m) with separate tile-drains f) central measuring station for the separate large plots.

30 cm below the soil surface to allow ploughing and harrowing over the lysimeters. The drainage discharge was collected in separate measuring stations for each lysimeter.

The measuring stations consisted of cylindrical cement cisterns (1.5 m diam., 2.5 m deep) dug into the ground (Fig. 2.16b). The water from the drainage pipes passed into tilting vessels. When one half was full the vessel tipped over and the other half came into position for filling. The number of tilts was registered with a mechanical counter. Water samples were collected with a vacuum hand pump directly from the incoming pipes.

In December 1981, six lysimeters made from cement rings (1.20 m diam., 0.75 m deep) were installed (Fig. 2.16c). A drainage pipe was placed at the bottom of each lysimeter and layers of drainage gravel (4–8 mm diam.) and filler (0.02–0.2 mm diam.) were placed on top of the pipe. The walls and bottom were sealed with rubber asphalt on the outside to prevent water movement through the concrete walls. The lysimeters were filled with sand at the bottom and top soil above. Grass turf from the grass ley was placed on top of the soil in four lysimeters, and the remaining two were sown with barley. The water discharge was collected in plastic bottles.

Twelve lysimeters made of plastic (PVC) pipes (0.30 m diam., 1.20 m deep) were installed in spring, 1982 (Fig. 2.16d). The soil in these lysimeters were undisturbed cores collected using a steel cylinder drill into which the empty plastic lysimeter casing was inserted. This technique was developed by L. Persson at the Dept. of Soil Sciences, Swedish Univ. of Agricultural Sciences. The drilling-cylinder rotated around the plastic casing and carved out a soil core that was gently pushed into the casing. Layers of filler, drainage gravel and gravel (8–12 mm diam.) were placed in the bottom after which the bottom plates were fitted. In each bottom plate two holes were drilled and fitted with plastic pipes. Meadow fescue was sown in 9 cylinders, and barley in the remaining three. Water was sampled with a vacuum hand-pump.

The four large plots (60×60 m each) with separate tile-drains (Fig. 2.16e) had tilting-vessels of the same type as those used in the measuring station for the rubber lysimeters. Each plot had one of the four cropping systems, i.e., B0, B120, GL or LL. Each plot had a separate drainage system which was connected to a central measuring station (Fig. 2.16f). Water samples were collected directly from the incoming pipes.

Meteorological and soil abiotic measurements

In July 1980 a meteorological station was established to measure air temperature, air humidity and precipitation (Tab. 2.8). In June 1981 measurements of global radiation, wind speed, soil temperature, soil water content, and soil water tension were included and recorded on a data logger (Solartron Compact Logger). A standard

Tab. 2.8. The meteorological and soil abiotic measurement program.

	Height or depth (m)	Location	Measurement period	Instrumentation	Sampling frequency
Air temperature	1.5	Research site	Jul 1980–Dec 1985	Thermohygrograph, thermistor, thermocouple	Continuously
Air humidity	1.5	Research site	Jul 1980–Dec 1985	Thermohygrograph	Continuously
Precipitation	1.5	Research site	Aug 1980–Dec 1985	Pluviograph	Continuously
Precipitation	1.5	One km from the field (Bergby)	May 1981–Dec 1985	Standard precipitation gauge	Daily
Global radiation	1.5	Research site	Jun 1981–Dec 1985	Pyranometer	Continuously
Wind speed	2.0, 8.0	Research site	Jun 1981–Dec 1985	Cup anemometer	Continuously
Wind direction	8.0	Research site	Jun 1981–Dec 1985	Wind vane	Continuously
Soil temperature	0.02–0.75	All treatments	Jun 1981–Dec 1985	Thermocouple, thermistor	Continuously
Snow cover depth		All treatments	Nov 1981–Apr 1985	Measuring rod	Once a week, winter
Snow water equivalence		All treatments	Feb 1982–Apr 1985	Spring balance	Several times, winter
Frost depth		All treatments	Nov 1981–Apr 1985	Frost indicator tube	Once a week, winter
Soil water content		Each plot	Sep 1981–Sep 1984	Neutron probe	Once a fortnight, summer
Soil water tension		Each plot	Jun 1981–Sep 1984	Tensiometer	Three times a week, summer

precipitation gauge for measuring daily precipitation was installed in 1981. Tensiometers and neutron probes were used for the soil moisture measurements. Measurements of snow and frost depth were started in the winter of 1981/82.

To measure artesian water levels and groundwater pressure and nitrogen content, piezometers were installed at three plots at two depths. Piezometer readings were taken once a week and groundwater samples for chemical analyses of nitrogen content were collected monthly.

The data, automatically or manually registered, were transferred to a VAX/VMS computer system. The program packages ECODATA and PGRA (Svensson 1979, Jansson and Christoffersson 1985) were used for data storage, retrieval, statistical treatment, and plotting. When meteorological data were missing, time series were supplemented with data from meteorological stations at Marsta (25 km S) and Uppsala (40 km S).

Management

The experimental field was managed according to normal agricultural practices, but additional measures had to be taken because of the experimental design. Before the experiment could start on a full scale some improvements of the field were necessary, especially drainage and weed control (Appendix 1).

The routine management consisted of soil preparation in spring in the barley plots, N- and PK-fertilization, sowing of barley in mid-May, harvesting the leys in late June, N-fertilization of the grass ley thereafter, harvesting the leys again in late August, harvesting barley in early September, removing straw and stubble and harrowing in the barley plots in late September, and ploughing them in mid-October. In January 1982 the open ditches surrounding the field were cleared.

To minimize the use of pesticides, applications were only made after inspections revealed a problem. When needed, the crops were treated with MCPA (2 kg ha^{-1}) or a mixture of dichlorprop, dicamba and MCPA (35:9:15, 3 kg ha^{-1}). Inspection for pests and diseases were carried out once or twice during the growing season. The barley plots were treated for aphids with etiofencarb (50%, 1 l ha^{-1}) in July 1980 and with pirimicarb (50%, 0.5 kg ha^{-1}) in July 1985. A detailed management calendar is given in Appendix 1.

3. Heat and water processes

Per-Erik Jansson, Holger Johnsson and Gunnel Alvenäs

3. Heat and water processes. P.-E. Jansson, H. Johnsson and G. Alvenäs

Introduction	33
Radiation and heat balance	33
Radiation balance	33
Heat balance	34
Soil temperature	34
Water conditions	35
Adaptation of the model	36
Annual water balance	37
Soil water-flow paths	38
Soil moisture conditions	38
Concluding remarks	40

Introduction

Ecosystems include both abiotic and biotic components. Heat and water conditions are important abiotic components in ecosystems because they influence the rate of all biological processes. In addition, water further regulates biological processes since it is a medium of transport for a number of important substances. However, biological processes also influence abiotic conditions, i.e., there is a complex interaction between abiotic and biotic components. This chapter describes abiotic processes, involving heat and water fluxes, with the focus on how biotic components influence abiotic conditions.

In contrast to many biotic transfers, the fundamental equations describing heat and water flows in soils are well-known, and this facilitates the formulation of mechanistic mathematical models. Provided that the soil physical properties can be estimated, modelling is a powerful tool for examining how systems behave under a wide range of conditions. Unfortunately independent and accurate quantifications of soil physical properties are difficult to obtain because of economic and methodological limitations. Thus the the possibility to thoroughly test the predictive capacity of models is often restricted. Within this project only basic soil properties such as textural composition, thickness of textural horizons, water retention and saturated hydraulic conductivity were measured (Chapter 2). Other important variables such as unsaturated conductivity and surface resistance of the different crops had to be estimated. A physically based water and heat model, named SOIL (Jansson and Halldin 1979; Fig. 3.1), was used to:

– Generate continuous time series of soil variables
– Generate driving variables to biologically oriented models
– Analyze how different crops and management practices affect soil temperature, soil moisture and water balance
– Analyze interactions between water and heat movements in partially frozen soil.

The model applications, with special emphasis on the differences between cropping systems, are summarized in this chapter. The heat studies were restricted to barley without N fertilization (B0) and grass ley (GL) cropping systems in block A and D, since they represented the total range of expected differences between cropping systems. However, the hydrological studies included all four cropping systems, i.e., also N-fertilized barley (B120) and the lucerne ley (LL), in the four different blocks A–D and in the large tile-drained plot F (Fig. 2.14).

Radiation and heat balance

Radiation balance is a fundamental determinant of heat and water flows in the biosphere, both between the atmosphere and crop canopy and at the soil surface. Global radiation from the sun acts as the driving force for biological processes which are dependent on assimilation of carbon by photosynthesis. It also acts indirectly as a driving force for micrometeorological conditions, and the following sections deal with this.

Radiation balance

The radiation that affects heat and water processes is commonly divided into short-wave (global radiation, 0.3–3.0 µm) and long-wave (terrestrial radiation, 30–50 µm). Net radiation is the sum of short-wave and long-wave radiation (Fig. 3.2). The radiation balance between atmosphere and crop canopy is mainly determined by meteorological conditions and, only to a minor extent, by the crop. Standard methods are often used for calculating net radiation from global radiation and cloudiness.

In contrast to the radiation above the crop canopy, the radiation at the soil surface is strongly influenced by the crop. However, it is difficult to estimate this influence. Detailed models for short-wave radiation within a crop canopy have been presented, for instance by Lemeur and Rosenberg (1979). The distribution of net radiation within crop canopies is difficult to model in detail because of the different scattering mechanisms of the two types of radiation. Therefore, most estimations of net radiation have been based on empirical knowledge in mechanistic micrometeorological models (e.g. Perrier 1979). In the model presented by Perrier the extinction of net radiation was calculated from Beers' law as:

$$R_n(z) = R_n(0)e^{(-a\ \text{LAI}(z))} \qquad (1:$$

where R_n is the net radiation above the crop cover and at different depths (z) below the reference height. The empirical coefficient a is commonly called the first-order extinction coefficient and the function LAI(z) represents the cumulative leaf area index. Attempts have been made to estimate this coefficient for net radiation

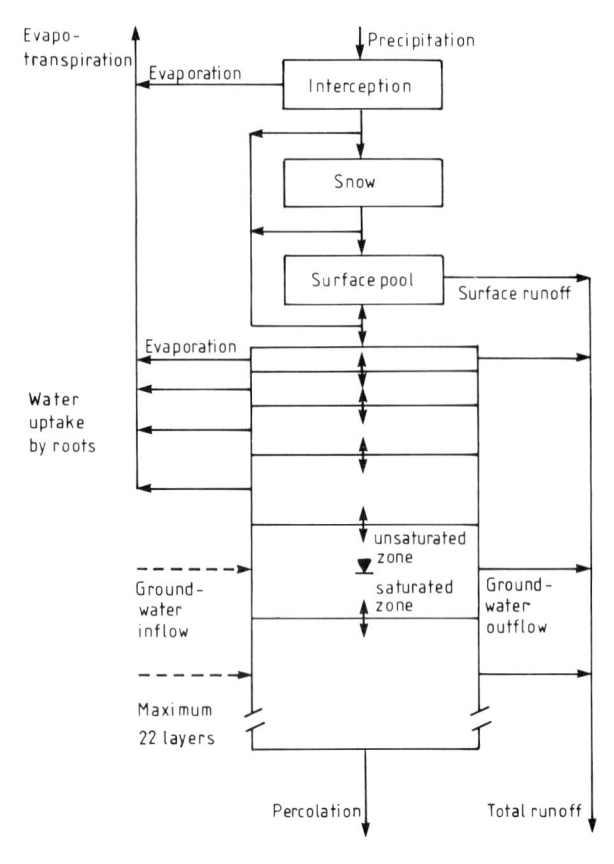

Fig. 3.1. Structure of the water and heat model SOIL. All flows in the soil profile except water uptake by roots represent both water and heat (from Jansson and Gustafson 1987).

(Perttu 1981) but most experimental data emanates from short-wave radiation measurements (Goudrian 1977).

Net radiation at the soil surface during 1982 (Fig. 3.3) was calculated based on equation (1) with $a = 0.3$ and using estimated annual LAI dynamics (Fig. 3.6). The daily net radiation balance was positive, i.e., net short wave radiation exceeded net long wave radiation during the periods selected for comparison. The grass ley (GL) was more efficient than the barley stand (B0) in intercepting radiation, even when the barley stand was fully developed. The greatest differences between B0 and GL occurred before emergence and after the harvest of barley when the soil was bare.

Heat balance

Assuming a daily sum of net radiation (R_n) available at the soil surface, the energy balance equation can be written as:

$$R_n = LE + H + S \quad (2)$$

where LE and H are the latent and sensible heat flows from soil surface to the air and S represents the heat flow into the soil. The latent heat flow corresponds to a mass flow of water vapour and the sensible heat flow corresponds to an exchange of air masses with different temperatures. The energy balance equation thus requires information on the gradients of temperature and vapour pressure in the atmosphere and temperature gradients in the soil. In addition, flow resistance must be known. If the heat flow to the soil is neglected, an analytical solution of Eq. (2) exists (Monteith 1981), but to solve the complete equation taking heat flow to the soil into account, numerical methods are required (van Bavel and Hillel 1976). In this study a numerical method, based on an iterative procedure to find unique solutions, was used. Technical details as well as all the equations are given by Alvenäs et al. (1986).

When comparing calculated heat flows during the 1982 growing season, substantial differences were found between barley (B0) and the grass ley (GL), both with respect to magnitude of and distribution between flows (Fig. 3.4). Both cropping systems had relatively high latent heat flows from the soil surface, with positive mean flows in all months. In B0 sensible heat flows were of the same magnitude as the latent heat flows. In GL the situation was quite different, with a negative sensible heat flow every month except June. This implies that a sensible heat flow from the air to the soil surface occurred simultaneously with a latent heat flow in the opposite direction in GL. This estimated contribution of sensible heat in GL, which counteracted the difference in soil temperature between B0 and GL, was of course influenced by the small plot size, which resulted in similar air temperature above the two different treatments. Heat flows into the soil were similar for the two cropping systems, but rates were lower in the grass ley.

Soil temperature

No independent direct measurements of heat flows were made within the project, so the quantification of these flows is uncertain. However, soil temperature measurements (Alvenäs et al. 1986) were used to indirectly test the calculated flows. The differences in soil

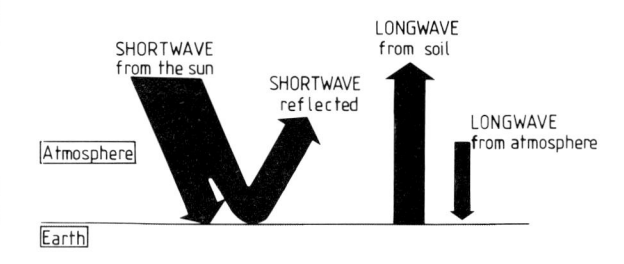

Fig. 3.2. The radiation balance between earth and atmosphere.

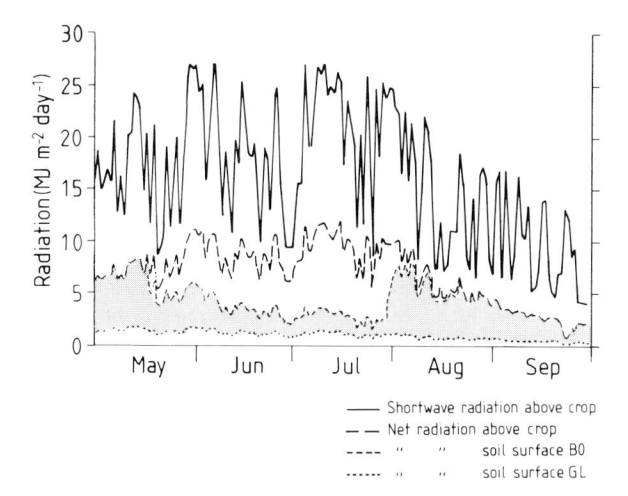

Fig. 3.3. Simulated radiation balance for 1982, including shortwave radiation, net radiation above crop or soil and net radiation at soil surface in B0 and GL. The difference between the two treatments is shaded.

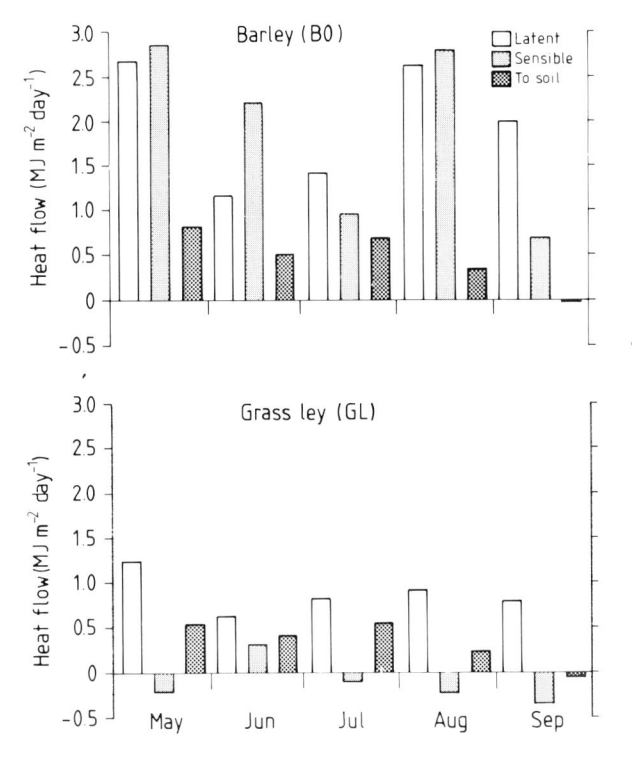

Fig. 3.4. Monthly mean values during 1982 of simulated latent and sensible heat flow to the air above the soil surface and of heat flow into the soil in B0 and GL.

temperature between the cropping systems were of special interest.

Both simulations and measurements showed the daily mean soil temperature in barley (B0) to be up to 5°C warmer during July than in the grass ley (Fig. 3.5). The difference between the grass ley (GL) and B0 in net radiation, i.e., available energy at the soil surface in July (Fig. 3.3), was greater than the corresponding soil temperature difference between the two cropping systems (Fig. 3.5b). The great difference between the two cropping systems, primarily in sensible heat flow but also in latent heat flow (evaporation) at the soil surface, leveled out the available energy for soil heating (Fig. 3.4). The soil temperature differences between cropping systems were due to differences in heat flow into the soil (Fig. 3.4) and the heat flows deeper within the soil (not shown). The deeper heat flows were delayed compared with the surface layers which caused the annual variation in temperature dynamics to be delayed 1.5 months when compared with the heat flow dynamics. This was the main reason why the maximum difference in temperature between the two treatments occurred much later than when the heat flows into the soil showed maximum difference.

Water conditions

The fundamental equation stating conservation of water can be written as:

$$P = E+R+D+S \tag{3}$$

where P is precipitation, E is evapotranspiration, R is surface runoff, D is drainage to pipes or ditches and

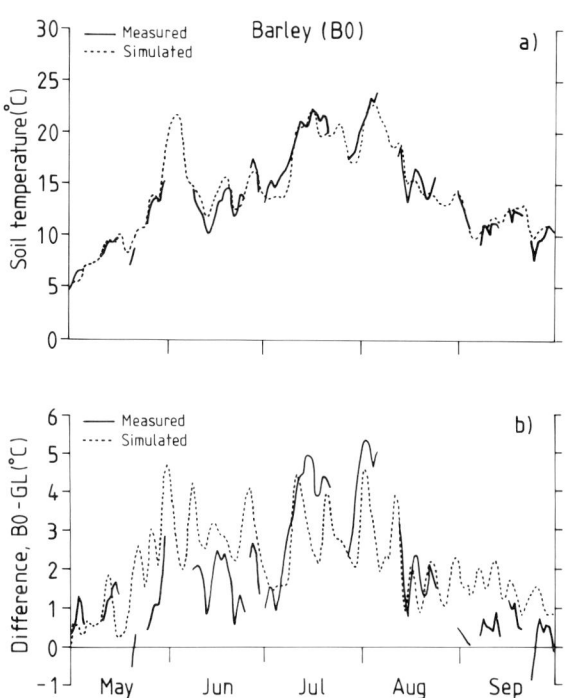

Fig. 3.5. Measured and simulated soil temperature during 1982 at a 5 cm depth in B0 (a) and measured and simulated soil temperature differences at a 5 cm depth between B0 and GL (b).

vertical flow to deep groundwater and S is the change of water storage in the system. When plots are small, as in this project, simple measurements of the components in the water balance equation are difficult. The water balance equation may instead be divided into separate equations for different subsystems, such as snow, soil and groundwater.

The distribution of water within a subsystem is important in ecological studies. An equation for water in a soil profile, which combines the Darcy equation and the law of mass conservation, can be written as:

$$\frac{\delta \Theta}{\delta t} = - \frac{\delta}{\delta z}\left[k(\Theta)\left(\frac{\delta \psi}{\delta z} + 1\right)\right] - s(t) \quad (4)$$

where the left side of the equation represents the temporal change of soil water content, Θ, and the right side represents the depth gradient of vertical water flow and the sink, $s(t)$. The water flow is calculated by the Darcy equation using the unsaturated conductivity ($k(\Theta)$) and the depth gradient of water tension ($\delta\psi/\delta z$) and the gravitational effect (= unity). Providing the appropriate boundary conditions to Eq. (4) are given, numerical methods can be used to solve the equation. The SOIL model (Fig. 3.1), as well as most other physically based soil-water models, are based on Eq. (4).

A detailed description of the SOIL model was presented by Jansson and Halldin (1980), and only a few important characteristics are described here. The water retention curve and the saturated and unsaturated conductivity, are treated with analytical functions according to Brooks and Corey (1964) and Mualem (1976). Calculation of potential transpiration, i.e., the sink term in Eq. (4), is based on the Penman combination formula as presented by Monteith (1965). Actual water uptake by roots from each layer is calculated according to an assumed depth distribution and an empirical reduction function accounting for water tension, rate of uptake demand and soil temperature.

Ground-water flow to the drainage pipes, like uptake of water by roots, is considered a sink. The flow rate, q_p, from each saturated layer above the depth of the drainage tiles is calculated as:

$$q_p(i) = z(i)\, k_s(i)\, (z_{gw} - z_p)/d \quad (5)$$

where z is the thickness of the layer, k_s is the saturated conductivity, z_p is the depth of the drainage tiles, z_{gw} is the depth of the ground water table and d is the distance between drainage tiles.

In addition to the water flow equation, the heat flow equation must be used for simulations during winter when water in the soil freezes. These interactions between water and heat flows are important for the redistribution of water in the soil profile during freezing and for the calculation of infiltration when the soil is partially frozen.

The applicability of a model based on Eq. (4) is totally dependent on the accuracy of the soil and plant properties in the equation, as well on both the upper and lower boundary conditions of the soil profile.

Adaptation of the model

Previous applications of the SOIL model for simulation of soil moisture conditions at the Kjettslinge field have been reported by Jansson and Thoms-Hjärpe (1986), Alvenäs et al. (1986), Jansson (1986a) and Johnsson et al. (1987). The present application of the SOIL model differs from the previous applications mainly with respect to its emphasis on crop properties and the inclusion of all four cropping systems in the simulations.

Meteorological data from the field site along with supplementary information from nearby meteorological stations (Alvenäs et al. 1986) were used as driving variables. The parameters concerning soil physical properties were chosen according to Jansson and Thoms-Hjärpe (1986) and adjusted to the areal mean values in thicknesses of the different textural horizons in the plots at Kjettslinge (Steen et al. 1984).

Differences between cropping systems were also taken into account in the calculations of evapotranspiration and in the vertical distribution of root water uptake.

Annual dynamics of surface resistance (Fig. 3.6), which control potential transpiration and soil evaporation, were based on crop development and were ad-

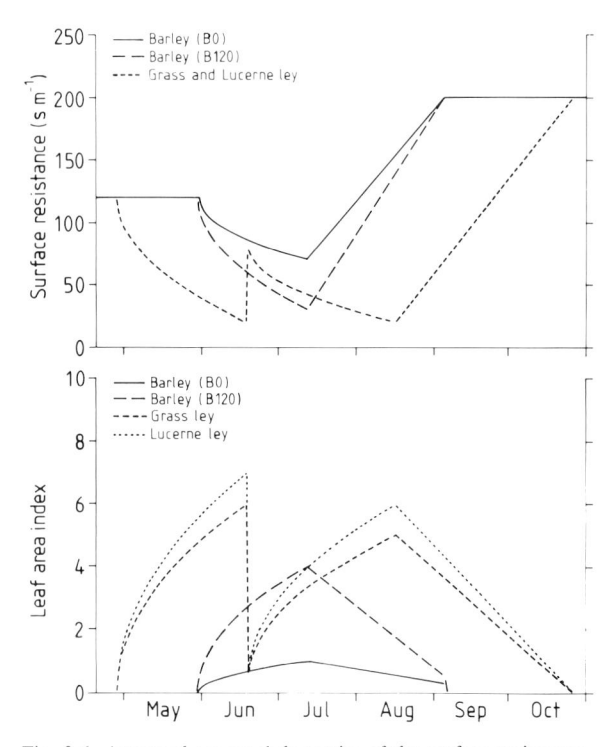

Fig. 3.6. Assumed seasonal dynamics of the surface resistance (top) and the leaf area index (bottom) for the four cropping systems.

justed to levels which resulted in an agreement with the relative differences found in observed drainage from the different cropping systems (Bergström 1987a). Evaporation of intercepted precipitation was treated separately with the same combination formula, using a low value for the surface resistance. The interception storage capacity was calculated from the leaf area index (Fig. 3.6), with a specific storage capacity of 0.5 mm per LAI unit as proposed by Jensen (1979). Aerodynamic resistance was calculated from the roughness length and the displacement height assuming the coefficients to be 10% and 70% of crop height, respectively. Measured above-ground biomass characteristics used for the calculations are reported in Chapter 4 and by Pettersson (1987) and Pettersson (1989).

Differences in the depth distribution of water uptake from the soil between crops were based on root biomass and distribution estimates reported by Hansson et al. (1987) and Hansson and Andrén (1987). Root development of barley started at sowing in the end of May, reached a maximum in the beginning of August and then decreased until the crop was harvested in the beginning of September. No seasonal differences in root depth were assumed for the leys. The vertical distribution of roots used in the calculations indicate shallower water uptake for the barley cropping systems and the grass ley than for the lucerne ley (Fig. 3.7).

Drainage was assumed to occur only through the drainage tiles. The fact that the subsoil below the tiles was a very heavy clay and that there was a slight tendency for an upward pressure gradient from the subsoil (Bergström 1987a) justifies this assumption.

Annual water balance

During 1981 to 1985, the mean annual precipitation (corrected for wind losses) was 610 mm. The simulations demonstrated interesting differences in the other water balance components between the four cropping systems.

The mean annual evapotranspiration showed a systematic difference between the cropping systems, with the highest value for the lucerne ley, followed by the grass ley, N fertilized barley and barley without N fertilization (Fig. 3.8). This was mainly due to differences in leaf-area-index and the corresponding differences in interception losses. The annual sum of transpiration and soil evaporation showed only slight, but similar, differences between cropping systems.

The mean annual runoff from the different cropping systems showed a reverse tendency, with the highest runoff from barley without N fertilization and the lowest from the lucerne ley (Fig. 3.8). The runoff was divided into surface runoff and drainage through the tiles, and the latter was measured enabling the model to be tested. The simulated differences between the cropping systems were similar to the measurements (Bergström 1987a) both from the lysimeters and the tile-drained plots. However, the simulated drainage was systematically higher than the measurements from tile-drained plots and, to a similar magnitude, they were lower than the measurements from the lysimeters (not shown).

A major uncertainty in the simulated runoff was the surface runoff, which was unsatisfactory when considering the high portion of water leaving the soil as surface runoff. The fact that little or no drainage was observed from the tiles during rapid snow melt was indirect evidence for the reliability of simulated surface runoff. There were no differences in simulated surface runoff between the cropping systems. However, it is not unreasonable to assume there were differences between the leys and the two barley cropping systems since barley was ploughed every autumn.

The differences in actual transpiration between years were quite small in spite of substantial differences in potential transpiration between years (Fig. 3.9a). The variation in potential transpiration had a minor influence on the annual sum of actual transpiration. The highest actual transpiration did not occur during the year with the highest potential transpiration, because of the differences in availability of soil water. During 1982 and 1983 soil moisture limited transpiration substantially, whereas the most important factor reducing transpiration during the other years was the high evaporation losses from intercepted water and the low soil temperatures, especially during winter and spring. The interception losses were high during years with wet growing seasons.

Compared with the relatively small differences in actual evapotranspiration between years, total runoff, especially drainage through the tiles, showed substantial variation between years (Fig. 3.9b). The variation in total runoff was mainly caused by differences in the amount of precipitation and snow melt between years.

The dynamics in the proportion of surface runoff versus drainage through tiles were complex. Simulated surface runoff only occurred during snow melt or precipitation in late autumn or early winter (Fig. 3.10). Differences between such events were substantial, depending on the degree of frost in the soil. During snow melt in the spring of 1982, when the thickest snowpack

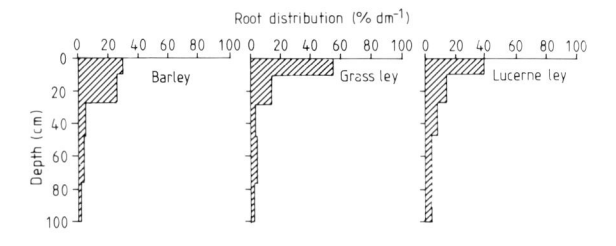

Fig. 3.7. Assumed vertical distribution of potential root uptake, at time of maximum root depth, for the three cropping systems. Note that the unit (% dm^{-1}) is necessary because of varying layer thickness.

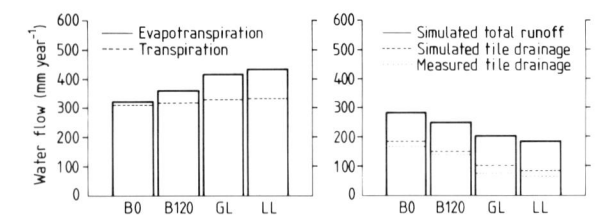

Fig. 3.8. Simulated mean values (1981–1985) of annual totals of transpiration and soil evaporation, evapotranspiration, tile drainage and total runoff. Measured drainage from the tile-drained plots is also included (Bergström 1987a).

existed, nearly all water infiltrated into the soil and resulted in drainage through the tiles. The situation was more or less the opposite in 1983 and 1984, when the soil was frozen and there was surface runoff of water. The simulated partitioning between surface runoff and infiltration into the soil was very sensitive to how the conductivity functions were defined for the different soil horizons in the model. Also the division of the soil profile into discrete layers was crucial when the soil was partially frozen, probably because of steep gradients in water tensions combined with large differences in hydraulic conductivity between layers. It is important to point out that even below 0°C soil water is only partially frozen and significant water flow can occur, especially when the temperature is close to 0°C.

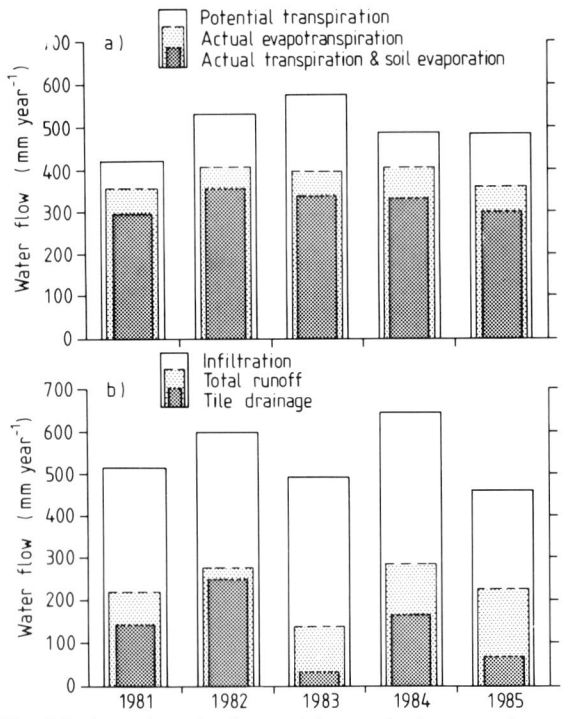

Fig. 3.9. Annual totals of potential transpiration, actual evapotranspiration, actual transpiration and soil evaporation (a), infiltration into soil, total runoff and drainage through the tiles (b). All variables are simulated means from all four cropping systems.

Soil water-flow paths

When simulated drainage was tested against drainage measurements, no information on the flow paths existed. The layered soil profile at Kjettslinge, with a mixture of sand and clay horizons (Chapter 2), makes flow paths a crucial consideration especially for leaching studies. The first assumption used was that a low saturated conductivity in the clay horizon below the sand resulted in a dominant role of the sand layer in determining the drainage of the entire soil profile (Fig. 3.11a). An analysis of the dynamics in soil nitrate storage (Johnsson et al. 1987; see also Ch. 7) indicated that the saturated conductivity in the clay was underestimated. When the possibility of macropores in the clay were accounted for by increasing the saturated conductivity from 2.4 to 144 mm d^{-1}, a substantially deeper flow path was simulated (Fig. 3.11b), although the total drainage was unchanged. The overall water balance was not influenced by the change in the saturated conductivity in the clay since only the conductivity of the macropores was altered. Unsaturated conductivities at tensions greater than 50 cm of water were not affected.

Soil moisture conditions

Two years, one wet (1981) and one dry (1983), were selected to illustrate typical differences in soil moisture conditions between the cropping systems. Previously, Jansson and Thoms-Hjärpe (1986) analyzed differences between fertilized and unfertilized barley for the same periods. In addition, Alvenäs et al. (1986) presented a comparison between one simulation and measurements made in the different plots and cropping systems from 1981 to 1984. All comparisons between simulated and measured tensions in the plough layer were limited to a 15 cm depth. Even though a number of systematic differences were indicated, it was difficult to quantify these differences between cropping systems because of the substantial spatial variability of soil properties between cropping systems and because of the short periods with reliable tensiometer measurements. Severe frost risk at Kjettlinge and a rapid drying out of the soil, causing tensions above the possible measurement range, reduced the duration of the tensiometer measurements. Neutron probe measurements were available some years but they had low measurement frequency and were limited in their ability to represent the uppermost horizons of the soil profile.

Some analyses were made of soil water contents from gravimetric measurements in the uppermost 10 cm of the top soil, in connection with the soil sampling for denitrification studies. Because of an irregular sampling program it was not possible to compare all cropping systems on the same dates. Instead, concurrent data from N fertilized barley and grass ley and from lucerne ley and grass ley were used (Fig. 3.12). The lowest water contents were found in the grass ley throughout

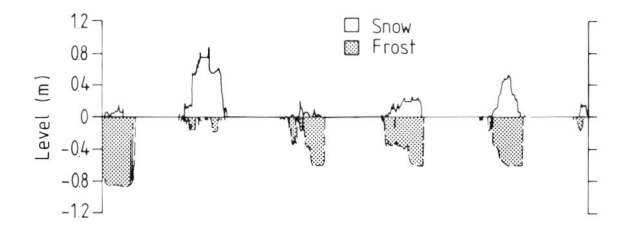

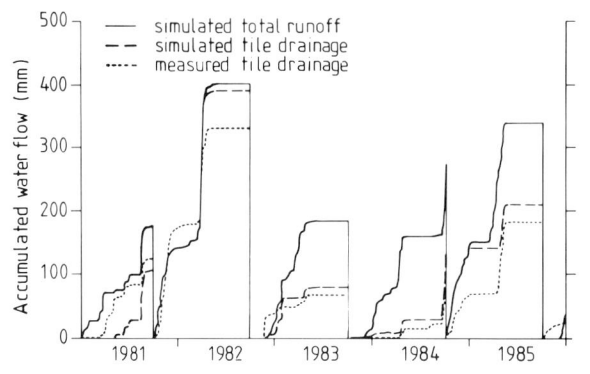

Fig. 3.10. Simulated snow and frost depths (top), simulated total runoff and tile drainage and measured tile drainage for B0 (bottom). The water flows are annual accumulations from 1 October to 30 September.

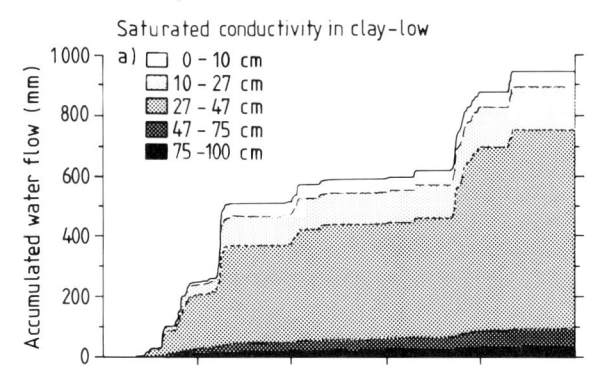

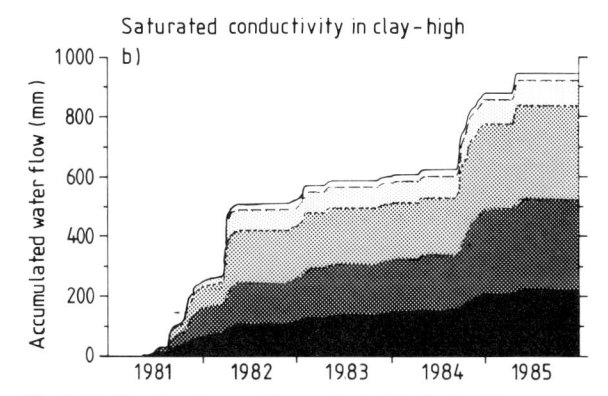

Fig. 3.11. Simulated water flow path, in B0 obtained by assuming a low (a) and a high (b) saturated hydraulic conductivity in the clay horizons from 47–100 cm depth. The uppermost solid line represents the total accumulated drainage from the entire soil profile, 0–100 cm depth.

the year. The lucerne ley was slightly moister than the grass ley in all periods, whereas N fertilized barley was slightly moister than the grass ley in summer and substantially wetter in spring and autumn.

The soil moisture tensions (Fig. 3.13) are the results of the adaptation of the model to agree with the observed annual amounts of drainage, and were thereby governed by the assumed properties of the different crops. The plough layer was divided into two layers, 0–10 and 10–27 cm deep, but no efforts were made to include the differences simulated at the soil surface using the energy balance approach (Figs 3.3 to 3.5). The results and especially the differences between the two horizons of the plough layer should therefore be interpreted with care.

In the first simulations, when the vertical distributions of water uptake-demand presented in Fig. 3.7 were used, there was a slight tendency for the lucerne ley to be drier than the grass ley. This should be compared with the opposite tendency in the measurements (Fig. 3.12). The discrepancy was probably caused by larger differences in root distributions and activities between the two leys than first assumed in the simulations. To force the model to simulate the drier conditions in the top 0–10 cm of grass ley, a uniform water uptake demand from all depths in the uppermost metre had to be used for the lucerne ley (Fig. 3.13). This does not correspond to the measurements of vertical belowground biomass distributions (Pettersson et al. 1986) but the root water uptake may not necessarily be directly coupled to below-ground biomass. First, it may be assumed that more than 50% of the biomass in the top 10 cm in lucerne ley consists of stem bases etc. (A.-C. Hansson pers. comm.). Second, numbers of roots, which is a better measure of root activity, were more evenly distributed throughout the profile down to 1 m with a mean depth of around 40 cm (Hansson and Andrén 1987).

The leys were substantially drier than the barley cropping systems especially during spring and autumn. During the dry summer of 1983 the tension became higher in the fertilized barley than in the grass ley. Otherwise tensions were always higher for the leys. The unusual situation in 1983, when tension became higher in fertilized barley than in the grass ley, was caused by the infrequent precipitation which resulted in a more important role for transpiration losses than for interception losses than was normal. Differences between the leys and the barley cropping systems were mainly the result of different interception losses (cf. leaf area index, Fig. 3.6), since only small differences were assumed in the surface resistance for transpiration during the height of the summer (cf. Fig. 3.6).

Periods when water tensions exceeded 800 cm water (−80 KPa) indicate periods when transpiration was reduced. The results indicated that transpiration from the leys was also reduced for the wet year of 1981 and that a substantial reduction occurred in the barley cropping systems during the dry summer of 1983. No reduction of transpiration, due to high moisture tension, occurred before the beginning of June or after the middle of September for any of the cropping systems.

Concluding remarks

Modelling has been an important part of the abiotic studies within the project. One of the most important outcomes of the modelling effort was realizing the need for more detailed experimental studies, as demonstrated by the lack of opportunities to thoroughly test different parts of the models. On the other hand, the models have been used as tools to extend information from the existing measurements to cover both new time periods and new variables. One example was the construction of continuous time series of soil temperatures using a combination of model output and measurements

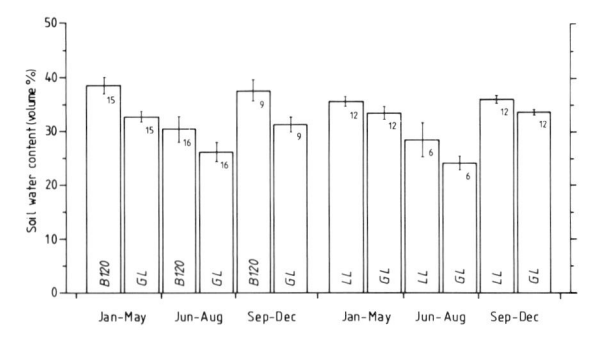

Fig. 3.12. Measured soil water content, at 0–10 cm depth, during different periods in 1983. Mean values were calculated from treatment means at occasions with concurrent sampling. At each occasion treatment means were normally calculated from the four blocks A–D. The standard error is shown by bars and the number of sampling occasions is listed for each cropping system and period.

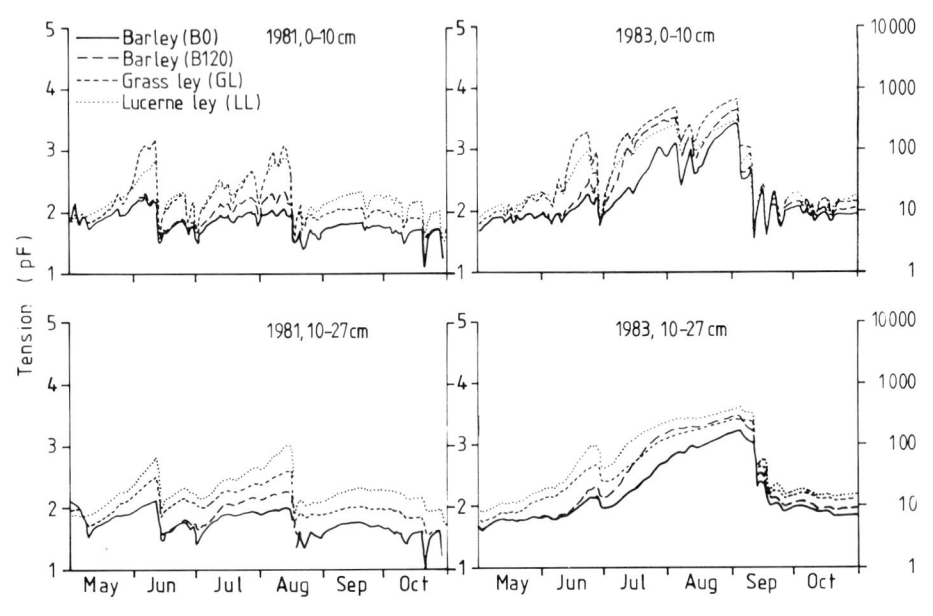

Fig. 3.13. Simulated soil water tension in the top soil (0–10 cm depth, upper figures and 10–27 cm depth, lower figures) during the wet growing season of 1981 and the dry growing season of 1983.

(see Alvenäs et al. 1986). Another example was the simulation of nitrogen dynamics which was based on continuous time series of both temperature and moisture variables generated with the water and heat model (Johnsson et al. 1987; Chapter 7). We have demonstrated how soil temperature measurements can be used to analyse evaporation and sensible heat flow at the soil surface and how drainage measurements can be used in the adaptation of a soil water model to the different cropping systems. The lack of detailed abiotic measurements caused the greatest problem when we were trying to evaluate differences between the cropping systems. The temporal behaviour of many variables was more easily simulated because of their strong correlation with meteorological conditions, which are comparatively easy to measure.

To analyse water movements in partially frozen soils, a new project was initiated involving detailed measurements of water in the experimental field during winter (Lundin 1989). Other pertinent studies for increasing our quantitative knowledge of abiotic processes are: water movements in macropores, seepage of water below the depth of the drainage tiles, combined water and heat flow in the air adjacent to the soil surface, moisture conditions close to the soil surface, mechanisms for water uptake by roots and the stomatal control of transpiration.

4. Structure of the agroecosystem

Ann-Charlotte Hansson, Olof Andrén, Sven Boström, Ullalena Boström, Marianne Clarholm, Jan Lagerlöf, Torbjörn Lindberg, Keith Paustian, Roger Pettersson and Björn Sohlenius

4. **Structure of the agroecosystem.** A.-C. Hansson. O. Andrén, S. Boström, U. Boström, M. Clarholm, J. Lagerlöf, T. Lindberg, K. Paustian, R. Pettersson and B. Sohlenius

Introduction	43
Methods	43
Components	47
The crop	47
Herbivores	51
Soil microorganisms, including protozoa	51
Nematodes	55
Microarthropods and enchytraeids	57
Soil and soil surface macroarthropods	60
Earthworms	61
Seasonal dynamics of the components	61
The crop	62
Herbivores	63
Litter	63
Decomposer organisms	64
Soil microorganisms, including protozoa	64
Nematodes, including root-feeders	70
Microarthropods and enchytraeids	71
Soil and surface macroarthropods	71
Earthworms	74
Long-term dynamics of the components	74
Litter	76
Herbivores	76
Decomposer organisms	76
Soil microorganisms, including protozoa	76
Nematodes, including root-feeders	78
Microarthropods and enchytraeids	79
Soil and surface macroarthropods	82
Earthworms	82
Summary	83

Introduction

Agroecosystems are composed of interacting biological and physical components. The components discussed in this chapter follow the conventional division into producers, consumers and decomposers (Fig. 4.1). Although this division is inadequate for describing the functional organization of soil organisms except at the broadest level of resolution (Heal and Dighton 1985), it provides a convenient basis to examine the structure of the agroecosystem. The next level of organization in this chapter is taxonomic and functional, which facilitates a quantitative description. Although individual species are not emphasized in the text, all species identified in the Kjettslinge field are listed in Tab. 4.1 for reference. Agroecosystems are generally described as depleted systems with a low diversity (Odum 1984), but they are still far from simple.

Characterizing the structure of an ecosystem requires estimates of abundance and types of organisms present and their spatial and temporal relationships (Richards 1974). In addition, information about the amounts and distribution of non-living organic as well as inorganic substances in the system is necessary. All systems, independent of treatment, are limited by their physical and chemical environment.

The systems studied in this project represent typical cropping systems of modern Swedish agriculture, i.e., perennial leys vs a spring-sown annual cereal and a fertilized grass ley vs a nitrogen-fixing legume ley. In addition to having different growth patterns, these crops also represent different agricultural practices, i.e., they are sown and cut at different times and tilled with different intensities.

The different cropping systems result in differing resource quality and quantity for decomposers, as well as differing physical environments. As a consequence, structural changes of the microflora and fauna were expected (Usher 1985, Heal and Dighton 1985). In other words, we hypothesized that different ecosystem "structures", in terms of number and biomass of organisms would develop within each system. To test this hypothesis, all organisms were sampled repeatedly during the growing seasons to follow fluctuations in number and biomass.

The field samplings were coordinated so that all organism groups in fertilized barley (B120) and meadow fescue (GL) were sampled monthly during the 1982 growing season. Furthermore, from 1980 to 1984 all cropping systems were sampled in September in order to follow long-term dynamics of all organism groups (Tab. 4.2). In addition, frequent samplings, concerning, e.g., seasonal dynamics of certain organisms, were performed within various subprojects.

The average values presented in this chapter are based on samplings from all years. Consequently, they differ in some cases from the results used in calculating carbon and nitrogen budgets (Chapters 5 and 7) which are based on measurements from 1982 and 1983 when the cropping system differences were more established and intensively sampled. The results in this chapter concerning seasonal dynamics are mainly from 1982 and 1983 if not otherwise stated in the text. Finally, the long-term dynamics are based on the September samplings 1980 to 1984.

If not otherwise stated, all results presented in this chapter are from sampling depths listed in Tab. 4.3. The results are normally expressed as dry mass m^{-2}, and exceptions are indicated in the text.

Methods

The descriptions of the methods used have been made as brief as possible and are presented in Tab. 4.3 and 4.4. For the interested reader, more detailed information on the methodology is available in the indicated references. Field sampling equipment, the number of samples per plot and the sampling depths are summarized in Tab. 4.3. Sample processing for each organism group is summarized in Tab. 4.4. When the development of a method has been a prerequisite for an experiment, the method is discussed in more depth in the relevant section.

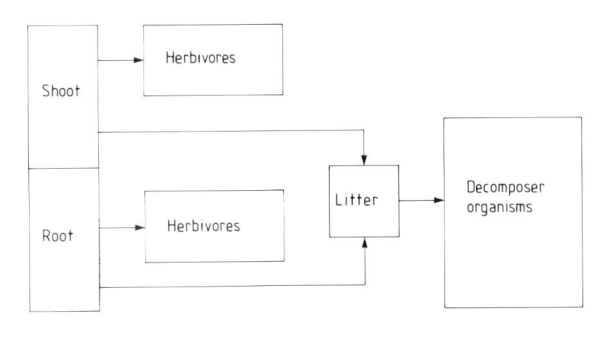

Fig. 4.1. The main components (boxes) and material flows (arrows) in an agroecosystem.

Tab. 4.1. List of identified heterotrophic organisms at Kjettslinge. All organisms excepting microorganisms and protozoa were, if possible, identified to species level. [1] = found in suction trapping (Curry 1986, J. Curry pers. comm.), [2] = found at soil surface or in soil using dry funnel extraction (Lagerlöf and Andrén 1988, Lagerlöf and Scheller 1989, J. Lagerlöf unpubl.).

BACTERIA
FUNGI

PROTOZOA
Rhizopoda
Flagellates
Ciliates

NEMATODA
TYLENCHIDA
Aphelenchoides spp.
Aphelenchus avenae
Boleodorus sp.
Cyst nematode sp.
Ditylenchus spp.
Helicotylenchus canadensis
H. sp.
Merlinius brevidens
M. microdorus
Neotylenchus sp.
Pratylenchus crenatus
P. fallax
P. sp.
Paratylenchus spp.
Psilenchus sp.
Tylenchorhynchus dubius
T. maximus
Tylenchus s.l. spp.
RHABDITIDA
Acrobeles ciliatus
Acrobeloides nanus
Acrobelophis minimus
Acrolobus emarginatus
Cephalobus persegnis
Cervidellus serratus
Chiloplacus sp.
Eucephalobus mucronatus
E. oxyuroides
E. striatus
Heterocephalobus sp.
Panagrolaimus rigidus
Rhabditis s.l. spp.
DORYLAIMIDA
Aporcelaimellus sp.
Clarkus papillatus
Diptherophora sp.
Dorylaimellus sp.
Eudorylaimus spp.
Mesodorylaimus spp.
Mylonchulus brachyuris
Oxydirus sp.
Pungentus spp.
Trichodorus primitivus
Tylencholaimellus striatus
OTHER NEMATODA
Achromadora sp.
Alaimus spp.
Bastiania sp.
Cylindrolaimus sp.
Monhystera spp.
Plectus sp.
Prismatolaimus sp.
Wilsonema sp.

INSECTA
HEMIPTERA
Aphididae
Acyrthosiphon pisum[1]

Corylobium sp.[1]
Metopolophium dirhodum[1]
Rhopalosiphum padi[1]
Sitobion avenae[1]
Cicadellidae
Arthaldeus pascuellus[1]
Diplocolenus abdominalis[1]
Empoasca sp.[1]
Macrosteles laevis[1]
Cicadellidae spp.[1]
Streptanus aemulans,[1]
Delphacidae
Dicranotropis hamata[1]
Javesella pellucida[1]
Lygaeidae
Taphropeltus sp.[1]
Miridae
Chlamydatus saltidans[1]
Lygocoris limbatus[1]
Lygus rugulipennis[1]
Neomecomma sp.[1]
Plagiognathus arbustorum[1]
P. chrysanthemi[1]
Trigonotylus sp.[1]
Tytthus sp.[1]
Nabidae
Dolichonabus limbatus[1]
Nabis ferus[1]
Myrmecoris gracilis[1]
Pemphigidae[2]
Piesmidae
Piesma maculatum[1]
Psyllidae
Aphalara polygoni[1]
Aphalaridae sp.[1]
Saldidae
Saldula saltatoria[1]
THYSANOPTERA[2]
Aelothripidae
Aelothrips sp.[1]
Phlaeothripidae
Haplothrips sp.[1]
Thripidae
Anaphothrips obscurus[1]
Chirothrips hamateus[1]
Frankliniella tenuicornis[1]
Limothrips denticornis[1]
Oxythrips sp.[1]
Thrips sp.[1]
HYMENOPTERA
Braconidae[1]
Cephidae[1]
Ceraphronidae[1]
Chalcidoidea[1]
Cynipoidea[1]
Ichneumonidae[1]
Myrmaridae[1]
Proctotrupoidea[1]
DIPTERA
Agromyzidae[1]
Anisopodidae[1]
Anthomyidae[1]
Anthomyzidae[1]
Asteiidae[1]
Bibionidae[2]
Calliphoridae[1]
Cecidomyidae[12]

Ceratopogonidae[12]
Chironomidae[12]
Smittia aterrima[2]
Bryophaenocladius cfr. *illimbatus*[2]
Gymnometriocnemus sp.[2]
Metriocnemus sp.[2]
Pseudosmittia sp.[2]
Limnophyes sp.[2]
Chloropidae[1]
Culicidae[1]
Dolichopodidae[12]
Drosophilidae[1]
Empididae[12]
Ephydridae[1]
Lauxaniidae[1]
Lonchaeidae[1]
Lonchopteridae[12]
Micropezidae[1]
Muscidae[12]
Opomyzidae
Opymyza sp.[1]
Geomyza tripunctata[1]
G. sp.[1]
Phoridae[1]
Psilidae[1]
Psychodidae[12]
Scatopsidae[12]
Sciaridae[12]
Sepsidae[12]
Sphaeroceridae[1]
Syrphidae[12]
Tabanidae[2]
Therevidae[2]
Tipulidae[12]
Trichoceridae[2]
COLEOPTERA
Cantharidae[2]
Malthodes sp.[1]
Psilothrix sp.[1]
Carabidae[2]
Agonum dorsale[2]
Bembidion lampros[12]
B. obtusum[1]
B. guttula[2]
B. quadrimaculatum[12]
B. spp.[2]
Clivina fossor[2]
Heliophorus sp.[2]
Tachyporus chrysomelinus[2]
T. obtusus[2]
Trechus quadristriatus[1]
T. secalis[1]
Chrysomelidae
Chaetocnema concinna[1]
Chrysolina sp.[1]
Lema melanopus[2]
Longitarsus sp.[1]
Phyllotreta flexuosa[1]
P. nemorum[1]
P. ochripes[1]
P. undulata[1]
P. vittula[1]
Coccinellidae
Coccidula rufa[1]
Coccinella quinquepunctata[1]
C. septempunctata[12]
Propylea quattordecimpunctata[1]

Cryptophagidae
Atomaria sp.[1]
Ephistemus sp.[1]
Curculionidae
Apion spp.[1]
Barypeithes sulcifrons[1]
Ceutorhynchus assimilis[1]
Ceutorhynchus sp.[1]
Phytobius sp.[1]
Sitona hispidulus[1]
S. lineatus[1]
Tychius sp.[1]
Elateridae
larvae[2]
Hydrophilidae
Helophorus sp.[2]
Lathridiidae
Corticaria sp.[1]
Enicmus histrio[1]
Lathridius sp.[1]
Nitidulidae
Kateretes pedicularis[1]
Meligethes aeneus[1]
Pselaphidae[2]
Ptiliidae[2]
Silphidae
Aclypea opaca[2]
Staphylinidae[2]
Acrulia sp.[1]
Aloconota insecta[1]
Amischa analis[1]
Anotylus rugosus[1]
Atheta fungi[1]
Lathrobium pallidum[1]
L. sp.[2]
Oligota sp.[1]
Othius laeviusculus[1]
Oxytelus sp.[2]
Philonthus sp.[1]
Quedius sp.[1]
Staphylinus sp.[1]
Stenus flavipes[1]
Tachinus sp.[1]
Tachyporus chrysomelinus[1]
T. hypnorum[1]
T. obtusus[1]
Trogophloeus sp.[1]
COLLEMBOLA
Entomobryidae
Entomobrya nicoleti[1]
Lepidocyrtus lanuginosus[1]
Pseudosinella alba[12]
P. decipiens[12]
P. immaculata[1]
Tomocerus sp.[1]
Willowsia platani[2]
Isotomidae
Folsomia fimetaria[12]
F. quadrioculata[2]
Isotoma notabilis[2]
I. viridis[12]
Isotomiella minor[12]
Proisotoma minima[2]
Onychiuridae
Onychiurus armatus s. lat.[2]

Tullbergia krausbaueri[2]
Poduridae
Friesea mirabilis[2]
Hypogastrura denticulata[12]
Sminthuridae
Arrhopalites caecus[2]
Bourletiella hortensis[12]
Deuterosminthurus sulphureus[1]
Neelus minimus[12]
Sminthurides malmgreni[1]
Sminthurinus aureus[1]
S. elegans[1]
Sminthurus nigromaculatus[1]
S. viridis[1]
S. spp.[2]
LEPIDOPTERA[12]
Geometridae[2]
Hepialidae[2]
Noctuidae[2]
Nymphalidae
Aglais urticae[1]
Inachis io[1]
Psychidae[2]
NEUROPTERA[1]
ORTHOPTERA[2]
PSOCOPTERA[12]
DERMAPTERA[12]
DIPLURA[12]
PROTURA[12]
TRICHOPTERA[2]
ACARI
Astigmata
Anoetidae sp.[2]
Schwiebia talpa[2]
Tyrophagus longior[1]
Tyrophagus palmarum[2]
Acaroidea spp.[2]
Mesostigmata
Gamasina
Alliphis siculus[12]
Amblygamasus stramenis[2]
Amblyseius cucumeris[1]
A. graminis[1]
A. obtusus[1]
Ameroseius echinatus[1]
A. sp.[2]
Androlaelaps sp.[1]
Arctoseius cetratus[12]
A. venustulus[2]
Dendrolaelaps strenzkei[2]
Holoparasitus sp.[1]
Hypoaspis aculeifer[2]
Iphidozercon sp.[1]
Lasioseius berlesei[12]
Paragamasus lapponicus[2]
P. runciger[1]
Parasitus sp.[1]
Pergamasus septentrionalis[1]
P. sp.[1]
Rhodacarellus silesiacus[2]
R. apophyseus[2]
R. sp.[2]
Rhodacarus coronatus[2]
R. sp.[2]
Sejus borealis[12]

Veigaia exigua[2]
V. nemorensis[2]
Gamasina sp.[2]
Uropodina
Nenteria breviunguiculata[2]
Olodiscus sp.[2]
Uropoda sp.[1]
Cryptostigmata
Brachychthonius sp.[2]
*Carabodes minusculus*1
Diaptoribates humeralis[1]
Hypochthonius sp.[2]
Liebstadia similis[1]
Nanhermannia sp.[2]
Oppia spp.[2]
Punctoribates punctum[1]
Tectocepheus sarekensis/velatus[2]
Cryptostigmata larvae/nymphs spp.[2]
Prostigmata
Alicorhagia sp.[2]
Anystis sp.[1]
Bryobia praetiosa[1]
Eriophyidae[12]
Erythraeidae[1]
Eupodes spp.[12]
Nanorchestidae[2]
Pachygnathidae[2]
Pygmephorus spp.[2]
Rhagidia sp.[12]
Raphignathidae[12]
Scutacaridae[12]
Siteroptes graminum[12]
Stigmaeidae[2]
Tarsonemidae[2]
Trombidiidae[1]
Thoribdella lapidaria[1]
Tydeus sp.[1]

ARANEAE[12]
OPILIONES[2]
CHILOPODA[2]
DIPLOPODA
PAUROPODA
Allopauropus cuenoti[2]
A. gracilis[2]
A. vulgaris[2]
A. verticillatus[2]

SYMPHYLA
Symphylella vulgaris[2]
Scolopendrellopsis subnuda[2]
OLIGOCHAETA
ENCHYTRAEIDAE
Enchytraeus buchholzi
Fridericia bulbosa
Henlea ventriculosa
LUMBRICIDAE
Allolobophora/Aporrectodea caliginosa
A. longa
A. rosea
Lumbricus terrestris
L. castaneus
L. rubellus
GASTROPODA
Limax spp.

Tab. 4.2. Sampling program for the organism groups at Kjettslinge. Sampled with soil cores (1), metal frames (2), formalin application (3), back-pack motor fan (4), scissors (5). A = B0, B = B120, C = GL and D = LL.

Year Organism group	April A B C D	May A B C D	June A B C D	July A B C D	Aug A B C D	Sept A B C D	Oct A B C D	Nov A B C D	Dec A B C D
1980 Bacteria, fungi, protozoa (1)						x x x x			
Nematodes (1)						x x x x	x x x x		
Enchytraeids, microarthropods (1)				x x x x		x x x x			
Earthworms (2) (3)									
Macroarthropods (2)		x x			x				
Above-ground arthropods (4)			x x x x	x x x x	x x x x	x x x x	x x		
Above-ground crop (5)			x x x x	x x x x	x x x x	x x x x	x x		
Below-ground crop (1)						x x x x			
1981 Bacteria, fungi, protozoa (1)						x x x x			
Nematodes (1)		x x	x x	x x		x x x x			
Enchytraeids, microarthropods (1)		x x x x	x x x x			x x x x			
Earthworms (2) (3)		x x							
Macroarthropods (2)		x x x			x x x		x x x x	x x x x
Above-ground arthropods (4)							x	x	
Above-ground crop (5)		x	x x x	x x x	x x x	x x x x	x		
Below-ground crop (1)		x	x x x x	x x x x x	x x	x x x x	x	x x	x x
1982 Bacteria, fungi, protozoa (1)	x x	x x	x x x x	x x x x x	x x	x x x x	x x	x x	x x
Fumigation biomass (1)		x x	x x	x x	x x	x x x x	x x	x x	x x
Nematodes (1)		x x	x x	x x		x x x x	x x		
Enchytraeids, microarthropods (1)	x x			x x	x x	x x x x		x x	
Earthworms (2) (3)		x x				x x x x			
Macroarthropods (2)	x x x x	x x x x x	x x x x x	x x x x x	x x x x x	x x x x	x x x x	x x x x x	x x x x x
Above-ground arthropods (4)		x x x x x	x x x x x	x x x x x	x x x x x	x x x x	x x	x x	
Above-ground crop (5)	x x	x		x	x	x x x x	x	x x	
Below-ground crop (1)	x x					x			
1983 Bacteria, fungi, protozoa (1)	x x	x x	x x	x x	x x	x x x x x	x x		
Fumigation biomass (1)						x			
Nematodes (1)						x x x x			
Enchytraeids, macroarthropods (1)						x x x x			
Earthworms (2) (3)			x x	x	x	x x x x	x		
Macroarthropods (2)	x x x x	x x x x	x x x x x	x x x x x	x x x x x	x x x x	x x x x	x x x x x	x x x x x
Above-ground arthropods (4)		x x x x	x x x x x	x x x x x	x x x x x	x x x x	x x	x x	
Above-ground crop (5)	x x	x		x	x	x x x x	x x	x x	
Below-ground crop (1)	x x	x		x	x	x	x	x	
1984 Nematodes (1)						x x x x	x		
Fumigation biomass (1)						x x	x x		
Earthworms (2) (3)		x x				x x	x x	x	
Above-ground crop (5)		x x x	x x x	x x x	x x x	x x x	x	x	
Below-ground crop (1)	x	x	x			x		x	
1985 Nematodes (1)	x		x	x					
Earthworms (2) (3)						x x			
Above-ground crop (5)		x x	x x	x x	x x	x x			
Below-ground crop (1)		x x	x x	x x	x x				

Tab. 4.3. Sampling methods used for the different groups of organisms at Kjettslinge.

	Above-ground crop	Below-ground crop	Bacteria, fungi, protozoa	Fumigation biomass	Nematodes	Micro-arthropods	Enchytraeids	Macro-arthropods	Earthworms	Herbivores
Equipment	Scissors	Soil corer	Soil corer	Soil corer	Soil corer	Soil corer	Soil corer	Metal frame	Formalin application/metal frames	Back-pack fan
Numbers of samples/plot	2–4	4–6	1	2–4	3	1–4	1–4	2	4/3	5
Diameter (cm)		7	3.2	3.2	2.3	3.35/6	6.5			
Area (m^2)	0.25							0.04	0.5/0.06 /0.25	0.46
Depth (m)		0.5–1	0.1	0.1	0.2	0.12–0.2	0.1	0.1		
References	Pettersson et al. 1986	Hansson and Andrén 1986	Schnürer et al. 1986a	Schnürer et al. 1985	Sohlenius et al. 1987	Lagerlöf 1987	Lagerlöf 1987	Lagerlöf unpubl.	Boström 1988a	Curry 1986

Components

The crop

The crop is the only primary producer of importance in normally managed agroecosystems. Consequently, the crop has a major influence as energy source for other organisms. It is also the major nutrient sink.

Energy becomes available to decomposers in the soil during the growing season by two main pathways: (1) via above-ground litter, produced from shedding of dead plant parts and spillage at harvest, which is either decomposed on the surface or transported below-ground by earthworms and (2) via carbon translocated to roots, which by rhizodeposition and decay after death deposit organic material in the soil. At ploughing root biomass and above-ground harvest residues (stubble and litter) are incorporated into the soil and become available for decomposition by soil organisms.

In this project, the cultivated plant was considered to be a treatment as well as an organism under study. The four crops in this experiment differ in many aspects, but the most important may be the differences in growth period, i.e., spring-sown annuals vs perennials.

All crops were sown in rows 12.5 cm apart, so their areal distribution was extremely regular in comparison with natural ecosystems. After emergence, a density of about 330 plants m^{-2} was recorded in barley, whereas single plants were not easily distinguished within the rows of meadow fescue or lucerne.

A barley plant consists of one main stem and several tillers. At harvest the main stem and some of the tillers have ears and the canopy is around 1 m high (Fig. 4.2). At normal harvest time each lucerne plant has several high stems with branches and leaves and is around 1 m high. Meadow fescue on the other hand, forms low tufts of grass which increase in diameter with time, and the soil surface becomes more densely covered than in barley and lucerne. The maximum height of the meadow fescue canopy was only around 60 cm.

In Sweden barley is a commonly grown crop on farms both with and without cattle. In a normal crop rotation barley is followed by a winter cereal or by a ley. Often leys are undersown in spring barley, i.e., the grass or legume seeds are sown a few days after sowing the barley. This is a normal practice, and reduces the occurrence of weeds in a young, open and slowly growing ley. The leys in this experiment were undersown in barley during the first experimental year (1980). Unfortunately, the leys were badly damaged during the first winter due to ice-burning and it was necessary to harrow and resow the ley plots the following spring (see Chapter 2 for details).

Phenology and canopy parameters such as density of tillers and ears, leaf area etc. are valuable when relating above-ground growth to environmental factors. Not until recently has comparable nomenclature been developed for root systems. Using this system individual roots can be identified, named and their appearance and branching related in time to morphological development above ground (Rickman et al. 1984, Belford et al. 1987). However, these measurements were taken infrequently within this project since the main objective of the root production studies was to estimate the amount of organic material supplied to the soil through the root system.

Grasses, e.g., barley and meadow fescue have two distinct root systems: the primary, or seminal system, which develops from the seed and the secondary, or nodal root system, which develops adventitiously from the lower nodes of the stem (Fig. 4.3 a, c). During growth several orders of root laterals develop from the axes, and seminal axes are sequentially replaced by new nodal axes eventually producing a highly-branched root system. The development of the nodal root system is correlated with above-ground tillering and since the

Fig. 4.2. Barley fertilized with 120 kg N ha^{-1} (B120) to the left and barley without nitrogen fertilization (B0) to the right in August 1984 (Photo J. Lagerlöf).

number of tillers is reduced by nitrogen deficiency (Gallagher and Biscoe 1978), the number of nodal roots is normally lower in an unfertilized crop than in a fertilized crop.

In lucerne the primary root evolves into a taproot (Fig. 4.3b), which thickens during growth as a result of cambial activity. The branching is normally less intensive in the top soil than in the subsoil and the upper part of the taproot has only a few short laterals (Weaver 1926, Kutschera 1960).

For root biomass determinations, soil samples were washed and separated into living and dead organic material. Flexible and light-coloured roots were considered living. The fraction "living roots" also included below-ground stem bases, seminal and nodal roots and *Rhizobium* sp. nodules. Dead organic material (DOM) consisted of dead roots and above-ground harvest residues, e.g., stubble and awns, incorporated into the soil at ploughing and more or less decomposed.

Roots were separated from soil by rinsing the root samples with water in sieves with 0.5 mm mesh screens, and organic and inorganic material with diameter <0.5 mm were removed with the washing water. On average the Kjettslinge soil contained 16 kg m^{-2} of soil organic

Tab. 4.4. Sample treatment and methods of analyses used for the different groups of organisms at Kjettslinge.

Organism group	Treatment	References
Above-ground crop	Sorting into living and dead parts, drying, weighing, combustion, nitrogen analysis	Pettersson et al. 1986
Below-ground crop	Separation from soil, sorting into below-ground biomass and dead organic material, drying, weighing, combustion, nitrogen analysis	Hansson and Andrén 1986
Bacteria	Counting and size estimation in acridine-orange stained soil smears. Conversion to biomass	Clarholm and Rosswall 1980
Fungi	Hyphal length measurement according to the agar film technique, fluorescein diacetate staining. Conversion to biomass.	Jones and Mollison 1948 Söderström 1977 Schnürer et al. 1986a
Protozoa	Counting, using a most probable number technique	Darbyshire et al. 1974 Clarholm 1981
Microbial biomass carbon and nitrogen	Chloroform fumigation incubation	Jenkinson and Powlson 1976 Schnürer et al. 1985 Lindberg et al. 1989
Nematodes	Wet funnel extraction, counting and biomass calculation.	Sohlenius 1979 Sohlenius et al. 1987 Andrássy 1956 Yeates 1979
Enchytraeids	Wet funnel extraction	O'Connor 1962
Microarthropods	Dry funnel extraction	Andrén 1985
Macroarthropods	Tullgren funnel extraction	Carter et al. 1985
Earthworms	Formalin watering, hand sorting	Boström 1988a

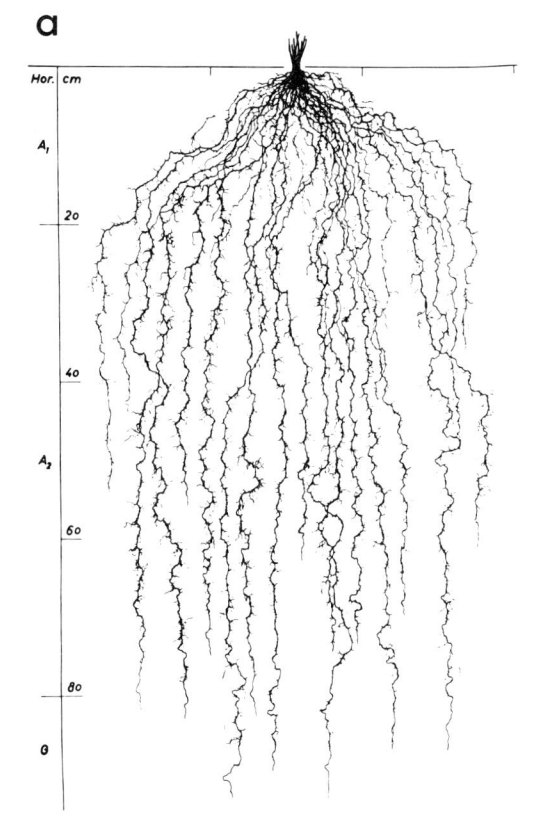

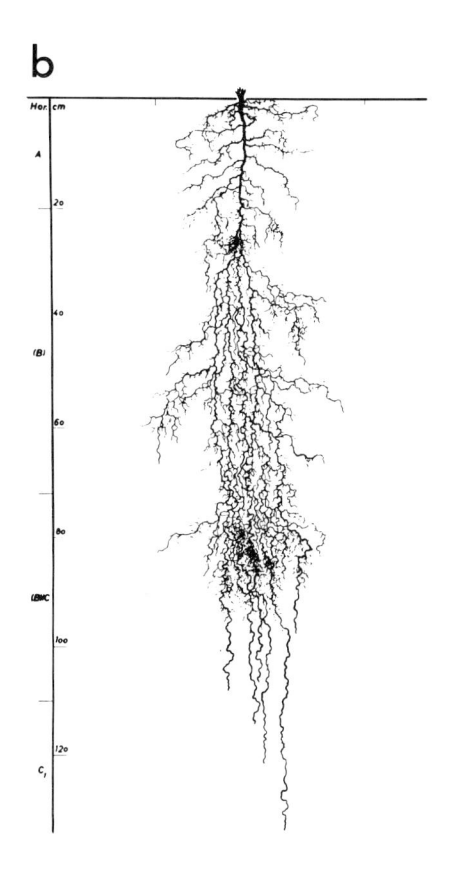

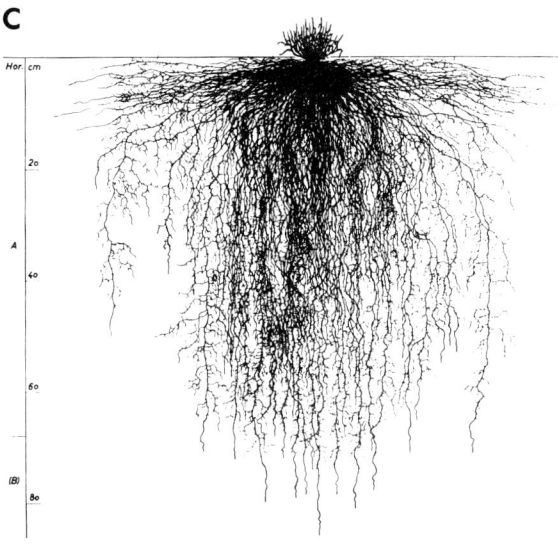

Fig. 4.3. Schematic illustrations of the root systems of a) barley, b) lucerne and c) perennial ryegrass (Kutschera 1960).

matter (SOM) in the top soil, and at the time of barley harvest in August it contained about 360 g m^{-2} of dead organic material that was coarse enough to be captured, i.e., the captured fraction was only about 3% of the total SOM (Fig. 4.4).

Roots of cereals and perennial grasses tend to be concentrated in the upper layers of the top soil (Troughton 1957). However, some cereal roots penetrate deep into the soil; for example, Kmoch et al. (1957) found roots of winter wheat at 4 m depth. The deepest cereal roots found in a Swedish study were 2 m below ground (Wiklert 1960). Lucerne roots also have the potential to grow into deep soil layers and they may penetrate to depths exceeding 9 m (Diekmahns 1972). However, most of the root biomass in lucerne is normally found in the upper 30 cm of the soil (Bolton 1962). In our study, the normal sampling depth was 0.5 m, but occasionally samples were taken down to 1 m, and some roots were found below 0.5 m (Hansson 1987). However, most of the root biomass (85% in barley, 85% in lucerne) was found in the top soil, 0–27 cm (Fig. 4.5). According to the results from a minirhizotron study at Kjettslinge, where transparent tubes were inserted into the soil for counting roots, mean root depth was around 30 cm for both B0 and B120, and 40 cm for lucerne. In the meadow fescue, however, the mean root depth was only around 20 cm (Hansson and

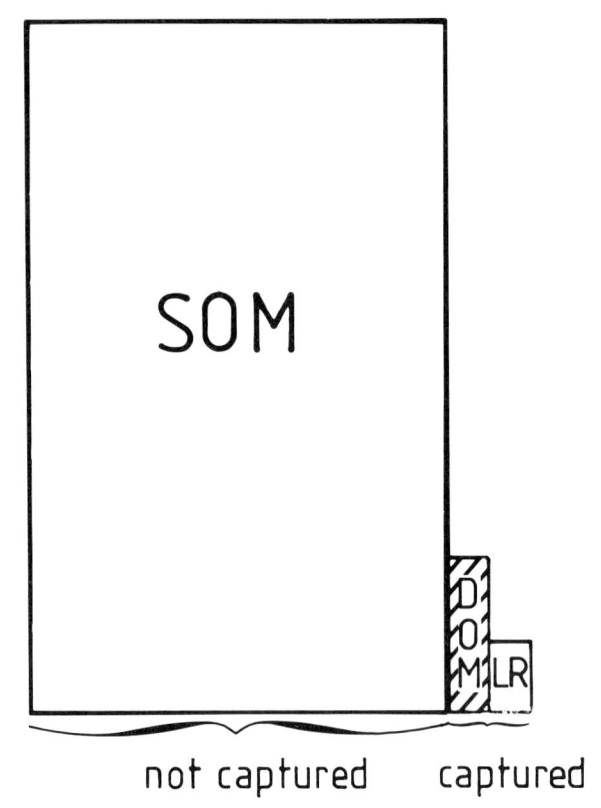

Fig. 4.4. The composition of soil organic matter in the top soil (0–27 cm) at time of harvest of barley at Kjettslinge. Amount of soil organic matter (SOM) calculated from 4% organic matter and 400 kg m^{-2} of soil down to ploughing depth. LR = living roots, DOM = dead organic material, consisting of aboveground harvest residues and dead roots captured on a sieve with 0.5 mm mesh size (Hansson 1987).

Andrén 1987). The highest number of roots in GL was at 10–20 cm depth, whereas the root biomass (77%) was concentrated in the upper 10 cm (Fig. 4.5). In all cropping systems the root biomass was concentrated in the top soil (0–27 cm). The sand layer or a plough pan probably restricted root growth into the subsoil.

At Kjettslinge, root development within and between rows of barley was studied in 1980 using mini-rhizotrons (Hansson and Andrén 1987). During that year barley was sown in all treatments and grass and lucerne were insown in two. During the first half of the growing season the insown leys were suppressed by the barley and they did not grow rapidly until the second half of July. The results from all four treatments, up to the second half of July, were therefore used to compare numbers of roots within and between rows. On average, more roots were recorded within than between barley rows, but due to large variations, the difference was not significant (Fig. 4.6). According to Andersen (1986) root length density did not differ in core samples taken within and between rows of barley, whereas in a study by Groth (1987) root biomass within rows, estimated by soil coring, was two times higher than that between rows. These contradictory results may be explained if stem bases were included in root biomass measurements within rows in the former investigation. When using rhizotrons, which record changes in those parts of the root system in contact with the observation window, stem bases are not included.

Nitrogen is the mineral element required in greatest amounts by plants. Nitrogen availability affects plant growth primarily by influencing the partitioning of carbon between photosynthetic and respiring biomass. Reduced soil nitrogen availability reduces the leaf area growth rate. As a consequence of lower total leaf area, canopy photosynthesis is reduced resulting in a lower total production. However, the photosynthetic rate in already existing leaves is only slightly affected (Gregory et al. 1981). Carbohydrates not used for leaf growth are allocated to the roots, i.e., when nitrogen availability is reduced a higher proportion of the total production is used for root production.

If nitrogen availability is very low, the photosynthetic rate is reduced, since it is a function of the nitrogen concentration in the plant. However, the details of this functional relationship are not well known (Ågren and Ingestad 1987).

Nitrogen increases main-stem leaf area and thus the supply of assimilates to the tiller buds, with more tillers and higher total leaf area as a result (Gallagher and Biscoe 1978, Pettersson 1987). Consequently, the structure of the canopy changes, i.e., number of stems, leaf size and number of leaves per stem is increased and more light is intercepted by the canopy. This can be expressed as an increase in the leaf area index (LAI). LAI is leaf area per unit area of land, and a high LAI indicates a dense canopy and an effective light interception (Hunt 1978). Further, a crop under nitrogen stress has a lower leaf area duration, i.e., leaf longevity is reduced. As a consequence of these conditions, unfertilized barley grew much slower (7 g m^{-2} d^{-1}) than the fertilized barley (20 g m^{-2} d^{-1}) during the vegetative growth phase (Pettersson 1987).

The two barley cropping systems in our experiment represent large differences in nitrogen availability, since one (B120) received 120 kg N ha^{-1} each year and the other (B0) did not receive any nitrogen fertilizer during seven years. This resulted in considerable differences in canopy structure and distribution of peak organic matter (Tabs 4.5, 4.6 and 4.7). Compared with the unfertilized barley, the fertilized barley had a higher and denser canopy, i.e., there were more stems. The ears had more grains. Finally, a lower proportion of the organic matter was allocated below ground but the rooting depth was greater.

In agronomic contexts the economic yield, i.e., the grain, is often of most interest. The harvest index expresses the economic yield as a proportion of total biological yield, which is the total above-ground dry matter at harvest (Beadle 1985). Harvest index is an impor-

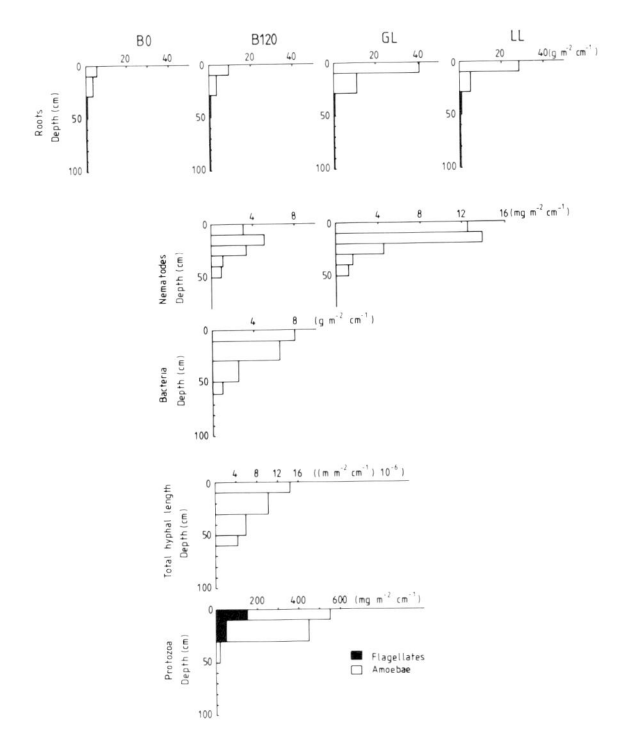

Fig. 4.5. Depth distribution of root biomass (LR) in July (B0 and B120), in September (GL) and in June (LL), nematode biomass in May (B120 and GL) total hyphal length, bacterial and protozoa biomass in September (Hansson 1987, Sohlenius and Sandor 1987, Schnürer et al. 1986a). Values are given as per m^2 for 1 cm depth intervals.

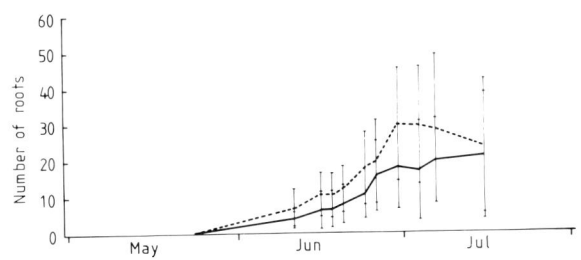

Fig. 4.6. Mean numbers of barley roots down to 100 cm depth as observed in 0.2 m^2 windows in mini-rhizotrons inserted within seed rows (broken line, n=7) and between seed rows (solid line, n=9). Bars indicate standard error (Hansson and Andrén 1987).

tant criterium in plant breeding, since cultivars with a high harvest index are usually selected. According to our results, the harvest index was not much affected by nitrogen availability, which is in accordance with results by Gallagher and Biscoe (1978).

Perennial crops grown for hay are almost never allowed to enter the reproductive phase because of frequent cuttings. This management also affects canopy structure and distribution of organic matter. Leaf area index (LAI) of pasture and forage crops often considerably exceed those of grain crops (Stoskopf 1981). At Kjettslinge we found the following sequence of LAI: grass ley > fertilized barley > unfertilized barley (lucerne was not measured). At its peak, above-ground organic matter had the following distribution in lucerne: above-ground biomass >> litter > attached dead, and in GL: above-ground biomass >> attached dead > litter (Pettersson 1987). In both the meadow fescue and lucerne leys, shoot biomass exceeded the amount of attached dead shoots during most of the growing season, and the amounts of surface litter were relatively small, with means of 50 and 70 g m^{-2} in GL and LL, respectively. This is different from most natural temperate grasslands, which are not disturbed by cutting and ploughing, where the amount of dead shoots can exceed shoot biomass by several times throughout most of the year, and the mean amounts of surface litter can range between 70–900 g m^{-2} (Sims and Coupland 1979). Finally, perennial crops allocate relatively more of their production below-ground than annual crops do and have a shoot/root ratio between 1 and 5. Grasslands which are not treated as croplands (i.e., sown, harvested and ploughed regularly) accumulate more biomass below-ground than managed grasslands do (Tab. 4.7).

Herbivores

Above-ground herbivorous arthropods reduce primary production by consuming or damaging the plant tissues, increasing water losses and transmitting diseases (Crawley 1983). Feeding is influenced by, e.g., fertilization which increases the vegetation mass, as well as the nutrient content and quality of the vegetation as food for herbivores (Andrzejewska 1976, Prestidge 1982).

Around 200 species or other taxonomic groups of herbivorous arthropods were recorded in the experimental field (Tab. 4.1). Collembola was a dominating group, in terms of biomass and abundance (Tab. 4.8). Other common groups were Hemiptera, Diptera and Coleoptera (Curry 1986). Arthropod population densities at Kjettslinge were similar to those recorded by Curry and O'Neill (1978) and Purvis and Curry (1980) from grass and clover leys in Ireland, by Vickerman (1978) from barley and grass fields in West Sussex, and by Persson and Lohm (1977) from an old grassland in Sweden.

Soil microorganisms, including protozoa

Microflora, i.e., bacteria, fungi and protozoa, constituted more than 90% of the Kjettslinge soil biomass, excluding plant roots (Chapter 7). Microorganisms fulfil three main functions in the soil ecosystem. Together with soil animals, they decompose organic matter, thereby releasing inorganic nutrients to the soil solution. Secondly, the microbial biomass serves as a tem-

Tab. 4.5. Dry matter in the total above-ground standing crop and grain at harvest, harvest index (HI), and other yield components for the barley cropping systems at Kjettslinge, compared with other temperate barley cropping systems. The Kjettslinge values are mean values for 1982–85. ND = not determined.

Place and N fertilization	Dry matter (kg 10^{-3} ha^{-1}) Total	Grain	Harvest index	Stems (no. m^{-2})	Ears (no. m^{-2})	Grains per ear (no.)	Mean weight per grain (mg)	Reference
Kjettslinge								
(B120)	8.6	4.1	0.49	868	753	19.6	35.5	Pettersson 1989
(B0)	3.8	1.7	0.44	502	424	15.7	33.8	
Southern Sweden								
with N (80 kg)	9.3	4.7	0.44	ND	ND	ND	ND	Ekman 1982
without N	7.1	3.7	0.44	ND	ND	ND	ND	
Sutton Bonington, England								
with N	7.6	3.5	0.46	ND	650	17.2	31.5	Gallagher and Biscoe 1978
Rothamsted, England								
with N	14.0	6.3	0.45	ND	850	22.1	32.3	Watson et al. 1958
without N	12.5	5.7	0.45	ND	744	21.0	35.1	

Tab. 4.6. Peak above-ground crop organic matter and peak canopy height in the cropping systems at Kjettslinge. Mean values for 1982–1983 in GL and LL, and for 1982–1985 in B0 and B120, and between-year variation (range) (Pettersson et al. 1986, Pettersson 1987). ND = not determined.

Cropping system	Peak above-ground crop organic matter (g m^{-2}) Living (LP) mean (range)	Dead (DP) mean (range)	Litter (AL) mean (range)	Total (LP+DP+AL) mean (range)	Peak canopy height (cm) mean (range)
B0	406 (377–432)	ND	<10	416 (387–442)	50 (42–56)
B120	844 (678–915)	ND	<10	854 (688–925)	77 (64–90)
GL					
1st harvest	570 (507–632)	97 (59–135)	51 (33–70)	718 (599–837)	59 (55–62)
2nd harvest	365 (244–487)	118 (0–236)	54 (33–74)	537 (520–554)	35 (25–45)
autumn	166 (156–176)	134 (95–167)	71 (22–50)	336 (274–393)	18 (16–21)
LL					
1st harvest	444 (251–638)	18 (12–25)	100 (62–139)	562 (325–802)	69 (48–90)
2nd harvest	528 (505–551)	32 (13–52)	107 (58–157)	667 (576–760)	80
autumn	197 (172–222)	27 (17–36)	66 (32–100)	290 (221–358)	ND

Tab. 4.7. The proportion of biomass allocated to the root system and root/shoot (R/S) ratios in the cropping systems at Kjettslinge. Means values for 1981 and 1985 (Hansson et al. in press), compared with two natural grasslands (FM = *Filipendula* meadow, Balsberg 1982; NM = *Nardus* meadow, Kotanska 1975).

	Cropping system					
	Cereals		Perennial crops		Natural grassland	
	B120	B0	LL	GL	FM	NM
Roots (%)	17	20	47	49	66	90
R/S ratios	0.20	0.26	0.95	1.00	1.9	9.83

porary sink and source for inorganic nutrients (Paul and Voroney 1980). Thirdly, microorganisms produce substances which build up and improve soil aggregate structure and stability (Tisdall and Oades 1982).

Bacteria in soil, with the exception of actinomycetes, grow as unicellular organisms with a diameter of 1 μm or less. Heterotrophic bacteria are mainly dependent on low molecular weight carbon sources such as sugars and organic acids for growth; consequently bacteria are most abundant and metabolically active close to root surfaces and fresh plant remains. In the mineral soil bacteria are concentrated on the surfaces of soil particles, where the main part of the population exists in a dormant state with low maintenance requirements (Lynch 1976).

Fungi grow in soil predominantly as hyphae with diameters between 1.5–6 μm. They are the soil organisms that are the best-equipped, enzymatically, to decompose the chemically complex and high molecular weight material characteristic of humified organic matter (Frankland 1982). The filamentous form also confers several advantages to fungi as compared with unicellular bacteria. They are not, as bacteria, restricted to growth in the water phase, but can bridge air-filled pore spaces. This is reflected in the much higher potential of fungi, compared with bacteria, to grow during dry conditions (Griffin 1981). Hyphae are able to actively penetrate and ramify through organic material and this, cou-

Tab. 4.8. Abundance and biomass of above-ground herbage arthropods at Kjettslinge. Mean values for 1982–1983 (Curry 1986).

Organism group	Abundance (no. m^{-2})	Biomass (mg m^{-2})
Collembola	849	45.2
Acari	294	1.4
Hemiptera	404	50.7
Thysanoptera	252	4.5
Coleoptera	110	17.9
Lepidoptera	83	0.5
Hymenoptera	109	5.8
Diptera	301	30.5
Araneae	2	21.7
Neuroptera	2	1.9
Psocoptera	2	<1
Opiliones	<1	<1
Chilopoda	<1	0.8
Dermaptera	<1	0.4
Diplura	3	<1
Total		180

pled with an ability to translocate cytoplasm within the mycelium, allows them to efficiently exploit localized areas of high substrate availability (Dowson et al. 1986). From theoretical analyses, the hyphal growth form was shown to be very energy and nutrient-efficient (Paustian 1985, Paustian and Schnürer 1987a,b). In growing hyphae, most of the energy (as ATP) used in biosynthesis is used to produce cytoplasm and much less is required to synthesize hyphal walls (Paustian and Schnürer 1987a). Thus translocation of cytoplasm from older hyphae into growing hyphal tips amounts to an internal recycling of the most energy and nutrient-rich parts of the fungi. Older hyphae become successively more vacuolated and no longer contribute to active metabolism, although they may remain intact for some time until lysis and decomposition of the cell wall occurs. Consequently, much of the total hyphal length in soil consists of empty or highly-vacuolated hyphae and a lesser fraction are cytoplasm-filled and capable of growing. Hyphae with cytoplasm can be metabolically active or inactive as judged by their ability to metabolize fluorescein-diacetate (FDA; Söderström 1977). Both total hyphal length and length of metabolically active hyphae were measured during the project.

Soil protozoa are unicellular, eucaryotic water-dependent organisms ranging in diameter from 2–1000 μm. Their main food source is bacteria, but many species also consume algae and spores. Some specialists attack fungal hyphae and feed on their cytoplasm (Pussard et al. 1979). Large protozoa are often predators on smaller members of the group. There are three main groups of protozoa in soil: flagellates and ciliates, both of which are free-swimming, and amoebae (Tab. 4.7) which live on soil particle and root surfaces. Average sizes for the respective groups are reported to be 50 μm^3, 3000 μm^3 and 400 μm^3 (Stout and Heal 1967).

When food is abundant, generation times for all groups of protozoa, except the shelled members of amoebae, are counted in hours. However, most of the time, protozoa in soil are present as non-growing cysts. If moisture conditions are suitable, the presence of actively growing bacteria will induce hatching of the cysts (Clarholm 1981, Chapter 5).

As bacterial consumers, soil protozoa may influence the size of bacterial populations. Field investigations, both older (Cutler et al. 1923) and more recent (Elliott and Coleman 1977) emphasize the dominating role of naked amoebae as bacterial consumers. As such, they also play an important role as regulators of nitrogen mineralization close to plant roots (Chapter 5). Flagellates can be as numerous as naked amoebae, but their feeding impact on the bacterial populations is much less, due to their 10–20 times smaller size. Ciliates are the largest protozoa but they are by far the least numerous in normal agricultural soils. Their importance could be expected to be greater in wet soils, where there is enough water-filled space for their feeding activities.

Microbial abundance was estimated by direct counts (bacteria, fungi) or by the most-probable-number (MPN) method (protozoa and nitrifying bacteria). Biomass was estimated from direct counts according to size class (bacteria) and hyphal diameter classes (fungi) and nominal values of biovolume, density, dry weight and C and N content (Schnürer et al. 1986a). Protozoa were classified into amoebae, flagellates and ciliates. The MPN-method used to enumerate protozoa was selective for bacteria-feeding individuals. No attempts were made to distinguish between different species or genera of bacteria or fungi, except for nitrifying bacteria (Chapter 6).

For many purposes it would be desirable to be able to measure the nutrients contained in different organism groups of the soil microbial biomass without having to rely on various factors for conversion from direct counts. At present this is not possible. However, with the introduction of the chloroform fumigation incubation method (CFIM) by Jenkinson and Powlson (1976) it is now considered feasible to quantify the C, N, P and S contents of the total soil microbial biomass (Jenkinson and Powlson 1976, Voroney and Paul 1984, Saggar et al. 1981, Brookes et al. 1982). The method has become widely used, partly because of the easy methodology as compared with direct counts and partly because it is well suited for use with isotope tracers. At Kjettslinge, fumigation biomass C was measured on several occasions in B120 (Schnürer 1985) and fumigation biomass N in field lysimeters with B120 and GL (Lindberg et al. 1989).

The abundance of bacteria and fungi in Kjettslinge were of the same magnitude as those found in other agricultural and grassland soils (Tab. 4.9). In the humus layer of forest soils, 10–30 times higher amounts of microorganisms per gram soil have been recorded (Bååth et al. 1981, Fogel and Hunt 1983, Lundgren 1981, Sundman and Sivelä 1978). However, on a per

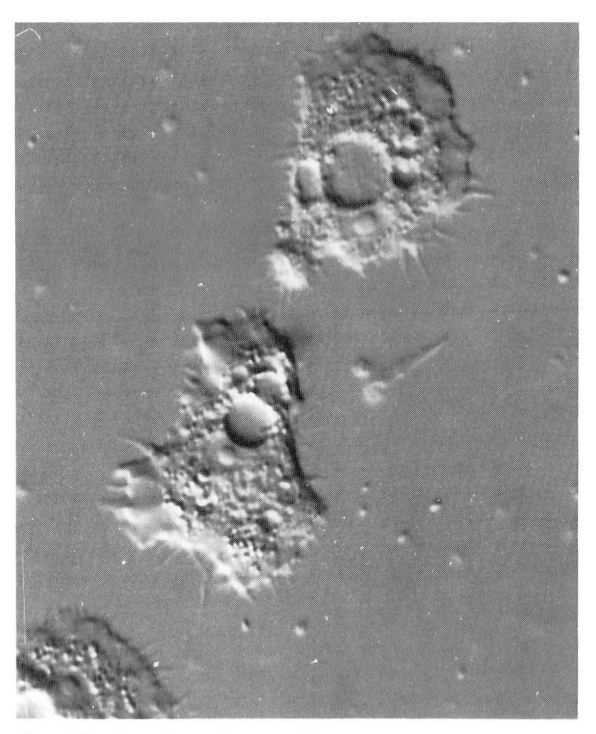

Fig. 4.7. *Acanthamoeba castellanii*, a bacteria-feeding soil amoeba common in soil. Magnification ×1000. (Page 1988).

10^2–10^3 g^{-1} dry soil (Stout and Heal 1967). Numbers of ciliates in the field at Kjettslinge only rarely reached the limit of detection, i.e., approximately 250 individuals g^{-1} dry soil. However, large numbers of ciliates were recorded in a microcosm experiment with Kjettslinge soil in treatments with constantly high water contents and C additions (Clarholm 1985a).

When crops are sown in rows, there is a horizontal stratification of root biomass, with lower root biomass occurring between rows than within rows. This in turn can influence microbial abundance and activity. The distribution of microorganisms within and between plant rows was investigated in B120 on two occasions (Tab. 4.11). The mean values of bacteria and bacteria-feeding naked amoebae were higher, but total and FDA-active fungi were lower within rows than between rows. However, the observed differences were not statistically significant, except for nitrite-oxidizers, which were more numerous within plant rows. The observed trends are in accordance with observations of a bacterial dominance in the rhizosphere and a fungal dominance in the bulk soil (Vancura and Kunc 1977).

Microorganism abundances in the auxillary bare fallow (Fig. 2.12) were compared with values in B 120 on one of the two occasions mentioned above (Tab. 4.11). The bare fallow had been devoid of fresh organic matter input for three years. Therefore, the C sources available for the microorganisms must have been quantitatively and qualitatively different to those in the cropped soil. Heterotroph oxygen consumption and FDA-active hyphae in the bare fallow had decreased to less than half of that in the cropped soil, while bacterial and protozoan abundances and total hyphal length were the same or only slightly lower (Tab. 4.11). Thus, the activity of the microbial biomass declined in the absence of fresh organic matter, although the size of the total microbial biomass was less affected.

unit area basis, forest values are quite similar to arable ones, due to the higher soil bulk density and greater microbial abundance that occurs deeper in arable soils than in forest soils. The abundance of flagellates and naked amoebae at Kjettslinge (Tab. 4.10) was similar to those reported elsewhere, i.e., 10^4–10^5 g^{-1} dry soil (Cutler et al. 1923, Elliott and Coleman 1977). Typical values for ciliates are much lower than for amoebae,

Tab. 4.9. Fungal hyphal lengths and bacterial abundance in temperate arable and grassland soils. Values per gram dry soil. ND = not determined.

Cropping system	Soil organic C (%)	Total hyphal length (m 10^{-3})	Bacterial abundance (no. 10^{-9})	References
Wheat	2.1	0.35	1.1	Shields et al. 1973
Arable	0.9	0.14	1.1	Jenkinson et al. 1976
Arable	1.2	0.16	2.0	–
Arable	2.7	0.21	3.7	–
Arable	2.8	0.34	5.5	–
Grassland	9.9	2.03	19.7	–
Arable	2.7	1.10	ND	Sundman and Sivelä 1978
Fallow	1.2	0.19	1.4	Domsch et al. 1979
Arable	35.4	1.08	ND	–
Grassland	4.7	0.80	3.3	–
Arable	2.0	1.77	ND	Bååth and Söderström 1980
Grassland	2.0	0.27	1.0	Nannipieri et al. 1978
Grassland	2.0	0.04	ND	Ingham and Klein 1984
Fallow	1.2	0.68	3.0	Schnürer et al. 1985
Cereals	1.3	1.03	3.5	–
Cereals	1.4	1.40	4.3	–
Cereals	1.7	1.79	6.6	–
Cereals	1.9	1.37	5.8	–

Tab. 4.10. Microorganisms in the top soil at Kjettslinge. Means per gram dry soil for the September samplings 1981–1983, and for 20 sampling occasions during the two field seasons (Schnürer et al. 1986a). The C content of the top soil was 2.2%.

Organism group	September samplings	Two field seasons
Fungi		
FDA-active hyphal length (m)	80	39
total hyphal length (m 10^{-3})	1.1	1.1
Bacteria		
abundance (no. 10^{-9})	5.5	5.7
biomass (mg)	0.44	0.46
Protozoa (no. 10^{-4})		
flagellates	7.9	4.2
amoebae	3.8	4.2

Bacterial and protozoan biomass and total hyphal lengths decreased with depth (Fig. 4.5), as has been found in other studies (Bååth and Söderström 1979, Fogel and Hunt 1979, Persson et al. 1980). Seventy-nine percent of the bacterial biomass and 73% of the total hyphal lengths were found in the top soil (Schnürer et al. 1986a).

Nematodes

Nematodes are small transparent animals, usually less than 2.5 mm long, which occupy pore spaces of the soil and need a film of water on the soil particles for their movements. They occur in high abundance in arable soils (Wasilewska 1979, Sohlenius et al. 1987).

The nematode fauna was classified into the following four feeding groups: plant feeders, fungal feeders, bacterial feeders and omnivores/predators according to Sohlenius et al. (1988). The orders Tylenchida and Aphelenchida have mouth parts in the form of a stylet, which can pierce plant cells or fungal hyphae and withdraw the contents. In this group we can recognize obligate root feeders, facultative root and fungal feeders, and fungal feeders. Some species belonging to the Tylenchida are known as serious agricultural pests. Although species of *Tylenchus* and *Ditylenchus* may be classified as facultative fungal feeders, we have, for practical reasons, classified them as true fungal feeders. Fungal feeders have, in vitro, been demonstrated to indirectly increase or decrease the rate of mineralization through their influence on fungal hyphae (Ingham et al. 1985).

Bacterial feeders are often the most abundant part of the nematode fauna (Fig. 4.8). This group may compete with protozoa for bacterial cells. By feeding on bacteria they have been shown to significantly increase the rate of mineralization in microcosm experiments (Ingham et al. 1985). Physiologically, bacteria-feeding nematodes may be better adapted than protozoa to dry conditions and to conditions with high concentrations of toxic metabolites. They also have an advantage over protozoa because of their greater ability to direct their move-

Tab. 4.11. Microorganisms and microbial processes in the top soil at Kjettslinge, sampled within and between plant rows of B120 and in an adjacent bare fallow in June 1982 (Schnürer et al. 1986a, Berg and Rosswall 1987) and July 1981 (M. Clarholm and J. Schnürer unpubl.). Means per gram dry soil of five samples from one plot.

	Year	Within plant rows	Between plant rows	Fallow
O_2 consumption (μl O_2 h^{-1})	1982	0.98[a]	0.83[a]	0.36[b]
FDA-active hyphal length (m)	1982	22[a]	27[a]	11[b]
	1981	50[a]	126[a]	
Total hyphal length (m 10^{-3})	1982	1.1[ab]	1.4[a]	0.9[b]
	1981	1.5[a]	1.8[a]	
Bacterial abundance (no. 10^{-9})	1982	5.0[a]	4.3[ab]	3.6[b]
	1981	4.7[a]	3.5[a]	
Bacterial biomass (mg)	1982	0.41[a]	0.35[ab]	0.33[b]
	1981	0.44[a]	0.33[a]	
Flagellates (no. 10^{-4})	1982	8.4[a]	12.2[a]	4.1[a]
	1981	3.8[a]	4.9[a]	
Amoebae (no. 10^{-4})	1982	3.8[a]	2.6[a]	3.9[a]
	1981	2.9[a]	2.7[a]	
Ammonium-oxidizers (no. 10^{-4})	1982	0.6[a]	0.5[ab]	0.2[b]
Nitrite oxidizers (no. 10^{-4})	1982	5.4[a]	1.8[b]	0.8[b]
Potential ammonification (ng N h^{-1})	1982	0.109[a]	0.101[a]	0.119[a]
Potential nitrite oxidation (ng N h^{-1})	1982	0.153[a]	0.159[a]	0.159[a]
Soil water potential (–kPa)	1982	115	121	121
	1981	72	67	

Values followed by the same letter do not differ significantly (P <0.05) between sampling locations.

ments towards microsites with high bacterial activity. By their movements inside and around bacterial colonies they increase the diffusion of nutrients and oxygen to the bacterial cells and may stimulate microbial activity (Bååth et al. 1981).

The fourth group of nematodes, the omnivores/predators belonging to the order Dorylaimida, feeds on various food sources, mainly algae and soil animals. They are usually less abundant than other nematodes but generally of large body size.

In comparison with other studies of agricultural soils (Wasilewska 1979, Hendrix et al. 1986), the total numbers of nematodes in this study were fairly high (Tab. 4.12). About 40 different taxa (species or genera) were identified (Sohlenius et al. 1987, see also Tab. 4.1). The true total number of species at Kjettslinge was probably around 70–100. The abundance of nematodes belonging to various orders and families is indicated in Tab. 4.13. The bacterial feeders were generally the most abundant group. The omnivores/predators, although an order of magnitude below the other feeding groups in density, contributed substantially to the total biomass. The plant feeders were strongly dominated by tylenchorhynchids, i.e., species of *Tylenchorhynchus* and *Merlinius,* in all cropping systems except for the meadow fescue where the genera *Paratylenchus* and *Helicotylenchus* were common (Paratylenchidae and Hoplolaimidae). The most common genera among the fungal feeders were *Tylenchus* s.l., *Ditylenchus*, *Aphelenchus* and *Aphelenchoides,* and the bacterial feeders were dominated by genera and species belonging to the Cephalobidae (i.e. *Acrobeles, Acrobeloides, Cephalobus, Cervidellus, Chiloplacus* and *Eucephalobus*). The genus *Rhabditis* s.l. was subdominant in all four crops. *Panagrolaimus* was abundant in the lucerne. The Dorylaimidae (i.e. *Dorylaimus* s.l., *Dorylaimellus, Eudorylaimus, Pungentus, Mesodorylaimus* and *Oxydirus*) were dominant among the omnivores/predators, and the predatory genera *Clarkus* and *Mylonchulus* were generally low in abundance.

The taxonomic composition of the fauna found in this study had many similarities to that found by Domurat (1970) in Polish barley fields, by Yeates (1984) in New Zealand pastures, by Wasilewska (1967) in Polish lucerne fields and for tylenchids only, to that found by Andersen (1979b) in Danish barley fields.

The largest nematode biomass was in the top soil (Fig. 4.5), but nematodes were abundant down to 40–50 cm, which was the greatest depth sampled (Sohlenius and Sandor 1987). The nematodes under meadow fescue had a more superficial distribution than those under barley. In meadow fescue only 22% of the nematode biomass was found below 20 cm, whereas under barley about 40% was found below this depth. These figures

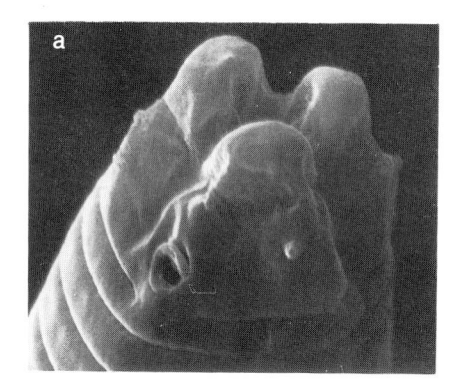

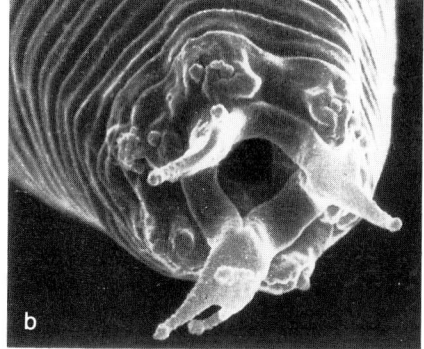

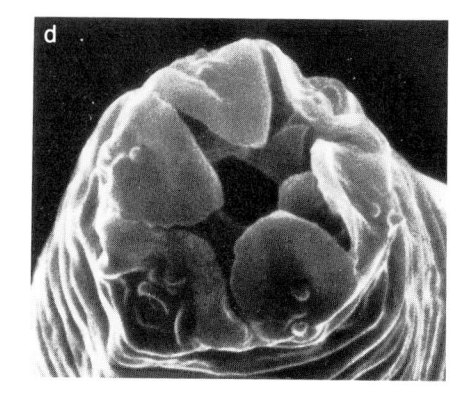

Fig. 4.8. SEM micrographs of the mouthparts of bacterial-feeding nematodes (Nematoda, Cephalobidae): a) *Cephalobus persegnis*, b) *Acrobelophis minimus*), c) *Acrobeles ciliatus*, d) *Acrolobus emarginatus* (Boström 1985, 1986a, 1988c).

Tab. 4.12. Abundance of nematodes (no. 10^{-6} m^{-2}) in the cropping systems at Kjettslinge in 1981 and 1982 compared to other arable and native grass ecosystem.

Ecosystem	Abundance (range)	References
Kjettslinge		
B0	3.5– 6.5	Sohlenius and Boström 1986
B120	4.0– 9.3	Boström and Sohlenius 1986, Sohlenius and Boström 1986
GL	6.9– 9.3	Boström and Sohlenius 1986
LL	4.3–12.8	Sohlenius et al. 1987
Arable land (barley)	3 –12	Andersen 1979a
Ungrazed grassland	1.8– 7.3	Freckman et al. 1979
Mixed-grass prairie	5 – 8	Ingham and Detling 1984
Arable land (grass ley)	20	Nielsen 1949
Pasture	1.0– 6.4	Yeates 1982
Arable land (cereals)	1.6– 9.7	Andrén and Lagerlöf 1983

also indicate the level of underestimation of the nematode biomass due to insufficient depth in the routine samplings (20 cm).

The plant feeders were thoroughly examined, to identify species which could be noxious to the crops. *Tylenchorhynchus dubius* (Bütschli), which at Kjettslinge was most abundant in GL (6.2×10^5 m^{-2}), was found by Sharma (1971) to be a noxious parasite which could cause damage at densities normal under field conditions. According to Whitehead and Fraser (1972) it can multiply in barley, but does not affect the growth of the crop. *Merlinius brevidens* (Allen) and *M. microdorus* (Geraert) were abundant in LL (3.2×10^6 m^{-2}) and in B120 (1.2×10^6 m^{-2}). *M. brevidens* was reported by Langdon et al. (1961) to cause general stunting, yellowing and reduced yields in barley. The populations of *Pratylenchus*, most dense in meadow fescue (1.3×10^5 m^{-2}) and B0 (7.7×10^4 m^{-2}), were a mixture of at least two species, *P. crenatus* Loof and *P. fallax* Seinhorst and possibly *P. flakkensis* Seinhorst. *P. crenatus* has been reported to cause damage to barley (Decker 1961) and *P. fallax*, to be a direct pathogen of barley (Corbett 1972).

Tab. 4.13. Abundance and biomass of nematodes at Kjettslinge. Mean values for the September samplings 1981–84. (B. Sohlenius, S. Boström, A. Sandor, unpubl.)

Feeding group/taxa	Abundance (no. 10^{-3} m^{-2})	Biomass (mg m^{-2})
Plant feeders		
Tylenchorhynchidae	1746	
Hoplolamidae	132	
Pratylenchidae	57	
Paratylenchidae	235	
Heteroderidae	2	
Trichodoridae	1	
Subtotal	2173	63
Fungal feeders		
Tylenchoidea	997	
Aphelenchoidea	871	
Neotylenchoidea	101	
Subtotal	1969	38
Bacterial feeders		
Rhabditidae	513	
Cephalobidae	2120	
Panagrolaimidae	161	
Monhysteridae	98	
Chromadorida	1	
Araeolaimida	52	
Alaimidae	16	
Subtotal	2961	84
Omnivore/Predators		
Dorylaimidae	375	
Belondiridae	2	
Leptonchidae	8	
Diphtherophoridae	68	
Mononchoidea	66	
Subtotal	519	82
Total	7622	267

Microarthropods and enchytraeids

Protozoa and nematodes are by far the most numerous soil animals, but the group with the largest number of species is the phylum Arthropoda. The most commonly occurring soil arthropods are mites (Acari) and springtails (Collembola). Together with a few less common groups of small arthropods they constitute the microarthropods. Taxonomically the microarthropods can be divided into:

a) Collembola (springtails), with the families Poduridae, Onychiuridae, Isotomidae, Entomobryidae and Sminthuridae. The first four families are comprised of species with elongated bodies, while Sminthuridae are more ball-shaped (Fig. 4.9).

b) Acari (mites). These are usually grouped into the suborders Mesostigmata, Prostigmata, Astigmata and Cryptostigmata. Mesostigmata consist of, among others, Gamasina, which are mainly predators, and the fungivorous or predatory Uropodina (Fig. 4.10).

c) Less common groups, such as Protura, Diplura, Symphyla and Pauropoda (Fig. 4.11). Their ecology is poorly known, but as far as known, the general ecology of Protura and Diplura is similar to that of Collembola (Nosek 1963, 1973). They are seldom found in large numbers. Symphyla and Pauropoda are small millipedes (Myriapoda), and both species of Symphyla found at Kjettslinge are considered to be microbivores.

Enchytraeids are related to earthworms (class Oligochaeta) and they look like small, white or transparent earthworms (Fig. 4.12). Microarthropods and enchytraeids together constitute the main part of the soil mesofauna, i.e., animals between 0.1 mm and 2 mm in body width (Swift et al. 1979) or between 0.2 mm and 10 mm in body length (Wallwork 1970).

Microarthropods live in air-filled soil pores, while enchytraeids need close association with free soil water. For movements both groups depend on soil pores and crevices, since they cannot make their own burrows, in contrast to earthworms and other larger soil animals. The mesofauna thus does not greatly influence soil macrostructure, but they can contribute to the formation of soil aggregates, i.e., they affect soil structure at a smaller scale. Both groups are also numercus in the soil surface litter.

Most microarthropods and enchytraeids are part of the decomposer community, and only few feed on living tissue of higher plants. Their grazing on plant litter leads to fragmentation, which increases the availability of the litter for microbial attack. The animals in general cannot metabolize plant structural polysaccharides and mainly assimilate microorganisms or their metabolic products when ingesting plant litter. Grazing on fungi and bacteria by the animals can increase the release of nutrients from microbial biomass into the soil, where they become available to plants. Microarthropod predation on protozoa, nematodes, enchytraeids and other microarthropods can regulate prey population levels as well as release nutrients locked up in prey biomass.

The activites of soil mesofauna are considered to be important control factors in decomposition processes (for review, see Anderson et al. 1981, Petersen and Luxton 1982, Parkinson 1983, Anderson and Ineson 1984, Seastedt 1984). However, some investigations indicate only small contributions by soil mesofauna to litter carbon or mass loss (Andrén and Schnürer 1985, Visser et al. 1981), while other indicate considerable effects (Hanlon and Anderson 1979, Setälä et al. 1988). Most field studies of soil animal participation in litter decomposition have been performed in natural ecosystems and comparatively few concern arable soil (Naglitsch 1966, Eitminaviciuté et al. 1976, Andrén and Lagerlöf 1983, Lagerlöf and Andrén 1985, Czarnecki 1983).

In conclusion, the large gaps in our knowledge about the specific roles of the soil mesofauna in agricultural soils make a thorough evaluation of their importance for litter decomposition, nutrient cycling and plant growth impossible. There are also gaps in our knowledge concerning the effects of agricultural measures on the soil mesofauna. Although practices such as soil cultivation, crop rotation, and the application of fertilizers, waste products and pesticides have been investigated, the results are often inconclusive or conflicting as regards their effects on soil mesofauna (Marshall 1977, Andrén and Lagerlöf 1983, Steen 1983, Edwards 1984, Ryszkowski 1985, Hendrix et al. 1986, Curry 1986).

A total of 19 taxonomic groups of Collembola were found in the Kjettslinge soil during the experimental period, and an additional six species were found at the soil surface or in the crop canopy (Tab. 4.1).

The numerically dominant Collembola in all crops were *Folsomia fimetaria* and *Tullbergia krausbaueri* s.l. followed by *Isotomiella minor*, *Folsomia quadrioculata* and *Friesea mirabilis*. *F. fimetaria* contributed to a relatively high proportion of the total biomass in contrast to *T. krausbaueri* s.l., which has a much lower individual biomass. Owing to their relatively large size, when present, *Isotoma viridis* constituted a major part of the biomass.

The number of collembolan species found in this study is similar to that reported from other studies in arable land, and the species identified are all widely distributed and have been recorded from many agricultural habitats (Müller 1959, Borg 1971, Krogh 1985).

A higher number of soil Collembola species were found in more natural habitats in the region. Persson and Lohm (1977) found 24 species in an old grassland 30 km from Kjettslinge, 12 of which were similar for both locations. Axelsson et al. (1984) found 24 species in a deciduous woodland 10 km from Kjettslinge, of which 8 species were also found at Kjettslinge.

The total abundance of Collembola found at Kjett-

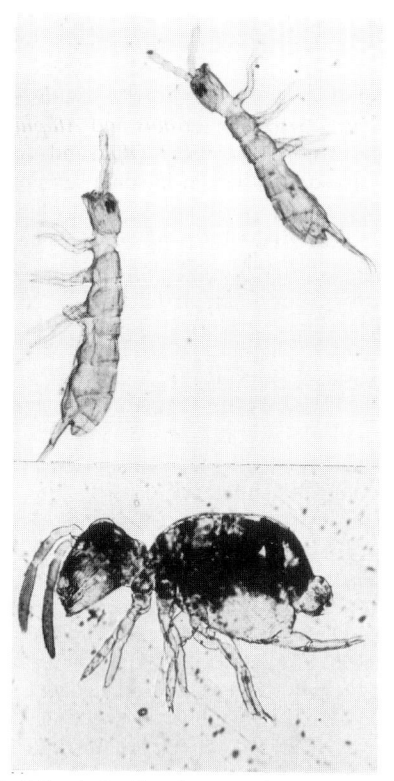

Fig. 4.9. Collembola. Top: *Isotoma olivacea* (Isotomidae). Bottom: *Sminthurinus elegans* (Sminthuridae). (Photo J. Lagerlöf).

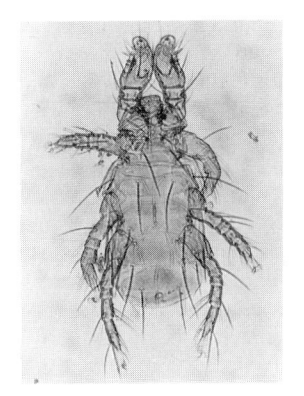

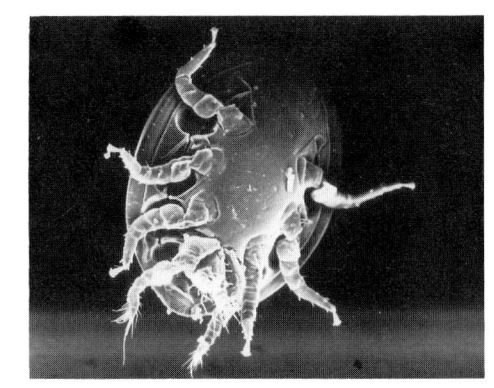

Fig. 4.10. Acari. Left: *Pygmephorus* sp. (Prostigmata), middle: *Hypoaseis nolli* (Mesostigmata, Gamasina), right: *Nenteria breviunguiculata* (Mesostigmata, Uropodina) (Photo J. Lagerlöf).

slinge (Tab. 4.14) is somewhat lower than the average found in a survey of different cropping systems in different parts of Sweden (Andrén and Lagerlöf 1983). In comparison with reports from arable soil in other countries, the Kjettslinge abundances were similar or higher. This difference may partly be due to high efficiency of extraction, since small extraction units and a carefully controlled extraction temperature regime were used (Andrén 1985).

Collembola abundances in native temperate soils are generally higher than in arable soils (Petersen and Luxton 1982). However, no differences in collembolan biomass or numbers were observed when cultivated fields were compared with natural meadows in Poland (Czarnecki 1969, cited in Ryszkowski 1985). On the other hand, Edwards and Lofty (1969) found that collembolan numbers were reduced by 40–70% when ploughing a 300-year-old grassland.

A total of 42 taxonomic categories of mites (Acari) were distinguished during the investigation (Tab. 4.1). The Prostigmata groups *Pygmephorus* spp., Tarsonemidae spp., and Alicorhagiidae spp. were the most numerous taxa. Within the predatory Mesostigmata, *Rhodacarellus silesiacus*, *Arctoseius cetratus* and *Alliphis siculus* were most common. *Schwiebia talpa* and Anoetidae spp. were common Astigmata in some samplings. The Mesostigmata contribution to total biomass was about 85% in all crops, due to their relatively high individual biomass and rather high numbers. Although the Prostigmata were the most abundant suborder they only constituted 7–15% of the biomass, owing to their generally low individual biomass. In all crops Astigmata and Cryptostigmata contributed little to the Acari biomass because they were both relatively small and uncommon.

The composition of the Acari fauna at Kjettslinge seems to be typical for agricultural soils, and most of the taxa listed have been reported in other European investigations (Karg 1967, Wasylik 1980, Emmanuel et al. 1985). In general, undisturbed sites have a more species-rich mite fauna than agricultural soils. In a grassland soil 30 km from Kjettslinge, Persson and Lohm (1977) recorded 72 species. The greatest difference between sites was in Cryptostigmata (20 species compared with 4 in Kjettslinge), while differences were less pronounced in the other groups. Karg (1967) found that a cultivated soil contained only 25 to 50% of the numbers of Acari species recorded in an adjacent woodland soil. In a 7-yr-old grass ley, species richness was still slightly lower than in a nearby woodland soil.

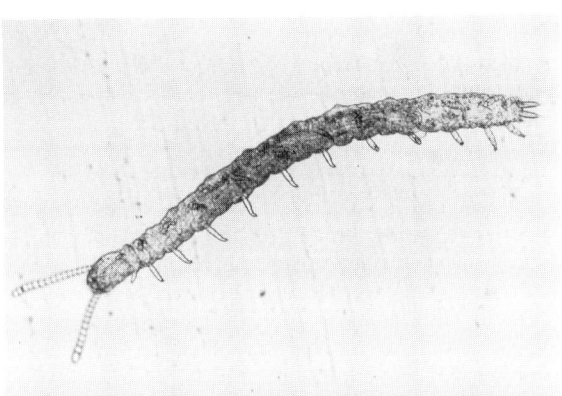

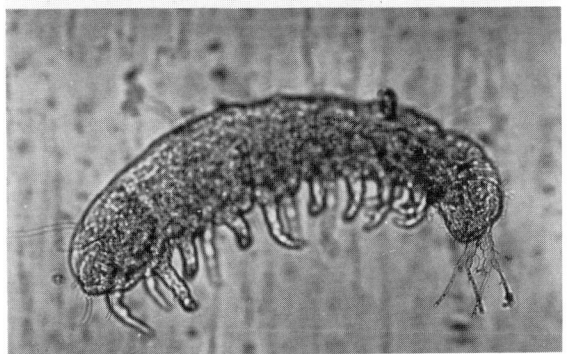

Fig. 4.11. Other microarthropods. Symphyla (top), Pauropoda (bottom). (Photo J. Lagerlöf).

Fig. 4.12. A large enchytraeid crawling in soil litter (Photo J. Lagerlöf).

However, certain groups are comparatively less affected and may even respond positively to agricultural practices. For example, Prostigmata, Astigmata and Rhodacaridae are often more abundant in arable than in native soils, even though the Acari fauna as a whole is reduced (Edwards and Lofty 1969).

Compared with other microarthropod groups, a greater proportion of mites are predatory and can have significant influence on nematode and microarthropod populations (Karg 1971, Edwards and Thompson 1973, Santos et al. 1981).

Numbers of Protura and Diplura at Kjettslinge were low, as also was reported from other agricultural land by Andrén and Lagerlöf (1983) and Edwards and Lofty (1969). Although the sampling intensity was too low to give a sensitive test of statistically significant differences between crops and between seasons, the highest numbers of Protura appeared to be in B0. Edwards and Lofty (1969) found that Protura were relatively unaffected by agricultural practices, probably because of their deep vertical distribution.

Pauropoda and Symphyla are minute myriapods living in the soil, and are often overlooked in soil fauna investigations. Symphyla have attracted some attention in agricultural soils, because this group includes species that can be economically important pests on field and greenhouse crops. The two species found at Kjettslinge are however considered to be microbivores, as are Pauropoda.

Both Pauropoda and Symphyla can be found at greater depths than most other soil microarthropods, probably due to their small size and slender body shape, effective for movement through narrow soil pores and crevices. Pauropods are susceptible to changes in the environment, and were considered by Edwards (1984) to be the group of soil animals that was most affected by agrochemicals. The mean abundance of pauropods ($1100-1900$ m^{-2}) and symphylids ($120-340$ m^{-2}) at Kjettslinge was similar to or exceeded abundances reported in most other cultivated or uncultivated soils (Peterson and Luxton 1982, Andrén and Lagerlöf 1983). In the September samplings at Kjettslinge, Pauropod abundance was higher in unfertilized barley than in the other crops, possibly due to the moist conditions under B0. The higher concentrations of N under the other crops may also have affected these fragile organisms (Lagerlöf and Scheller 1989).

Like most soil animals, enchytraeids are often found in lower numbers in arable soil than in comparable native soils, but their relative contribution to total biomass and respiration of the soil biota can be considerable even in arable soils (Kasprzak 1982, Andrén and Lagerlöf 1983, Lagerlöf et al. 1989). Enchytraeids are microbial feeders and detritivores (O'Connor 1967) and are part of the decomposer system. Their role in soil-forming processes is less obvious than that of earthworms, but their influence on soil aggregate formation is considered to be important in some systems (Kubiëna 1955). Compared with earthworms, their ecology is not well known and few publications have been devoted to their occurrence and activity in arable soils (Nielsen 1955, Ryl 1977, 1980, Andrén and Lagerlöf 1983, Lagerlöf et al. 1989). The genera and species found at Kjettslinge (*Enchytraeus buchholzi, Fridericia bulbosa, Henlea ventriculosa*) are all widely distributed (Lohm 1979) and reproduce sexually (Christensen 1961). They produce large numbers of egg cocoons which probably can survive for long periods and hatch when moisture and temperature conditions are favourable.

Soil and surface macroarthropods

This diverse group contains arthropods usually larger than those classified as microarthropods. However, there are great differences in body size between juveniles and adults in many species, and between species in this group. Some macroarthropods spend their entire life cycle in the soil or litter, e.g., Chilopoda and Diplopoda. Others only utilize the soil for overwintering or pupal development, e.g., some Lepidoptera, Thysanoptera and Coleoptera, and spend their active life stages in the vegetation.

Dominant macroarthropods at Kjettslinge were Coleoptera (beetles), especially Carabidae and Staphylinidae, and Diptera larvae, especially Chironomidae, Fungivoridae, Dolichopodidae and Cecidomyidae. Other prominent groups were Araneae (spiders) and Homoptera.

Macroarthropods are more affected by agricultural practices, especially soil cultivation, than most other soil animals. At Kjettslinge significantly higher abundances were recorded in the leys than in barley for most macroarthropod groups. This difference was greatest

Tab. 4.14. Abundance and biomass of soil mesofauna at Kjettslinge. Mean values for the September samplings 1980–84 (Lagerlöf 1987).

Organism group	Abundance (no. 10^{-3} m^{-2})	Biomass (mg m^{-2})
Collembola	18.2	22
Acari	12.4	32
Myriapoda	1.6	1.4
Enchytraeids	8.4	515
Total	40.5	570

immediately after ploughing in autumn (Carter et al. 1985, J. Lagerlöf unpublished).

Earthworms

Earthworms contribute to the decomposition of organic matter by bringing together mineral soil and plant residues. Some species feed on crop residues which are drawn into the burrows from the soil surface. Their ingestion of mineral soil is thereby comparatively low, while other species feed on organic matter, i.e., dead soil fauna, microflora, roots, root hairs and other plant residues mixed with a higher proportion of soil (Piearce 1978). The fragmentation of the organic matter that occurs during its passage through the earthworm gut increases particle surface area available for microorganisms. Whether earthworms take nourishment directly from the organic matter or from microorganisms attached to it is not known. Earthworm faeces, called casts, are deposited under as well as on the soil surface. Graff (1971) estimated that 25% of the upper 10 cm soil layer passes through earthworms yearly. In this way litter is gradually covered by surface-deposited casts and plant residues are mixed with the mineral soil.

The burrowing and casting activities of earthworms influence the soil structure. Where earthworms are present aggregate stability and water permeability are enhanced (Guild 1955). In the absence of surface-casting earthworms, water infiltration rates can be reduced, which together with litter accumulation, increases the surface runoff of ammonium and nitrate (Sharpley et al. 1979).

Aporrectodea caliginosa Sav. was the dominant species in all cropping systems at Kjettslinge and constituted more than 80% of the total earthworm biomass (Fig. 4.13). Other species found frequently were *A. longa* Ude, *Lumbricus terrestris* L. and *A. rosea* Sav., while *L. castaneus* Sav. and *L. rubellus* Hoffm. were found only sporadically (Boström 1988a). The three most common earthworm species in Kjettslinge soil: *A. caliginosa*, *A. longa* and *L. terrestris* often occur together in deciduous woodlands or in cultivated soils (Nordström and Rundgren 1973).

According to the classification into ecological categories developed by Bouché (1977), *A. caliginosa* and *A. rosea* can be characterized as endogeics, i.e., while burrowing through the soil the species consume large amounts of mineral soil together with organic matter. *A. caliginosa* is favoured by soil cultivation and is usually more abundant in annually cultivated soils than in pastures or leys (Evans and Guild 1948a, Edwards and Lofty 1975). *A. longa* can be classified as an anecic species which lives in a burrow system and comes to the soil surface to feed on litter. *L. terrestris* is an epianecic species, which in contrast to true anecics does not have an aestival diapause, i.e., the ability to survive unfavourable conditions by entering a dormant stage. The species is often more negatively affected by soil cultivation than other species (Edwards and Lofty 1977). Both *L. rubellus* and *L. castaneus* can be classified as epigeics, i.e., species that feed on litter and live above the mineral soil, although *L. rubellus* also burrow in the soil (Rundgren 1975). According to their vertical distribution, Nordström and Rundgren (1973) considered *L. castaneus* to be a surface living species, *L. rubellus* an intermediary species, while *A. caliginosa*, *A. longa*, *A. rosea* and *L. terrestris* were considered to be deep-burrowing species.

Earthworms which had developed clitellum were classified as adults, while others were classified as juveniles. As a mean for all 4 cropping systems, the annual earthworm biomass was 46 g m^{-2} (fresh weight) (Tab. 4.15), which is within the range 1–50 g m^{-2}, estimated by Andersen (1987) and Jensen (1985) for arable fields in Denmark. Graff (1953b) reported a mean of 20 g m^{-2} of earthworm biomass in arable land in Germany, while in Scotland the biomass in 5-yr-old arable fields varied between 7 and 14 g m^{-2} (Evans and Guild 1948a).

Seasonal dynamics of the components

Seasonal growth dynamics of a crop is a function of phenology, management and climatic conditions. The crop influences other organisms by competing for nutrients and water, but also by supplying food for herbi-

Fig. 4.13. *Aporrectodea caliginosa* with cocoons (Photo A. Lofs-Holmin).

Tab. 4.15. Abundance and biomass (fresh weight) of adult and juvenile *A. caliginosa* and of combined earthworm species at Kjettslinge. Mean values for the September samplings 1981–1983 (Boström 1988b).

	Abundance (no. m^{-2})	Biomass (g m^{-2})
Adults	24.9	30.2
Juveniles	48.8	9.0
All species	81.4	45.6

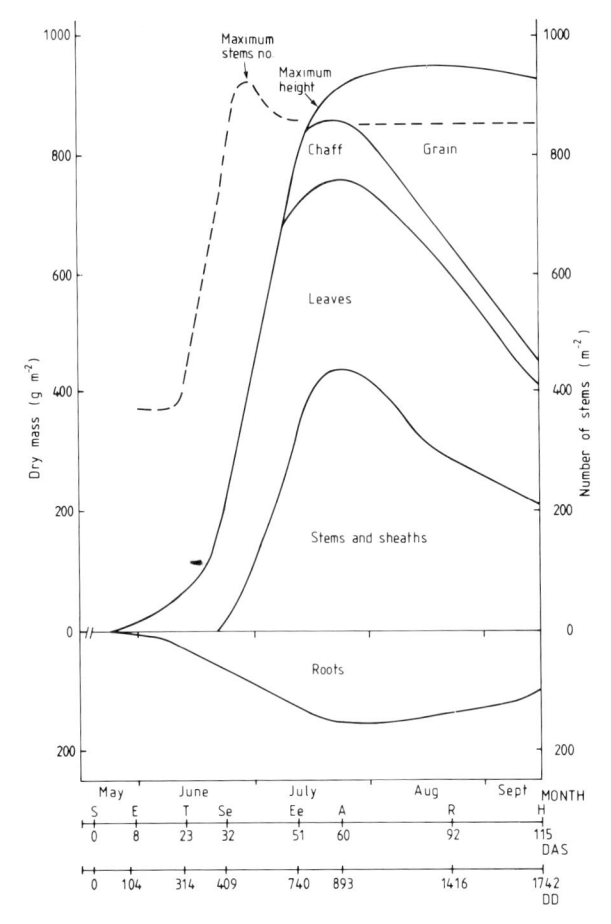

Fig. 4.14. Development and distribution of plant biomass in B120. DAS = days after sowing, DD = accumulated day-degrees using days with a mean temperature ≥ 0 °C. S = sowing, E = emergence, T = tillering, Se = stem elongation, Ee = ear emergence, A = anthesis, R = ripening, and H = harvest. Based on results from 1981 and 1985 (Pettersson 1987).

vores and fresh organic material to the soil organism community.

In general terms, the abundance of animals is determined by the difference between birth and death rates. Any factor which influences these processes may affect population density. Such factors are food availability and quality, competition and predation, environmental factors such as soil humidity and temperature, and agricultural practices. Inevitably, numbers of organisms and biomass only represent the net result of complex interactions between these and other factors. Nevertheless, comparisons of the seasonal dynamics of the abundance and biomass for the different organisms represent a necessary first step in an ecosystem study.

In this section emphasis is put on parallel presentations of the dynamics of the different organism groups. Most of the results have been published in separate papers during a period of several years. With all the results now available, it is possible to discuss how the components interact and how they are affected by abiotic factors.

The abundance, the spatial and the temporal distribution of the investigated organisms, as well as their relationships during one growing season are discussed. Most organisms were sampled intensively, i.e., several times per month, during 1982 and/or 1983. The sampling program was concentrated in B120 and GL.

The crop

Plant growth can be divided into several genetically governed phenological stages such as tillering, stem elongation, flowering etc. Different phenological scales have been developed for different purposes. We have used the decimal code (Tottman and Makpeace 1979) which is a modification of Zadoks' scale (Zadoks et al. 1974). Phenological development in general is influenced by temperature and day length, whereas lack of nitrogen mainly affects timing of senescence. Different phenological stages imply different patterns of carbon allocation, e.g., after flowering the carbon allocation is switched from vegetative parts to reproductive parts. Before seed formation, phenological stages in an annual and a perennial crop are similar, although the duration of the stages differ. However, after seed formation annual crops wilt and die, whereas a perennial crop adapted to a cold climate stores carbohydrates, becomes dormant and then continues to grow the following spring.

The spring barley at Kjettslinge was sown in May and nitrogen fertilizer was applied immediately after sowing. The plants emerged after 8 to 10 d (Fig. 4.14). The vegetative growth phase lasted about 5 wk and the reproductive phase about 8 to 9 wk. Rapid vegetative growth commenced in mid-June, when tillers started to emerge, and lasted until about anthesis, when also the root biomass reached its maximum (Hansson 1987). Tiller numbers peaked in late June, and ear emergence started in early July. After a growth period of 110 d the barley was harvested in late August or early September. At harvest, grain accounted for about 50% of the total above-ground standing crop. Total biomass and its distribution between different parts of the plant in relation to phenological development in B120 and B0 differed in magnitude but not in temporal distribution (Pettersson 1987).

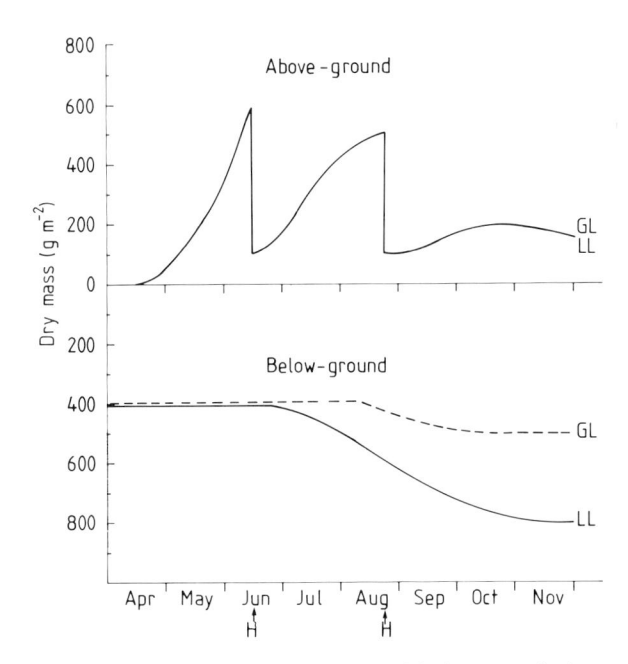

Fig. 4.15. Development of above- and below-ground plant biomass in GL and LL. H = harvest. Based on results from 1982 and 1983 (Pettersson et al. 1986, Hansson and Andrén 1986, Hansson and Pettersson 1989).

Beginning in 1981, when both leys were established (See Chapter 2), they were cut twice a year at time of flowering, resulting in fairly identical seasonal patterns of above-ground biomass. In GL and LL below-ground biomass did not change until the second half of the growing season, when there was a slight increase in GL and a pronounced increase of living roots in LL (Fig. 4.15). At the latitude of the project site, growth ceases in October–November, and snow normally covers the crop for about 100 d, between December and March (Rodskjer and Tuvesson 1975).

Relative growth rate (RGR) expresses the increase of biomass per unit of biomass per unit of time. RGR is a good measure for comparisons between treatments, and it is usually better than absolute growth rate, especially when the biomass in the treatments differ. Both in cereals and perennial hay crops, RGR normally decreases more or less linearly from start of growth to harvest. At least a part of this decline can be caused by lack of nitrogen, i.e., with adequate nitrogen availability RGR should remain at a higher level during the vegetative growth phase. During 1985 maximum RGR was the same in both barley crops, but it remained at this high value longer in B120 than in B0 (Hansson et al. in press). During the second vegetative growth period in hay crops, i.e., after the first harvest and before the second harvest, changes in RGR are similar to those during the spring. The maximum (start) value is usually lower than that during the spring. This phenomenon is often referred to as "summer dormancy" and may partly be genetically determined (Walton 1983). However, alternative explanations such as lack of carbohydrate reserves and high respiratory costs for the root system have been proposed. Growth also continues during the autumn after the second harvest, but RGR is low, mainly due to low temperature and light levels.

Low temperature in spring delays the development of leaf area, whereas low temperature in autumn may impair carbohydrate storage, which is needed the following spring for initial development of leaves. In 1982, production in the lucerne during spring was unusually low, probably both due to the unusually cold autumn in 1981 and the cold spring in 1982 (Pettersson et al. 1986).

Normal agricultural practice implies that a high amount of fertilizer is applied to the soil at the beginning of the growing season. This high concentration of nutrients in the soil results in a high internal concentration in the plant, which stimulates the development of new meristems. The resulting demand for more nutrients is partly met by the expanding root system, exploring previously untapped soil volumes. However, the supply through mineralization does not normally keep pace with the uptake and immobilization processes. Consequently, assimilates are produced faster than nutrients are absorbed and the internal nutrient concentration decreases. To be able to develop new tissues during these circumstances, nutrients from older leaves are recycled. Consequently, nitrogen concentrations differ between older and younger, vegetative and reproductive plant parts (Milthorpe and Moorby 1979) (Fig. 4.16 a,b).

Herbivores

Vertebrate consumers, such as voles and hares, were not observed in the experimental field and only a few droppings were found. Thus they were not included in this study. Instead, herbivorous insects, especially aphids are referred to under this heading. These insects can consume considerable amounts of plant biomass and consequently reduce the final yield. Even during years when pest organisms do not reach high enough numbers to significantly lower yields, their consumption of plant biomass is of interest when studying C and N flows. From an agricultural viewpoint, the transmission of plant pathogens by the herbivores is of great importance, but studies of herbivores as disease vectors were not included in the present project.

Population densities and biomass of most groups tended to be highest in July and August although this trend was less pronounced in the perennial ley crops where many groups were abundant in June 1983 (Curry 1986).

Litter

The seasonal dynamics of leaf death and litter-fall in agricultural crops is greatly influenced by agricultural practices, and thus differs considerably from that in a natural system. In barley all leaves died before harvest. In spite of this litterfall during the growing season was negligible, <10 g m^{-2} (Hansson et al. 1987), since most of the dead leaves were still attached to the stems. After the harvest, however, stubble and harvest residues were left in the field and incorporated in the soil at ploughing. Consequently, almost the entire input of above-ground organic material into the soil occurred as a pulse in September.

Since the leys were cut just when entering the reproductive phase, very little above-ground vegetation senesced during spring and summer, except for during very dry periods. In both leys the amount of litter on the ground was rather stable during the growing season, ranging between 20 and 75 g m^{-2} (Pettersson 1987). During autumn, there was usually an accumulation of dead leaves in GL, which remained on the stems until next spring. During spring the attached dead leaves fell to the ground as litter (Fig. 4.17). In LL, however, above-ground biomass senesced later during winter and most of the standing dead fell to the ground before the first sampling in the following spring. A small amount, about 3–35 g m^{-2}, of green shoots persisted under the snow in both leys and continued to grow during the new season (Pettersson et al. 1986, Pettersson 1987).

Another source of litter are roots. In this section we are only dealing with macroscopic roots, i.e., "macroroots", visible to the naked eye. Rhizodeposition, i.e., turnover of finer roots and exudation of carbohydrates, is discussed in Chapter 5. Roots may die during the growing season even during periods when root biomass increases. However, death of roots in barley was limited and occurred mainly after maximum root biomass at anthesis. In the meadow fescue, parallel studies using soil coring, mesh bags and ^{15}N-labelling indicated that more than 50% of the over-wintering roots died before the second cut (Fig. 4.18). Between the first and second cut, the rate of root death decreased and after the second cut mainly roots produced the current year died (Hansson and Andrén 1986, Hansson and Pettersson 1989). In an experiment with ingrowth cores, roots produced the current year died during autumn (Steen 1984). According to Stuckey (1941) most roots of meadow fescue do not survive more than one year.

Decomposer organisms

Under this heading all organisms in the decomposer food chain will be discussed, i.e., microorganisms and saprovorous animals as well as the predators that feed on them. For convenience, the faunal community is separated into taxonomic groups, which are discussed separately. Below-ground herbivores, i.e., plant-feeding nematodes, are discussed together with the saprovorous nematodes.

Soil microorganisms, including protozoa
The most important factor regulating the growth and activities of soil microorganisms is the supply of carbon

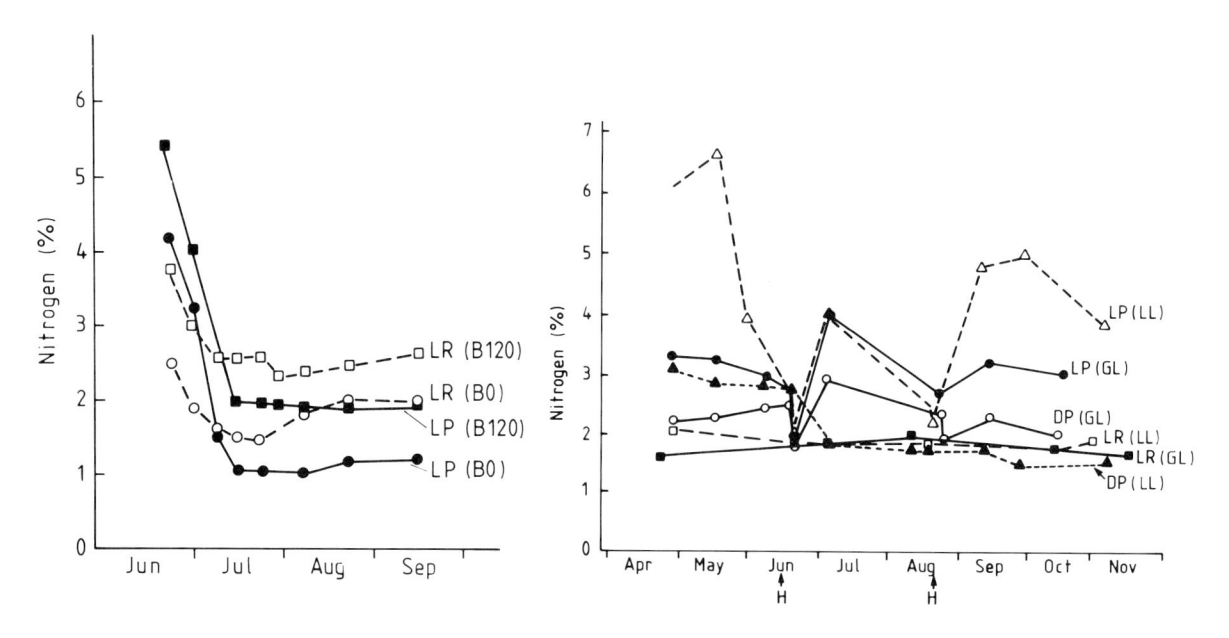

Fig. 4.16a. Nitrogen concentration in above- and below-ground plant biomass in B0 and B120. Based on results from 1981 and 1985 (Hansson et al. 1987, Pettersson 1987, Hansson et al. in press). LP = living plant parts above ground, LR = living roots.

Fig. 4.16b. Nitrogen concentration in above- and below-ground plant biomass in GL and LL. Based on results from 1982 and 1983 (Pettersson et al. 1986, Hansson and Pettersson 1989, Hansson et al. in press). LP = living plant parts above ground, DP = dead plant parts above ground, LR = living roots, H = harvest.

(Stotzky and Norman 1964). Growth rates are modified by soil temperature and soil moisture conditions. Calculations based on the annual carbon inputs to the microbial biomass of temperate arable soils, indicate that, on average, less than one new microbial generation can be produced per year (Tab. 4.16). These calculations are of course uncertain, both with regard to determinations of substrate inputs and to biomass estimates. Average values provide a very simplified and coarse picture of microbial life in soil. For example, the nutrient and energy situation in the rhizosphere is different to the situation in the bulk soil. Similarly, the situation in the upper soil layers near the soil surface is more dynamic than in deeper soil layers, due to greater litter inputs, as well as wider fluctuations in temperature and moisture. Nevertheless, it is evident that energy inputs only allow limited microbial growth in soil.

In Kjettslinge, abundance of bacteria and protozoa, length of fungal hyphae, and chloroform fumigation biomass, were determined several times during the growing season. There were no clear seasonal trends in length of fungal hyphae (Tab. 4.17). This contrasts with results from forest soils where seasonal variations have been reported, with high spring and autumn values and lower summer values (Bååth and Söderström 1982, Hunt and Fogel 1983). In forests, this pattern could be attributed to input of leaf litter during discrete periods of the year, coupled with very dry conditions in the litter layer during summer. At Kjettslinge, bacterial biomass tended to be higher during the growing season than during the rest of the year (Schnürer et al. 1986a).

Fumigation biomass carbon increased by 25% between early June and the middle of July in B120 (Fig. 4.19), which coincided with the period of maximum root production (Fig. 4.18). Fumigation biomass carbon then declined gradually until October. Direct counts of microbial biomass (Schnürer et al. 1986a), suggested that the increase in fumigation biomass was mainly due to increased bacterial biomass. A greenhouse experiment with nitrogen-fertilized barley grown in Kjettslinge soil gave further indication that active roots are an important carbon source for the microbial biomass. During five weeks of root growth, fumigation microbial biomass increased by approximately 100 µg C g^{-1} dry soil (Schnürer and Rosswall 1987). This increase in biomass carbon was close to that registered in the field (Fig. 4.19). On the other hand no seasonal pattern was observed in fumigation biomass nitrogen in a field lysimeter experiment with B120 and GL (Lindberg et al. 1989).

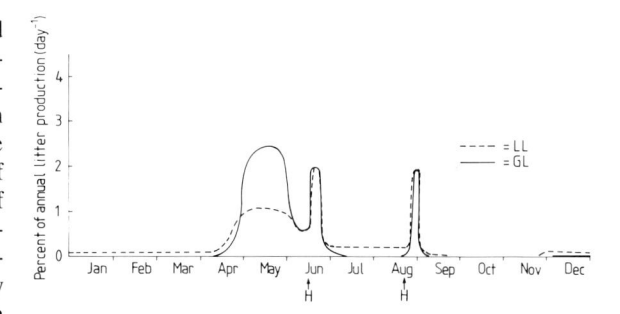

Fig. 4.17. Dynamics of litter production in GL and LL (Pettersson et al. 1986, Pettersson and Hansson submitted).

While it is not possible to determine the relative contribution of fungi, bacteria and protozoa to the fumigation biomass, the fumigation method may approximate the amount of cytoplasm-containing microbial biomass (Schnürer 1985). Thus, the sum of bacteria, protozoa and FDA-active fungal hyphae may be the best direct count value to compare with fumigation biomass, since much of the non-FDA hyphal lengths may be devoid of cytoplasm. Of the average total microbial biomass (measured by direct counts) in the Kjettslinge soil, fungi comprised about 75%, of which 1–10% were FDA-active, and bacteria constituted the remaining 25% (Schnürer 1985). Protozoa were of little importance in terms of total soil microbial biomass, with an amount equivalent to about 10% of the bacterial biomass. Based on these figures, the cytoplasm-containing fraction comprised approximately 30–35% of the microbial biomass, i.e., 230–280 µg C g^{-1} dry soil. If the assumptions made above are reasonable, this fraction should be roughly equivalent to the fumigation biomass, which amounted to approximately 400 µg C g^{-1} dry soil

Tab. 4.16. Microbial biomass, measured with the chloroform fumigation incubation method, and substrate input in B120, compared with two other temperate agroecosystems.

	Kjettslinge[a] (Sweden)	Rothamsted[b] (England)	Saskatchewan[b] (Canada)
Cropping system	B120	Wheat	Wheat-Fallow
Soil C (%)	2.6	1.2	2.4
Plant-C input[c] (µg C g^{-1} dry soil yr^{-1})	7 10^2	5 10^2	6 10^2
Microbial biomass-C[c] (µg C g^{-1} dry soil)	4 10^2	3 10^2	6 10^2
Microbial generations per year[d]	0.9	0.8	0.5

[a] Recalculated from Andrén et al. (1987) and Paustian et al. (in press).
[b] Recalculated from Voroney (1983).
[c] Sampling depth (cm): Kjettslinge 0–27, Rothamsted 0–23, Saskatchewan 0–24.
[d] Assuming 50% metabolic efficiency and no maintenance requirements.

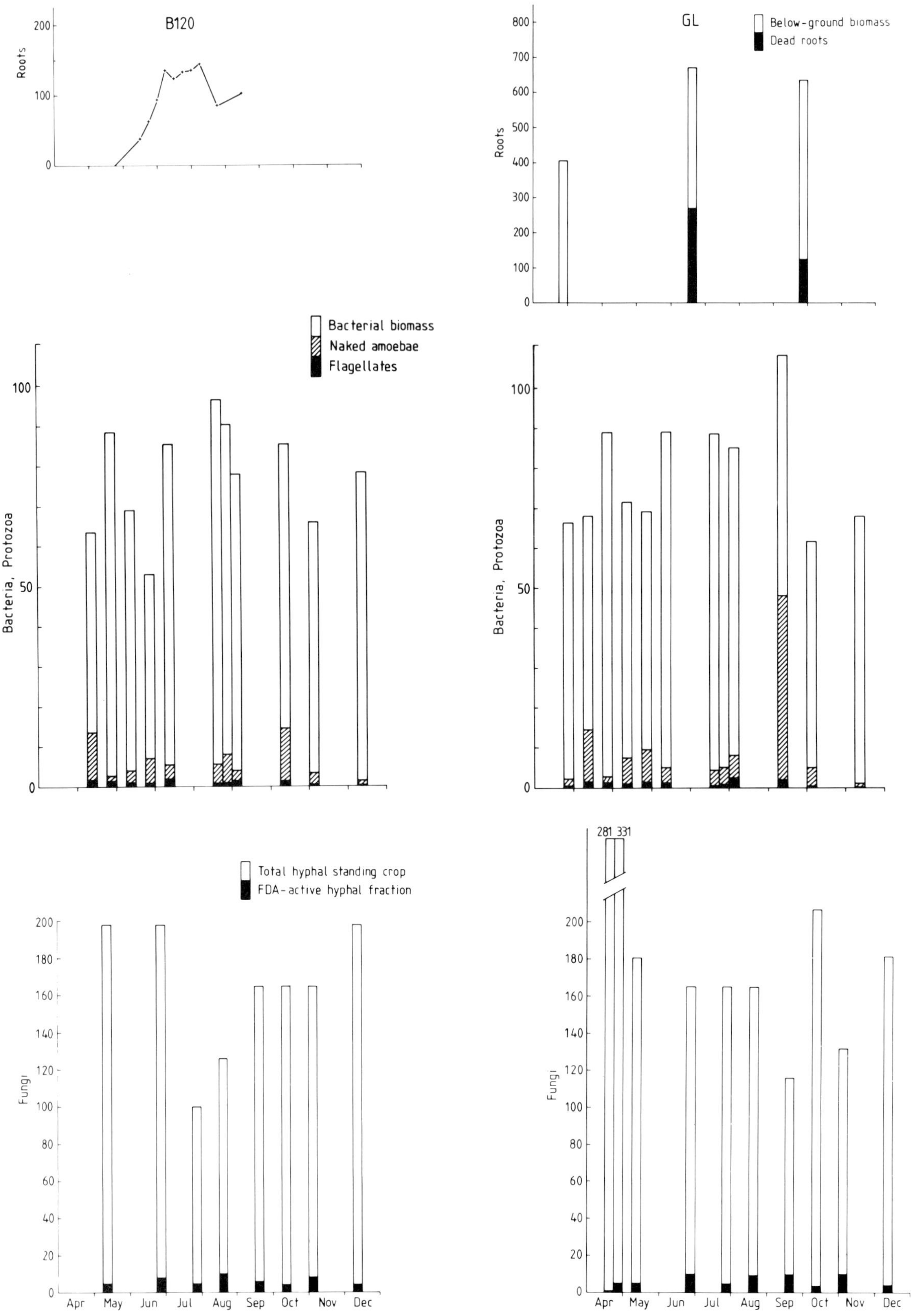
B120
Roots
GL
Below-ground biomass
Dead roots
Bacterial biomass
Naked amoebae
Flagellates
Bacteria, Protozoa
Total hyphal standing crop
FDA-active hyphal fraction
Fungi
281 331
Apr May Jun Jul Aug Sep Oct Nov Dec

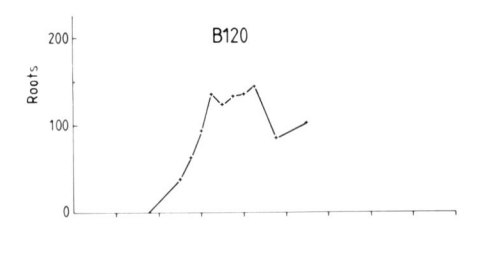

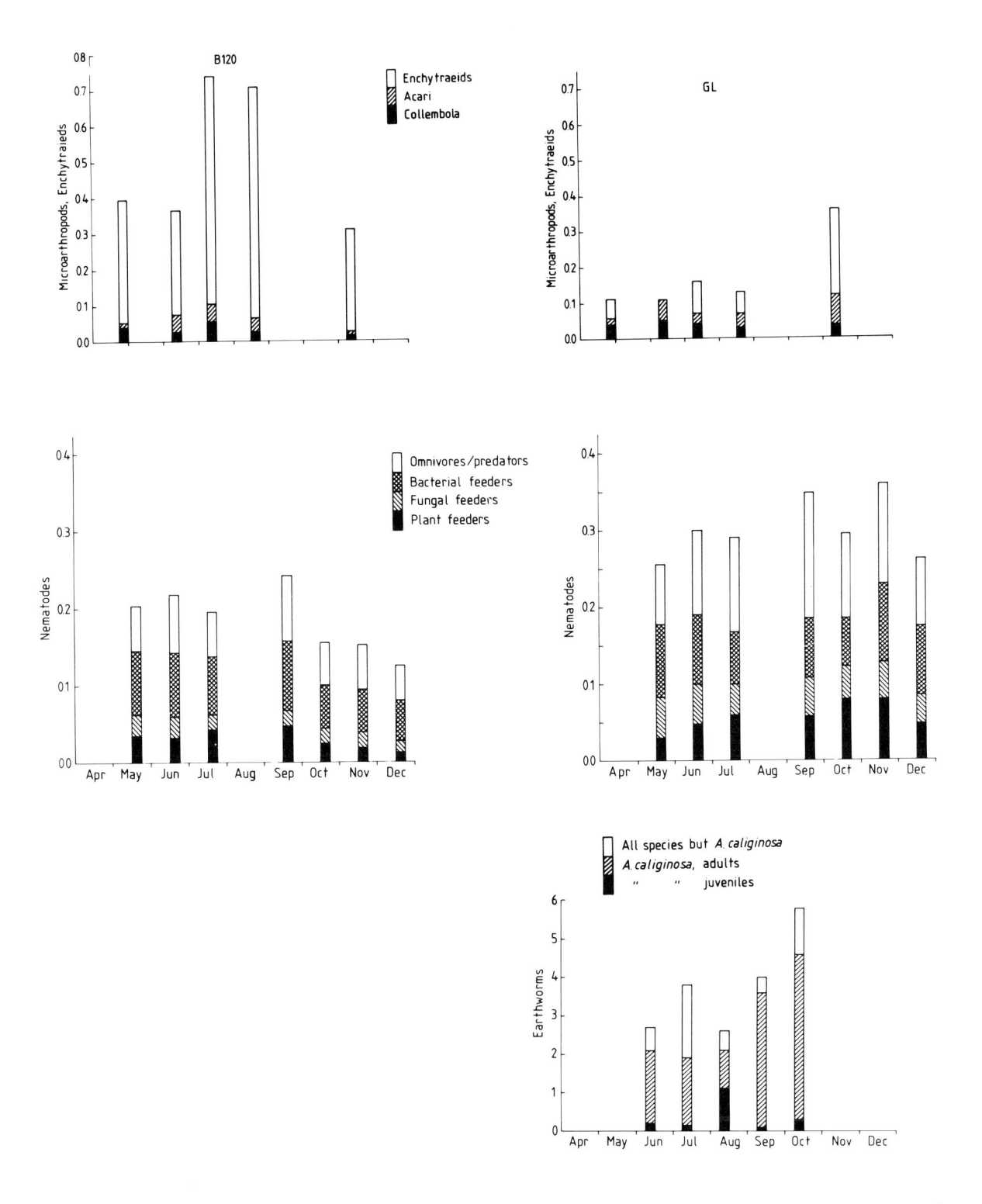

Fig. 4.18. Seasonal biomass dynamics of roots, microorganisms and soil animals in B120 and GL. Values in g dry weight m^{-2} (Hansson 1987, Hansson and Andrén 1986, Lagerlöf and Andrén in press, Lagerlöf and Andrén 1988, Lagerlöf et al. 1989, Schnürer et al. 1986a, B. Sohlenius and S. Boström unpubl., Boström 1988a). Conversion factors from length and numbers to weight were as follows: total and FDA-active fungi 1.14 10^{-6} g m^{-1}, flagellates 0.2 10^{-9} g, amoebae 0.8 10^{-9} g. Conversion factor for calculating dry weight excluding gut content for earthworms was 0.131. See Chapter 5 for conversion factor for calculating nematode dry weight.

Tab. 4.17. Seasonal variation in abundance of microorganisms and soil animals in B120 and GL during 1982 (1983 for earthworms). Standard errors are shown in parentheses under mean values. Values for microorganisms are per gram dry top soil. ND = not determined.

Organism group	April B120	April GL	May B120	May GL	June B120	June GL	July B120	July GL	August B120	August GL	September B120	September GL	October B120	October GL	November B120	November GL	December B120	December GL
Fungi[a]																		
FDA-active hyphal Length (m)	ND	8 (2) 1.7 (0.2)	59 (6) 1.2 (0.1)	54 (6) 1.1 (0.1)	47 (5) 1.2 (0.1)	59 (5) 1.0 (0.1)	32 (2) 0.6 (0.1)	28 (3) 1.0 (0.2)	57 (3) 0.8 (0.1)	57 (4) 1.0 (0.1)	3 (9) 1.0 (0.1)	59 (13) 0.7 (1.2)	22 (2) 1.0 (0.2)	21 (6) 1.3 (0.1)	49 (4) 1.0 (0.1)	55 (4) 0.8 (0.1)	26 (3) 1.2 (0.1)	24 (3) 1.1 (0.1)
Total hyphal length (m 10⁻³)	ND	6.3 (0.4)	5.4 (0.4)	5.5 (0.3)	4.6 (0.0)	5.7 (0.6)	7.4 (0.4)	7.9 (0.6)	7.6 (0.4)	7.5 (0.7)	5.8 (0.4)	6.4 (0.4)	5.5 (0.4)	4.7 (0.2)	4.8 (0.0)	4.2 (0.3)	6.0 (0.1)	5.1 (0.4)
Bacteria[a] (no. 10⁻⁹)																		
Protozoa[a] (no. 10⁻⁴)																		
Flagellates	ND	2.1 (0.1)	5.9 (1.0)	4.8 (1.2)	3.5 (0.4)	2.6 (0.3)	6.3 (0.6)	5.1 (0.8)	2.6 (0.3)	2.1 (0.4)	5.1 (0.9)	8.3 (2.0)	4.6 (1.3)	6.0 (1.6)	1.7 (0.3)	1.7 (0.4)	1.0 (0.4)	1.6 (0.4)
Amoebae	ND	1.1 0.0	9.4 (1.8)	7.8 (1.9)	1.9 (0.1)	2.5 (0.5)	2.3 (0.6)	5.5 (1.3)	3.2 (0.5)	2.4 (0.5)	1.6 (0.4)	3.8 (0.3)	9.1 (2.4)	31.8 (5.6)	2.3 (0.4)	3.2 (0.8)	0.6 (0.5)	0.2 (0.2)
Enchytraeids[b] (no. m⁻² 10⁻⁴)	ND	ND	ND	ND	1.2	0.1	0.9	0.2	1.8	0.2	1.0	0.4	ND	ND	0.4	0.4		
Microarthropods[c] (no. m⁻² 10⁻⁴)																		
Collembola	2.7	3.4	ND	ND	1.9	2.0	3.3	1.7	2.2	2.1	ND	ND	ND	ND	0.8	2.2		
Acari	0.4	1.6	ND	ND	1.7	1.5	2.1	1.4	2.6	3.8	ND	ND	ND	ND	2.3	4.0		
Nematodes[d] (no. m⁻² 10⁻⁶)																		
Plant feeders	ND	ND	7.2 (0.3) 1.8 (0.6)	7.4 (0.3) 0.8 (0.5)	7.2 (0.4) 1.6 (0.3)	7.9 (0.3) 1.3 (0.4)	6.8 (0.7) 2.1 (0.1)	6.9 (0.4) 1.6 (0.6)	ND	ND	8.0 (0.4) 2.4 (0.2)	7.9 (0.6) 1.6 (0.5)	5.2 (0.2) 1.3 (0.2)	7.2 (1.2) 2.2 (0.4)	4.9 (0.3) 0.9 (0.2)	9.0 (0.7) 2.2 (0.6)	4.0 (0.2) 0.6 (0.1)	7.0 (0.4) 1.3 (0.2)
Fungal feeders	ND	ND	1.5 (0.6)	2.8 (0.5)	1.6 (0.3)	2.9 (0.4)	1.2 (0.1)	2.2 (0.6)	ND	ND	1.1 (0.2)	2.7 (0.4)	1.2 (0.2)	2.2 (0.4)	1.2 (0.2)	2.5 (0.6)	0.8 (0.1)	2.0 (0.2)
Bacterial feeders	ND	ND	3.5 (0.5)	3.3 (0.3)	3.4 (0.5)	3.1 (0.4)	3.0 (0.5)	2.4 (0.2)	ND	ND	3.9 (0.2)	2.7 (0.4)	2.3 (0.4)	2.2 (0.3)	2.3 (0.2)	3.6 (1.2)	2.2 (0.3)	3.1 (0.8)
Omnivores/Predators	ND	ND	0.4 (0.1)	0.4 (0.1)	0.6 (0.2)	0.6 (0.2)	0.4 (0.1)	0.7 (0.2)	ND	ND	0.6 (0.1)	0.9 (0.1)	0.4 (0.1)	0.6 (0.1)	0.4 (0.1)	0.7 (0.2)	0.3 (0.0)	0.5 (0.1)
Earthworms[e] (no. m⁻²)																		
Adults	ND	ND	ND	ND	ND	17.7 (8.4)	ND	22.7 (5.6)	ND	9.3 (3.4)	ND	25.6 (4.8)	ND	29.3 (9.9)	ND	ND	ND	ND
Juveniles	ND	ND	ND	ND	ND	30.0 (4.4)	ND	50.0 (22.2)	ND	30.7 (7.0)	ND	20.0 (5.4)	50.7	ND (18.1)	ND	ND	ND	ND

[a] Schnürer et al. 1986a.
[b] Lagerlöf et al. 1989.
[c] Lagerlöf and Andrén 1988, Lagerlöf and Andrén, in press.
[d] Boström and Sohlenius 1986.
[e] Boström 1988a.

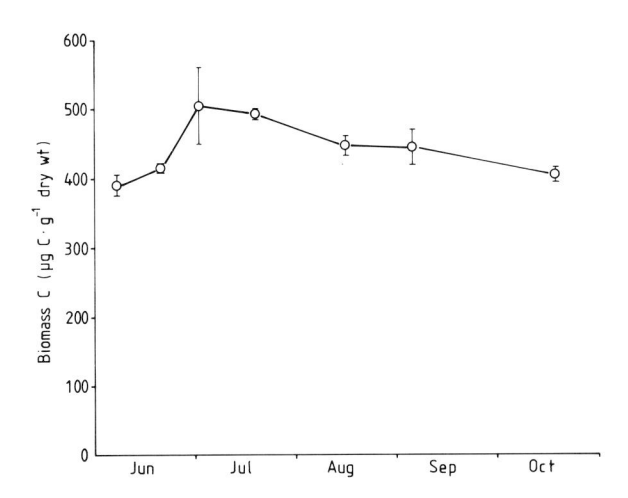

Fig. 4.19. Microbial biomass carbon dynamics in B120 top soil. Determined by the chloroform fumigation incubation method and calculated without subtraction of the control. Bars indicate standard errors (n=5) (Schnürer 1985).

(Tab. 4.16). However, FDA-active hyphal length may underestimate the total amounts of cytoplasm-filled hyphae. If the proportion of cytoplasm-filled hyphae is taken to be 15–37% of the total hyphal length, as observed by phase contrast microscopy in a forest soil (Frankland 1975), the cytoplasm-containing microbial biomass becomes 310–442 µg C g^{-1} dry soil, which is close to the fumigation biomass. The results support the view that the fumigation biomass measures a fraction of the total microbial biomass, presumably the cytoplasm-filled one.

Occasional short-term variations in direct counts of microorganisms were observed during the growing season. In B120, short-term variations in bacterial biomass seemed to be positively correlated with soil moisture contents (Schnürer et al. 1986a). During dry periods, a sudden rain can synchronize microbial activities and lead to large, though temporary, increases in microbial abundance and activity. The effects of soil moisture on microbial biomass dynamics and activity were studied in an irrigation experiment in B120, where bacteria, protozoa, fungi and oxygen consumption were measured frequently during a two-week period (Schnürer et al. 1986b). After an initial period when the soil was dry, one treatment received daily irrigation while the other treatment received rainfall on only two occasions. Oxygen consumption reacted most rapidly to the changes in water availability, and preceded increases in the soil microbial biomass. After an initial flush in bacteria numbers following wetting of the dry soil, values in both treatments declined (Fig. 4.20). A likely reason for the bacterial growth pattern is that the rainfall released readily-decomposable material, but after a few days the bacteria became carbon-limited and numbers decreased.

After wetting of dry soil, FDA-active hyphae reacted in a similar fashion to bacteria, however, total hyphal length development differed substantially. A near constant hyphal length was maintained in the rain treatment, while in the irrigation treatment there was an increase from 700 to 1700 m g^{-1} dry soil during the same period (Fig. 4.20). In simulations of the fungal biomass dynamics, the differences between treatments could be explained by the interaction between soil moisture, substrate availability and growth rates (Paustian and Schnürer 1987a, b). In the rain-only treatment active hyphae initially responded to wetting of the soil but the rapid drying of the soil inhibited hyphal length increases until the second rainfall. In the irrigation treatment there was a similar initial response to wetting the soil. However, favourable moisture conditions allowed growth to continue under C-limited conditions and there was a continuous increase in total hyphal length, largely subsidized by translocation of cytoplasm from older hyphae into the new growth.

As suggested earlier in this chapter, a large amount of the bacterial production in soil may be confined to the near-root rhizosphere. Most of the bacteria in bulk soil are not actively growing except for brief periods after organic matter addition. Similarly, only a small fraction of the total fungal hyphae, both in the rhizosphere and in the bulk soil, are in an active or potentially active state. However, when favourable conditions occur substantial net production of hyphae can occur and fungi are able to exploit a large volume of soil. The significance of the hyphal growth form can be exemplified by considering the doubling in total hyphal lengths observed in the irrigated treatment of the experiment described above. If this growth had represented complete synthesis of both walls and cytoplasm, i.e., no net translocation, then a supply of over 800 µg C g^{-1} dry soil would have been required; this is about equal to the total annual input of carbon, expressed on a per gram soil basis (Schnürer and Paustian 1986). However, if growth mainly involved synthesis of new cell walls together with translocation of cytoplasm from existing biomass, then C requirements for the doubling of total hyphal length would be only 150 µg C g^{-1} dry soil, which is more in line with the C supply to soil. It should be added that irrigation, combined with high soil temperatures, provided a near optimal abiotic environment, so that microbial production rates measured in this experiment were undoubtedly higher than normal.

It has been suggested that a soil has a 'protection capacity' for a certain amount of microbial biomass and that this capacity is related to soil texture. Biomass formed in excess of the protection capacity is assumed to die at a relatively high rate because of exposure to predators or unfavourable microclimatic conditions. The protection capacity is higher in clay soils than in sandy soils, which is believed to be a result of better environmental conditions and/or a greater number of soil pore spaces which are inaccessible to predators

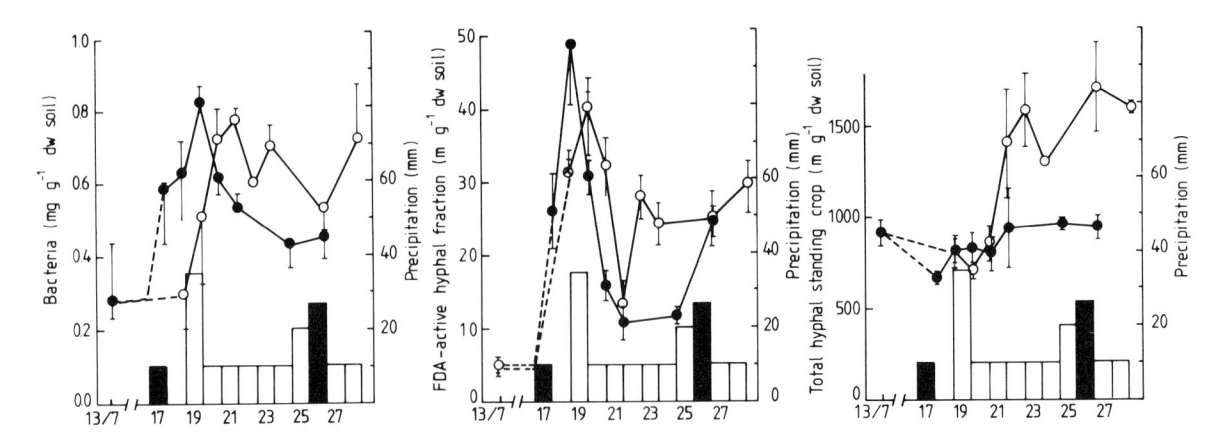

Fig. 4.20. Bacterial biomass, FDA-active hyphal length and total hyphal length in an irrigation experiment in B120. Open symbols = irrigated treatment. Filled symbols = rain-fed treatment. Bars show water input in mm. Open bars irrigation, filled bars rainfall. (Means ± standard error, n=5) (Schnürer et al. 1986b).

(Elliott et al. 1980, van Veen et al. 1984). Microbial measurements at Kjettslinge, both by direct counts and fumigation, appeared to fit this concept. Biomass tended to be maintained around a base level, with occasional increases followed by a return to the base level. In addition, the base levels were roughly the same for the two intensively studied cropping systems, B120 and GL, which would be expected if soil textural properties (which were the same for all treatments) are a main factor determining the protection capacity.

The pattern of biomass fluctuations above the protection capacity of the soil will vary for different predator – prey combinations. Bacteria are consumed by bacteriovores like protozoa and nematodes, which in turn are grazed upon by many other soil animals (Chapter 5). Bursts of bacterial growth cause protozoan cysts to excyst and to grow exponentially for a short period until bacteria are grazed down to the protection capacity level. At that time protozoan numbers decrease and some of them encyst. The whole sequence from cyst to cyst takes about 10 d (Pussard and Delay 1985). Other bacterivores and protozoan predators (such as nematodes) increase their populations (and thus grazing pressure) more slowly as compared with protozoa grazing on bacteria. It is therefore likely that large populations of protozoa could build-up locally and temporarily under favourable conditions. This was occasionally observed in the Kjettslinge field (Schnürer et al. 1986a).

At Kjettslinge, amounts of FDA-active hyphae were always small in relation to total hyphal lengths. Under long periods of favorable growth conditions (high moisture or surplus of available C), total hyphae accumulated (Schnürer et al. 1986b, Wessén and Berg 1986). Since fungal predators like Collembola and nematodes multiply comparably slowly, and since fungal hyphal walls (the dominant structural component of the fungal biomass) may be more persistent than bacteria, total fungal hyphae will normally show less pronounced short-term variation.

Methodological problems associated with microbial biomass measurements in soil make it difficult to distinguish real changes in biomass from random variation. The precision of a microbial parameter estimation is greatly dependent on sample size. The soil samples for direct counts were small (1–5 g fresh weight) and further diluted before the actual count. This makes the estimates extremely sensitive to the heterogeneity of the soil. The variation between replicates is fairly small at times of low microbial activity, i.e., when the microorganisms are at protection capacity levels. A much larger variation is recorded under periods of growth, due to the heterogeneity of the substrate distribution. Samples for fumigation biomass determinations were larger (50 g fresh weight), and therefore less sensitive to spatial heterogeneity. The coefficient of variation is generally around 4–10% for fumigation biomass, 15–30% for direct counts of bacteria, but often >100% for FDA-active hyphae and protozoa.

In conclusion, occasional fluctuations in soil microbial biomass were induced mainly by favorable moisture conditions, which created a favourable abiotic environment and temporarily increased the supply of energy-rich material. The resulting increases in active microbial biomass were subject to rapid consumption by microbivores, most clearly illustrated in the cases of bacteria and protozoa. The C budget estimates of the Kjettslinge cropping systems indicated that a major part of the seasonal microbial production was consumed by microbivores (Chapter 7). The influence of roots and microbial grazing on C and N flows are treated in more detail in Chapter 5.

Nematodes, including root-feeders
Important factors determining the potential abundance

response for a species to seasonal variation of, e.g., food availability, are generation time and fecundity. The generation times of nematode species vary widely (Nicholas 1984). Some species of bacterial feeders, such as *Rhabditis* s.l., reach maturity after 3–4 d at field temperatures. Quite a few members of Tylenchida and Cephalobidae have generation times of 10–20 d. The large dorylaims have much longer generation times, sometimes up to one year. Fecundity also varies widely between nematode species and may range from >500 eggs/female in rhabditids down to about 10 eggs/female in dorylaims (Nicholas 1984).

Due to their relatively short generation times and high fecundity, the abundance of some nematode feeding groups may follow short-term changes in the abundance of their food sources. If this is the case, the number of microbial feeders can be an indicator of microbial production. Similarly, the development of root systems of host plants and the abundance of plant-feeding nematodes are frequently correlated (Andersen 1979b, Ferris and Ferris 1974). Covariation of nematode abundance and root biomass was observed in B120 where both abundance and biomass decreased after harvest, cultivation and ploughing (Tab. 4.17, Fig. 4.18). Most of the decrease in abundance and biomass was caused by a decrease in root- and bacterial-feeders. This indicates that the bacterial feeders utilize the rapid flux of nutrients in the food chain: roots – root exudates – bacteria – bacterial feeding animals. Increases of abundance of plant and bacterial feeders, although not statistically significant ($p > 0.05$), occurred in GL during the autumn (Tab. 4.17). In addition to an increase in root biomass, soil moisture was higher in the autumn than during the summer months (Chapter 3). In forest soils nematode abundance generally increases during the autumn, due to high water contents and nutrient input (Sohlenius 1979).

The population densities of fungal feeders were remarkably stable in both B120 and GL, with a higher abundance in meadow fescue than in barley. Omnivores/predators were much less abundant than other feeding groups. Generally the highest densities were found under GL. For further details see Boström and Sohlenius (1986).

Microarthropods and enchytraeids
The dominant microarthropod and enchytraeid species found in Kjettslinge are able to reproduce during most of the growing season, and a population decline during an unfavourable period can rapidly be compensated for by an increase during a favourable one. This flexibility in reproductive timing is characteristic for euedaphic and hemiedaphic species (Petersen 1980) – to which all the dominant species in the soil under barley belonged. The epedaphic species, i.e., surface and litter dwellers, have a more defined reproduction season; and therefore, they cannot compensate as rapidly for an unsuccessful reproductive period. It is not surprising that such species were more common in the more environmentally stable leys than in barley.

Apart from population dynamics induced by biotic factors, environmental factors have a great influence on seasonal dynamics of the population. For example, from late autumn 1980 to spring 1981 heavy rain and waterlogging followed by a hard frost without protecting snow cover resulted in low numbers of Collembola. Dry conditions can also reduce populations, e.g., the superficially living *F. fimetaria* may have suffered from the dry, warm weather during the summer of 1982 (Tab. 4.17).

Agricultural practices may drastically change environmental conditions in the soil, e.g., ploughing mechanically destroys some of the fauna while also trapping animals in deep soil. Thus the overall effect of ploughing is to cause a rapid decrease in the abundance and often the diversity of the fauna. Both Collembola and Acari decreased considerably during the autumn after ploughing in B120, whereas the conditions in the unploughed GL were more favourable, and the litter layer and uppermost part of the soil was moist (Tab. 4.17, Fig. 4.18).

From the results of this investigation it is hard to generalize about the yearly dynamics of enchytraeids, but mortality seemed to be very high during dry periods of the growing season, and hatching and growth was rapid when moisture and temperature were suitable. Evidently the time for these events changed from year to year. For example, during 1981 and 1982 the population dynamics of enchytraeids in B120 and GL showed different trends. GL had a summer peak in 1981 with significantly higher densities than B120 ($p < 0.001$). In 1982 low numbers of enchytraeids were found in GL throughout the growing season, while B120 had significantly higher densities. In November 1982 the total abundance in B120 had declined to the same levels as in GL (Tab. 4.17). Biomass fluctuations were similar to the abundance values.

Soil and soil surface macroarthropods
In May 1981 all macroarthropod groups were found in low densities (Carter et al. 1985). Ice cover and waterlogging during the winter 1980–81 resulted in high mortality of the ley crops, and apparently the macroarthropod populations also suffered. During the following summer and autumn an increase of most groups was observed. This was especially obvious for Diptera larvae, especially Chironomidae, which reached a population density of over 9 000 ind m^{-2} in the lucerne ley in late autumn/early winter. The abundance of all other Diptera families in lucerne was, at the same time, about 500 ind m^{-2}. Diptera abundance in the grass ley was much lower than in lucerne, while it was similar in the other groups. The soil and surface active Collembola species *Isotoma viridis* was numerous in the leys in December 1981.

In 1982, when macroarthropods were sampled

monthly from April to December, the considerably higher abundances of most macroarthropods in the leys than in the barley crops could be clearly seen (Fig. 4.21). The dominant groups were Diptera larvae, *Isotoma viridis* (Collembola) and Coleoptera, followed by spiders (Araneae), thrips (Thysanoptera) and several other groups such as Homoptera, Hymenoptera, Lepidoptera, Psocoptera and Myriapoda.

Chironomidae and most other Diptera larvae were most abundant during spring, autumn and winter. This is due to their life cycle consisting of egg laying during summer and autumn, larval development during autumn, overwintering in the larval stage, pupation in the spring and hatching from pupae during spring and early summer. This type of univoltine life cycle is interrupted by autumn ploughing in annual crops, such as spring barley.

Isotoma viridis hibernated as adults or subadults in the leys. The population increased during the spring and summer and by analysis of numbers and size class distribution it could be concluded that two generations were produced during the season, one in May-June and one in August-September. Only low numbers of *I. viridis* were found in the barley cropping systems throughout the year.

Among the dominant beetle families, Carabidae and Staphylinidae, the highest abundances were also found in the leys. Spring and late autumn populations of larvae and adults were negligible in the barley cropping systems, but they increased during the summer.

Up to 400 ind m^{-2} of adult Staphylinidae were found in the grass ley during summer, when about 150 ind m^{-2} were found in LL and B120 and 70 ind m^{-2} were recorded in B0. A peak in numbers of Staphylinidae larvae was found in all cropping systems in June, between 400 ind m^{-2} in GL and 100 ind m^{-2} in B0. This indicates that adult Staphylinidae immigrated to the experimental plots during spring and early summer and laid eggs, which gave a peak of adults in late summer and autumn. In the ploughed barley only few adult or larval Staphylinidae were found in November and December while the abundance remained at a high level in the leys. The tendency for higher numbers of both adults and larvae in B120 than in B0 may have been a positive response to denser vegetation cover in B120.

Carabidae larvae and adults were found in low numbers in all crops during late autumn 1981. By immigration and reproduction, larval and adult populations increased during spring and early summer 1982. Larval abundances peaked in LL in June when 425 ind m^{-2} were found. In the other crops the abundances remained at a lower level, around 25–50 ind m^{-2}. Autumn and winter abundances in LL and GL were about 50 ind m^{-2} while in barley they decreased to near zero. Carabidae adult numbers were similar in all cropping systems during the growing season and between 20 and 60 ind m^{-2} were found, with high temporal and spatial variations. In late autumn, abundances in barley decreased to very low levels while they remained at summer levels in the leys.

Thrips (Thysanoptera) are small insects with a body length up to 3 mm. The wings of the adults consist of narrow strips with broad fringes of hair on each side. They have sucking mouth-parts and many species are plant feeders, feeding on above- and below-ground plant parts. Other are fungal feeders or predators, and mainly live in the soil or litter layer. Many species hibernate in soil and litter as adults. During the summer the highest numbers of thrips were found in the barley cropping systems, up to 1000 ind m^{-2} in B120 and slightly fewer in B0, while numbers during spring and autumn were the highest in the leys, up to 500 ind m^{-2} in GL. An increase was also observed in the leys during summer, but the abundances did not become as high as those in the barley. Spiders (Araneae) occurred in low numbers in the barley during the whole sampling period, ≤25 ind m^{-2}, while in the leys up to 250 ind m^{-2} were recorded during summer, and 50–75 ind m^{-2} were recorded in December. The most common family was the small Linyphiidae, which construct horizontal webs in the lower part of the vegetation or over open soil.

Less common arthropod groups such as Lepidoptera larvae, Hymenoptera, Chilopoda and Diplopoda showed similar distribution and dynamics patterns as the groups mentioned above, i.e., they occurred in low numbers or were absent in the barley during spring and autumn and increased during the summer. In the leys they were present all year, most often with the highest abundances occurring in summer.

All macroarthropod groups had a shallow depth distribution and most specimens were extracted from the upper 5 cm layer of the soil.

In Fig. 4.22 the abundance of the dominant macroarthropod groups in late November-early December in 1981–1983 is shown. The higher abundances of wintering macroarthropods in the leys than in the barley cropping systems is clearly shown. The between-year variation was great for most groups. Diptera larvae occurred in the highest numbers in 1981, while most other groups were more common during 1982–1983. The ley crops, LL and GL, in general had the highest abundances of macroarthropods.

The sampling technique used with large samples gave low extraction efficiency for the small Diptera larvae, Coleoptera larvae and Collembola. Extraction efficiency may also have been lower for ploughed soil than for undisturbed ley profiles, since animals might be locked up in pockets of surface litter that were ploughed under.

Soil and surface macroarthropods showed greater differences in abundance between the annual barley and the perennial leys than any other animal group studied at Kjettslinge. The major factor inducing this difference was the annual ploughing in autumn, and resultant lack of vegetation during winter, and the ploughing under of surface and litter-dwelling species to depths from where

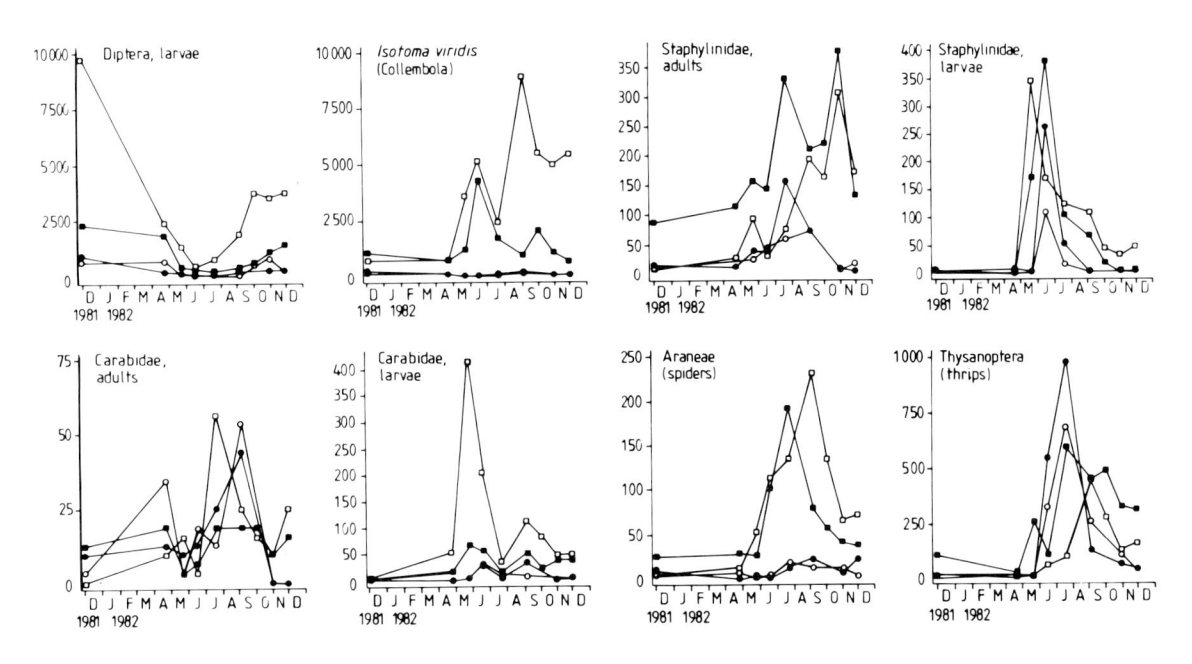

Fig. 4.21. Dynamics of major groups of soil and surface macroarthropods (no. m^{-2}) during December 1981–December 1982 in Kjettslinge. Open circles = B0, filled circles = B120, filled squares = GL, open squares = LL. (J. Lagerlöf unpubl.).

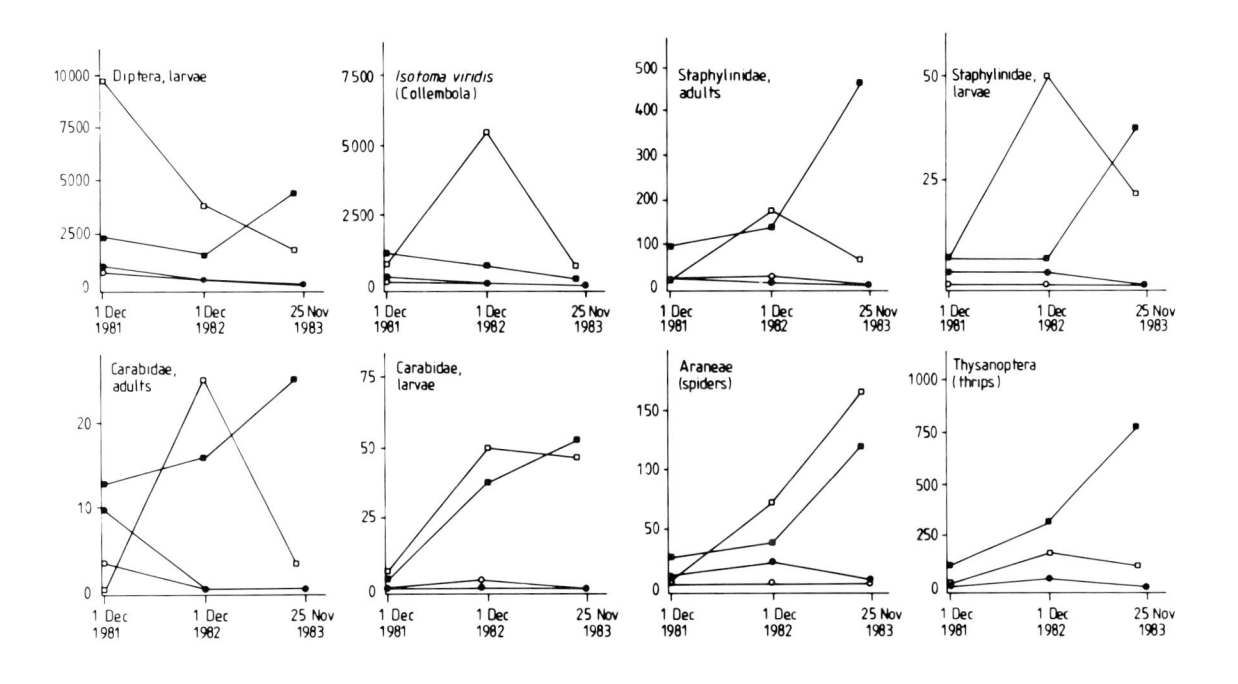

Fig. 4.22. Winter populations of major soil and surface macroarthropod groups (no. m^{-2}) during 1981–1983 in Kjettslinge. Open circles = B0, filled circles = B120, filled squares = GL, open squares = LL. (J. Lagerlöf unpubl.).

they could not come up to the surface again. Whether LL or GL showed the highest abundances varied from year to year, probably due to how favourable the weather was during the season for the growth of each crop.

The higher numbers of Staphylinidae and thrips in B120 than in B0 during summer indicated that thrips, like aphids, prefer dense vegetation that provides more plant juice (see the section "Herbivores"). Their higher abundances can result in increased abundances of Staphylinidae which prey on aphids and thrips.

Earthworms
Soil temperature and humidity influence the amount of time needed for earthworms to reach sexual maturity. Thus the age for maturation may vary considerably under field conditions, from 6 months to 1.5 yr for *A. caliginosa* (Gerard 1967, Nordström 1976, Bengtson et al. 1979). The breeding life of most lumbricids has been estimated to be only a few months long while the longevity of adult *A. caliginosa* was found to be at least 12–14 months (Satchell 1967, Bengtson et al. 1979). The average lifespan of *L. terrestris* in the field was estimated to about 15 months, although a few individuals survived for nine years (Satchell 1967).

Several earthworm species have been reported to enter diapause, a resting stage when the worm empties its intestines and coils itself inside a small chamber constructed at the end of a burrow. Earthworms can also enter quiescence, when the earthworm ceases feeding and goes into a torpid stage without excavating a resting chamber (Olive and Clarc 1978). Facultative diapause and quiescence are initiated by deteriorating environmental conditions. The type of resting stage varies between, as well as within, earthworm species. The occurrence of resting stages prolongs earthworm generation time and causes it to vary between different populations of the same species.

The number of cocoons produced annually by one adult varies with earthworm species, food supply, soil temperature and soil humidity (Evans and Guild 1948a, Boström 1987). *L. rubellus* produces 79–106 cocoons annually and *A. caliginosa*, 12–35, while *A. rosea* produces only 3–8 and *A. longa,* 8 cocoons (Boström 1988a, Evans and Guild 1948b, Graff 1953b).

The incubation time for cocoons varies with soil climate, being prolonged due to high or low temperatures and humidity. The incubation time for cocoons produced by *A. caliginosa* is 3–11 months, for *A. rosea* and *L. rubellus,* four months and for *A. longa,* three months (Boström 1988a, Evans and Guild 1948b, Nowak 1975). Juveniles emerging early in the spring in climates with cold winters most certainly originate from cocoons produced the preceding year.

A. caliginosa cocoons are yellow when newly produced, but turn reddish as the embryo develops and the time of hatching approaches. In June 1983 a large number of yellow cocoons and a low number of juveniles were found in GL. There were few red cocoons, which indicated that hatching probably did not commence before the end of June. In July the number of newly hatched juveniles as well as adults had increased (Tab. 4.17). At the beginning of August the sampled numbers of juveniles and adults had decreased drastically due to drought, and all worms found were inactive, i.e., their alimentary canals were empty. A portion of the population had probably moved downward and some individuals had probably died due to low soil humidity and high soil temperatures. Due to increased precipitation the number of adults increased in late September while the number of juveniles still decreased. The abundance of juveniles first began to increase in October after cocoon hatching had commenced. The number of both red and yellow cocoons decreased as juvenile abundance increased, which indicated a new peak of hatching. However, none of these changes were statistically significant. Earthworm biomass fluctuated in the same way as abundance although the changes were less pronounced (Fig. 4.18). Emergence periods for *A. caliginosa* in autumn and spring have also been recorded from southern Sweden (Rundgren 1977).

Considering the temperature, the number of adults and the cocoon production rates, it can be estimated that about 260 cocoons m^{-2} were produced in the meadow fescue ley during 1983 (Boström 1988a). Since the individual cocoon mass is positively correlated to the mass of adults producing them (Lofs-Holmin 1983, Phillipson and Bolton 1977), the high mean mass of the cocoons indicated that it was mainly the larger adults in the field that were reproducing.

Long-term dynamics of the components

Five years may seem a short time to obtain results usable for a discussion of long-term dynamics. The time span certainly is short if the intention is to discuss long-term dynamics of total soil nitrogen or carbon. However, if the purpose is to study soil organisms with comparatively short generation times, whose community structure and function are influenced by the quality and quantity of the annual food resource (Heal and Dighton 1985), five years of study can be sufficient for drawing some conclusions.

It should not be expected that the cropping systems developed into different equilibrium states during the five-year-long experimental period, since a constant management during many decades is required for an agroecosystem to reach steady-state (Jenkinson et al. 1987). However, due to continuous changes of cultivation practices, crops, crop rotations, a transitional state may be characteristic of agroecosystems. In this context five years with constant management may be considered "long-term".

Agroecosystems are often described as highly disturbed systems, i.e., organism communities in arable land are exposed to repeated management measures.

The types of disturbances differ between annual and perennial crops. In one sense the perennial system can be described as less disturbed than the annual, since no annual soil cultivation is performed. However, managed perennial crops are disturbed several times during the growing season by traffic in connection with fertilization and harvest. At harvest the phenological stage of the crop is suddenly set back, which implies a changed pattern of carbon allocation within the plant, and the microclimatic conditions just above the soil surface are changed (Curry 1986).

Besides direct influences of management practices, perennial and annual crops have different indirect effects on soil organisms. For example, a perennial crop covers the soil all year round, thus creating different soil moisture and temperature conditions as compared to an annual crop, which only covers the soil for a part of the year. In general, at Kjettslinge the meadow fescue ley had the lowest soil water content, while the fertilized barley was slightly moister than the grass ley during the growing season and substantially wetter during spring and autumn. The unfertilized barley had the highest soil water content of all cropping systems during the growing season (See Chapter 3).

Further, both the quality and quantity of organic material supplied to the soil differs between crops. There were lower nitrogen concentrations in B0 roots compared with roots in the other cropping systems. Nitrogen concentration in stubble, however, did not differ between the barley cropping systems.

Although the total supply of carbon and nitrogen in harvest residues is discussed in detail in Chapters 5 and 7, it is important when discussing long-term changes in structure to consider that the differences in carbon input to the soil were less than would be expected from the differences in total production. In other words, the differences in production were mainly manifested as differences in the exported harvests.

Following changes in structure, i.e., spatial distribution, species composition and abundance of organisms, over a longer period of time requires dealing with several specific problems. It is extremely resource-consum-

Tab. 4.18. Abundance (no. m^{-2}) and biomass (mg m^{-2}) of above-ground herbage arthropods in the cropping systems at Kjettslinge. Mean values are from 1982–1983 (Curry 1986).

Organism group	B0 Abundance	B0 Biomass	B120 Abundance	B120 Biomass	GL Abundance	GL Biomass	LL Abundance	LL Biomass
Collembola	322	12	328	9.6	830	38.4	1915	120.8
Acari	220	0.8	372	1.3	144	1.5	440	2.0
Hemiptera	96	17.8	1088	76.4	266	40.2	168	68.4
Thysanoptera	224	4.2	432	8.0	228	3.9	126	2.0
Coleoptera	57	9.0	122	22.5	126	17.5	134	22.6
Lepidoptera	<1	<1	3	0.7	1	0.2	4	5.0
Hymenoptera	36	1.6	60	2.9	83	6.0	256	12.4
Diptera	80	8.6	193	17.1	370	41.1	514	55.2
Araneae	52	13.8	64	16.9	91	23.9	123	32.4
Neuroptera	1	1.2	3	3.2	2	2.1	1	1.3
Psocoptera	0	0	7	<1	<1	<1	2	<1
Opiliones	0	0	<1	<1	<1	<1	1	<1
Chilopoda	0	0.6	<1	1.7	<1	0.4	<1	0.8
Dermaptera	0	0	0	0	0	0	1	1.8
Diplura	10	<1	2	0	<1	<1	<1	<1
Total	1098	69.6	2674	160.3	2141	175.2	3685	324.7

Tab. 4.19. Microorganisms in the top soil in the cropping systems at Kjettslinge. Mean values per gram dry soil for the September samplings 1981–83 and for 20 sampling occasions during two field seasons (Schnürer et al. 1986a).

Organism group	September samplings B0	B120	GL	LL	Two field seasons[1] B120	GL
Fungi						
FDA-active hyphal length (m)	74	58	90	96	39[a]	39[a]
Total hyphal length (m 10^{-3})	1.2	1.1	1.2	1.1	1.0[a]	1.2[a]
Bacteria						
Abundance (no. 10^{-9})	4.6	5.7	6.0	5.7	5.7[a]	5.7[a]
Biomass (mg)	0.38	0.45	0.49	0.45	0.46[a]	0.47[a]
Protozoa (no. 10^{-4})						
Flagellates	8.6	5.9	10.3	7.2	3.7[a]	4.6[a]
Amoebae	7.2	1.7	3.5	2.8	3.2[a]	5.1[b]

[1]Values followed by the same letter do not differ significantly (p <0.05) between cropping systems.

ing to sample intensively for several years; one or a few occasions must be chosen. Organisms with short generation times are likely to be influenced by events immediately before the sampling occasion and not only by the integrated effects of the overall cropping system. Organisms with longer generation times, on the other hand, are not as influenced by, e.g., occasional changes in weather. However, the seasonal maximum of different species may not coincide. In this study, yearly samplings of all soil organisms groups in September after harvest of the barley was agreed upon as a reasonable compromise.

Litter

As the lucerne ley became older, litter accumulated on the soil surface. The woody stems of lucerne decomposed slowly (Andrén 1987) and a litter layer was gradually built up, whereas in the grass annual ley litter production and disappearance were more or less equal. Still, a litter layer existed during large parts of the growing season in GL. In barley, however, litterfall was negligible, and harvest residues were ploughed under each autumn.

Herbivores

The lowest mean population densities and biomass of above-ground arthropods generally occurred in B0 and the highest occurred in the leys, particularly LL (Tab. 4.18). Herbivores contributed from 29% of total arthropod biomass in LL to 52% in B120. The range for detritivores/microbivores was from 20% in B120 to 51% in LL, and predators/parasites accounted for between 20% of arthropod biomass in LL and 37% in B0 (Curry 1986). These biomass values are comparatively high, and particularly in the leys they were comparable with those of undisturbed natural grasslands (Kajak 1980), indicating that the effects of cultivation on arthropods are not always as drastic as is sometimes suggested.

Decomposer organisms
Soil microorganisms. including protozoa

Perennial crops will, over a 5–12 yr period, develop a higher soil microbial biomass than annual crops (Jenkinson and Powlson 1976, Lynch and Panting 1980, Adams and Laughlin 1981). These increases in microbial biomass are mainly related to increases in organic matter content (Jenkinson et al. 1976, Adams and Laughlin 1981, Schnürer et al. 1985).

The build-up of soil organic matter is a very slow process. Carbon accumulates temporarily in roots under a perennial crop. The C-content in the top soil in GL increased from 2.3 to 2.6%, equivalent to 140 g C m^{-2} yr^{-1}, within five years (Tab. 2.5), while there was no change in C-content in the other cropping systems. A similar accumulation of soil organic C was also indicated

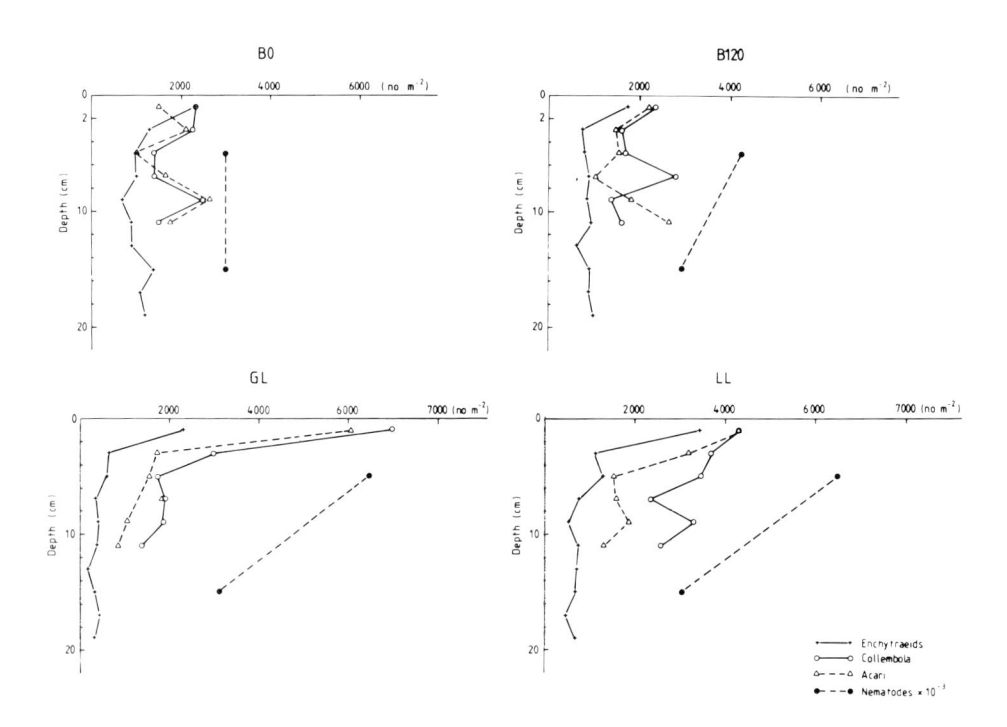

Fig. 4.23. Depth distribution of microarthropod, enchytraeid and nematode abundance in B0 and B120. Means from the September samplings 1981–1984 (Lagerlöf and Andrén in press, Lagerlöf and Andrén 1988, Lagerlöf et al. 1989, Sohlenius et al. 1987).

Tab. 4.20. Microorganisms and soil animals in the cropping systems at Kjettslinge in September during 1981–1984. Standard errors are shown in parentheses under mean values. Microbial values are per gram dry top soil. ND = not determined.

Organism group	1981 B0	1981 B120	1981 GL	1981 LL	1982 B0	1982 B120	1982 GL	1982 LL	1983 B0	1983 B120	1983 GL	1983 LL	1984 B0	1984 B120	1984 GL	1984 LL
Fungi[a]																
FDA-active hyphal length (m)	110 (10)	75 (8)	120 (6)	140 (8)	39 (4)	41 (4)	59 (13)	52 (11)	ND	ND	ND	ND	ND	ND	ND	ND
Total hyphal length (m 10^{-3})	1.7 (0.2)	1.1 (0.1)	2.0 (0.1)	1.4 (0.2)	0.7 (0.1)	1.1 (0.2)	0.7 (0.2)	0.8 (0.2)	ND (0.1)	1.0	0.8 (0.1)	1.0 (0.1)	ND (0.3)	ND	ND	ND
Bacterial abundance[a] (no. 10^{-9})	4.4 (0.2)	4.8 (0.5)	5.3 (0.1)	3.5 (0.2)	4.8 (0.0)	5.9 (0.4)	6.3 (0.4)	6.4 (0.5)	ND	6.4 (0.3)	6.5 (0.2)	7.2 (0.3)	ND	ND	ND	ND
Protozoa[a] (no. 10^{-4})																
Flagellates	10.1 (1.1)	9.1 (5.9)	20.8 (5.7)	13.4 (6.0)	7.1 (2.2)	5.1 (0.9)	8.4 (2.0)	5.7 (1.8)	ND	3.5 (0.6)	1.8 (0.9)	2.4 (0.5)	ND	ND	ND	ND
Amoebae	7.4 (1.2)	2.5 (0.3)	5.3 (1.2)	4.0 (0.3)	7.1 (2.2)	1.7 (0.3)	3.8 (0.3)	2.2 (0.4)	ND	1.0 (0.7)	1.4 (0.7)	2.2 (0.5)	ND	ND	ND	ND
Enchytraeids[b] (no. m^{-2} 10^{-4})	0.6	0.7	0.5	0.9	1.0	1.0	0.4	0.7	1.6	1.0	0.3	1.0	1.5	0.8	1.3	1.6
Microarthropods[c] (no. m^{-2} 10^{-4})																
Acari	1.0	1.3	1.1	1.4	0.8	0.9	1.0	1.3	1.3	1.1	1.6	2.3	0.5	0.4	1.2	1.4
Collembola	1.8	1.7	1.4	1.2	1.2	1.1	1.6	6.0	1.0	1.0	2.1	2.3	0.9	0.9	2.0	3.5
Nematodes[d] (no. m^{-2} 10^{-6})	5.3 (0.7)	8.8 (1.4)	7.8 (1.5)	7.3 (0.7)	6.3 (0.5)	8.0 (0.6)	7.9 (1.2)	12.8 (2.7)	5.9 (0.6)	6.2 (0.6)	9.3 (3.7)	12.5 (1.8)	4.6 (0.7)	6.1 (1.3)	9.0 (0.9)	4.3 (0.5)
Earthworms[e] (no. m^{-2})	28.8 (8.3)	43.5 (4.3)	33.6 (9.4)	52.6 (8.9)	35.3 (7.1)	29.5 (7.6)	34.9 (8.8)	82.0 (74.8)	25.9 (7.0)	18.8 (2.2)	50.8 (11.8)	289.3 (44.8)	ND	ND	38.4 (8.5)	55.0 (11.0)

[a] Schnürer et al. 1986a.
[b] Lagerlöf et al. 1989.
[c] Lagerlöf and Andrén 1988, Lagerlöf and Scheller 1989, Lagerlöf and Andrén, in press.
[d] Sohlenius et al. 1987.
[e] Boström 1988a.

in the C-budget for GL (Chapter 7). Although the increase in C-content in GL was mainly caused by a build-up of roots, slower decomposition in GL compared with the other cropping systems might have also contributed to the increase (Chapter 5).

According to the protection capacity concept, in a short time-perspective a given soil should have a 'typical' abundance for each group of microorganism, regardless of the cropping system. Our observations support the existence of such typical values, although occasional fluctuations may occur. Bacterial and protozoan numbers, as well as total and FDA-active hyphal length in the top soil of B120 and GL were measured at 20 occasions during two years (Schnürer et al. 1986a). Mean values from these samplings and means from all cropping systems sampled in September 1981–1983 are presented in Tab. 4.10. There were only occasional differences in microbial abundance between cropping systems, and these occurred during 1981 and 1982 (Schnürer et al. 1986a). For example, only the abundance of naked amoebae were significantly higher in GL than in B120 (Tab. 4.19), but the difference was created by an extremely high count at one sampling occasion (Schnürer et al. 1986a). In September 1983, when the leys were more developed, there were no significant differences (Tab. 4.20). The stable size of the soil microbial biomass was confirmed for fumigation biomass C and N (Lindberg et al. 1989, Paustian et al. 1990).

Nematodes, including root-feeders
The different cropping systems had little influence on the number of species and genera. Possibly the nematode communities in the perennial systems did not stabilize during the five year project. Greater differences between cropping systems may have emerged after a longer time.

In the annual crops nematode abundance in the two sampled layers (0–10 and 10–20 cm below ground) was similar, while in the perennial crops the abundance in the top layer was usually twice that in the lower layer (Fig. 4.23). This is probably due to cultivation practices, that even out vertical differences in the upper soil (to about a 25 cm depth) of the annual crops. Similar depth distributions were also recorded for several other organisms, as can be seen in the following.

The total number of nematodes, sampled in September each year, was fairly constant in all crops except LL (Tab. 4.20). In LL the abundances in 1982–83 were higher than in the other years. In particular plant feeding nematodes fluctuated more in the perennial than in the annual crops (Tabs 4.21, 4.22, Sohlenius et al. 1987). The relative contribution of various feeding groups of nematodes to total nematode numbers fluctuated more in the perennial than in the annual cropping systems. This is in contrast to the suggestion (Norton 1978) that stability should be greater in perennial than in annual cropping systems.

The proportions of feeding groups under barley differed from Danish arable soils (Andersen 1979a,b). In the Kjettslinge study a larger part of the nematode fauna consisted of fungal feeders and dorylaims than in Andersen's study. Nematodes belonging to the Dorylaimida are sometimes considered to be particularly sensitive to soil disturbances. Fields with annual crops in Poland had a lower proportion of dorylaims than pastures and fields with perennial crops (Wasilewska 1979). A similar tendency was found in our study, where GL had a higher abundance of dorylaims than the barley did (Boström and Sohlenius 1986).

The relative abundance of trophic groups under meadow fescue was quite similar to that in an ungrazed North American prairie soil (Freckman et al. 1979). Compared with a Polish lucerne field (Wasilewska

Tab. 4.21. Abundance and biomass of nematode feeding groups in the cropping systems at Kjettslinge. Mean values for the September samplings during 1981–84. Index letters indicate significant (p <0.05) differences, i.e. "a" indicates significant differences with cropping system A, etc. Index letters in italics indicate significant differences between cropping systems in any of the sampling years (B. Sohlenius, S. Boström, A. Sandor, unpubl.).

Feeding group	Cropping system			
	B0 (A)	B120 (B)	GL (C)	LL (D)
Abundance (no. 10^{-3} m^{-2})				
Plant feeders	924[bd]	2375[ad]	2037[d]	3351[abc]
Fungal feeders	1919[cb]	1313[cad]	2731[abd]	1912[bc]
Bacterial feeders	2235[cd]	3111	3072[ad]	3430[ac]
Omnivores/predators	448[c]	483	628[a]	520
Total	5526[cbd]	7282[ad]	8468[ad]	9213[abc]
Biomass (mg m^{-2})				
Plant feeders	30	48	76	97
Fungal feeders	36	22	50	46
Bacterial feeders	60	76	89	113
Omnivores/predators	86	63	113	67
Total	212	209	328	323

1967), the Swedish lucerne had a larger proportion of bacterial feeders.

Fungal feeders constituted a greater proportion of the fauna in B0 and GL than in B120 and LL. Bacterial feeders were relatively more abundant than fungal feeders in B120 and LL (Tab. 4.22). However, there was no direct relationship between number of nematodes and biomass of microorganisms (Schnürer et al. 1986b).

Microarthropods and enchytraeids
Abundance of Collembola increased in GL and especially in LL, whereas it remained the same in B0 and B120 throughout the experimental period (Tab. 4.20).

The abundance of Acari varied considerably between years and cropping systems. Fluctuations between years were greater than within-year differences between specific cropping systems. Still, GL and LL tended to have the highest abundances. Similar results have also been reported earlier (van de Bund 1970, Andrén and Lagerlöf 1980, Andrén and Lagerlöf 1983, Emmanuel et al. 1985, Mallow et al. 1985).

The small Prostigmata was the most numerous suborder and was largely responsible for the variation in total numbers of Acari. Although Astigmata were generally uncommon, they were unusually abundant in GL during 1984. Cryptostigmata was the least common suborder.

Mean abundance of enchytraeids in September samplings during 1980–84 did not differ significantly between the crops. LL and GL showed parallel abundance dynamics, but enchytraeid abundances were significantly higher in LL than in GL in three of five years.

Tab. 4.22. Mean abundance (no. 10^{-3} m^{-2}) of nematodes in the cropping systems at Kjettslinge in the September samplings during 1981–84. Standard errors in parentheses (B. Sohlenius, S. Boström, A. Sandor, unpubl.).

Feeding group/Taxa	B0		B120		GL		LL	
Plant feeders								
Tylenchorhynchidae	711	(69)	2082	(274)	913	(323)	3278	(1226)
Hoplolaimidae	66	(20)	45	(21)	378	(151)	38	(15)
Pratylenchidae	76	(32)	28	(14)	103	(39)	20	(6)
Paratylenchidae	68	(43)	217	(75)	640	(436)	15	(9)
Heteroderidae	3	(3)	3	(3)	0	0	0	0
Trichodoridae	0	0	0	0	3	(3)	0	0
Fungal feeders								
Tylenchoidea	961	(138)	739	(208)	1610	(186)	677	(116)
Aphelenchoidea	827	(102)	533	(40)	1083	(259)	1041	(205)
Neotylenchodea	131	(73)	41	(18)	38	(9)	194	(95)
Bacterial feeders								
Rhabditidae	418	(36)	338	(78)	468	(121)	827	(325)
Cephalobidae	1683	(171)	2654	(239)	2276	(318)	1869	(384)
Panagrolaimidae	10	(4)	12	(6)	104	(33)	517	(253)
Monhysteridae	66	(23)	66	(22)	114	(39)	147	(78)
Chromadorida	0	0	0	0	0	0	5	(5)
Araeolaimida	42	(19)	31	(5)	82	(26)	55	(30)
Alaimidae	16	(9)	10	(8)	28	(16)	10	(5)
Omnivore/predators								
Dorylaimidae	340	(36)	384	(65)	418	(80)	359	(42)
Belondiridae	2	(2)	2	(2)	4	(4)	2	(2)
Leptonchidae	12	(12)	5	(5)	10	(10)	6	(6)
Diphtherophoridae	43	(16)	10	(4)	115	(46)	105	(63)
Mononchoidea	51	(13)	82	(34)	81	(26)	48	(15)

Tab. 4.23. Mean abundance (no. m^{-2}) and fresh weight biomass (g m^{-2}) of adult and juvenile *A. caliginosa* and of combined earthworm species in the cropping systems at Kjettslinge in the September samplings during 1982–1983 (Boström 1988a).

Organism group		B0	B120	GL	LL
A. caliginosa					
Adults	Abundance	13.2[bc]	15.6[bc]	26.7[b]	44.0[a]
	Biomass	14.4[bc]	16.0[bc]	30.3[b]	60.0[a]
Juveniles	Abundance	20.5[a]	11.5[a]	21.9[a]	141.4[a]
	Biomass	5.3[a]	3.4[a]	5.1[a]	22.3[a]
All species	Abundance	35.2[a]	27.8[a]	49.2[a]	213.5[a]
	Biomass	21.4[bc]	20.6[bc]	38.2[b]	102.2[a]

Values followed by the same letter do not differ significantly (p >0.05) between cropping systems.

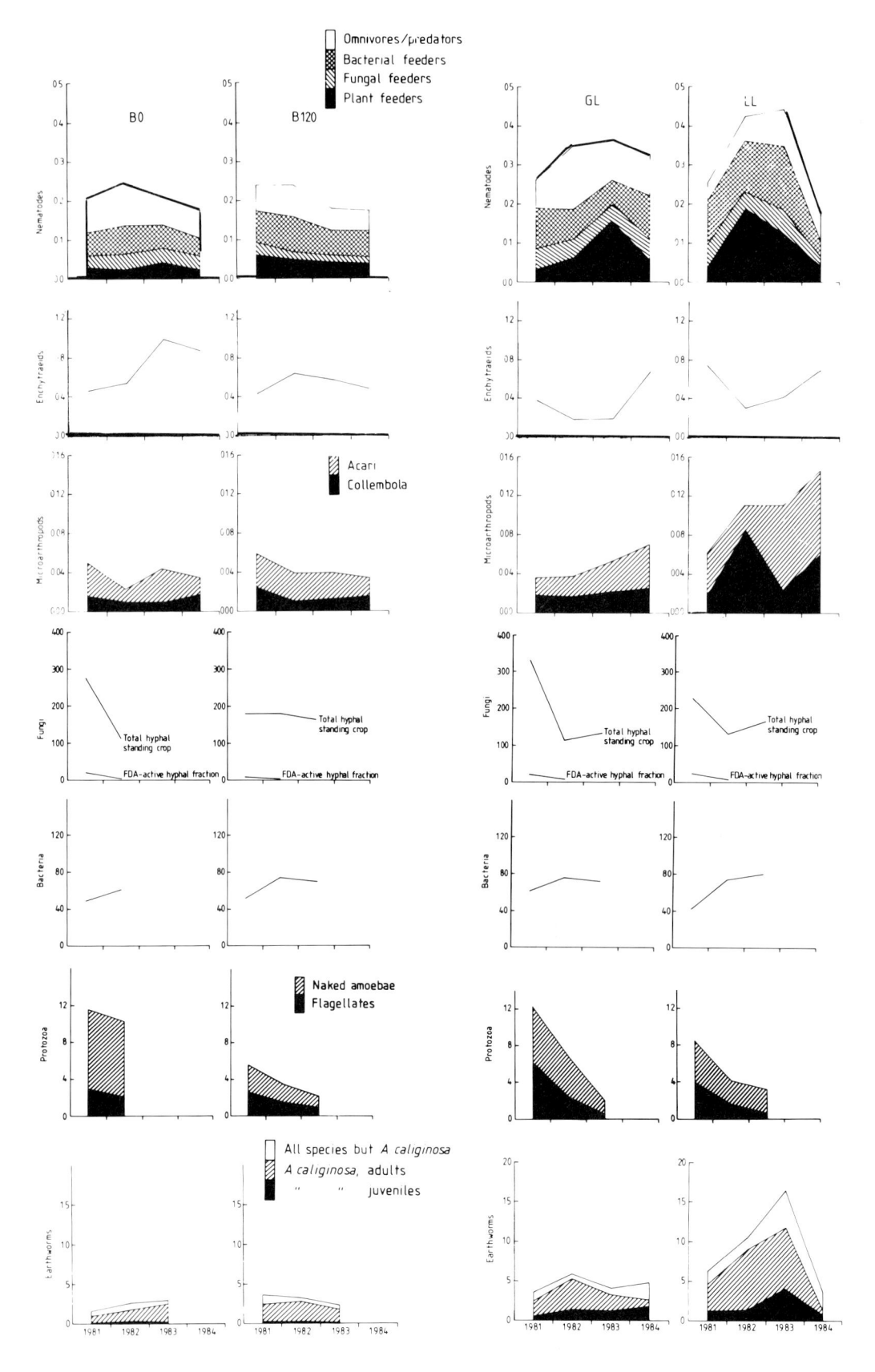

Fig. 4.24. Biomass of microorganisms and soil animals sampled in September 1981–1984 in Kjettslinge. Values in g dry weight m^{-2}. (Lagerlöf and Andrén in press, Lagerlöf and Andrén 1988, Lagerlöf et al. 1989, Schnürer et al. 1986a, Sohlenius et al. 1987, Boström 1988a). For conversion factors see text to Fig. 4.18.

Abundances of enchytraeids in B0 and B120 were similar during 1980–82 but later on numbers in B0 increased relative to those in B120.

In general, biomass in September samplings was proportional to abundance for all groups (Fig. 4.24).

Although most Collembola and Acari became more abundant in the leys than in barley, the number of species remained similar in all crops during the five-year period. Several earlier studies have reported the number of species and diversity increase over time in perennial crops, as the soil environment becomes more like that in a natural grassland (Karg 1967, Curry 1986). A partial explanation for the similarity in the number of species found in the barley and ley crops may be that the abundance of large and active litter-dwelling species, such as certain Mesostigmata and predatory Prostigmata, were underestimated due to the small sampling units. In addition, considering the disturbances associated with crop management (See Chapter 2), the perennials in this experiment could hardly develop into something that resembles natural grassland.

The differences in species richness and abundance between the Kjettslinge arable system and an old grassland in the region investigated by Persson and Lohm (1977), were especially great for Cryptostigmata mites. For other groups, including Mesostigmata, Pauropoda, Collembola, and Enchytraeidae, the differences were less pronounced. A uniform depth distribution is a consequence of ploughing and harrowing, since crop residues are incorporated into the soil, creating a more porous structure. Consequently, there was a highly significant difference in the depth distribution of both Collembola and Acari between GL and B120 during 1982–1983. During this period the greatest part of the Collembola fauna in B120 was found in the deeper part of the profile, while in GL it was concentrated near the surface. Acari were near the surface in GL, and uniformly distributed in B120. This pattern was consistent throughout the study for all groups except Astigmata. For enchytraeids, the positive effects of ploughing seemed to outweigh the negative effects of mechanical damage and disturbance (Fig. 4.23). Vertical distribution of enchytraeids was even throughout the sampling depth in B120. In GL the vertical distribution was shallow during 1981, while equally low numbers were found at all levels during 1982.

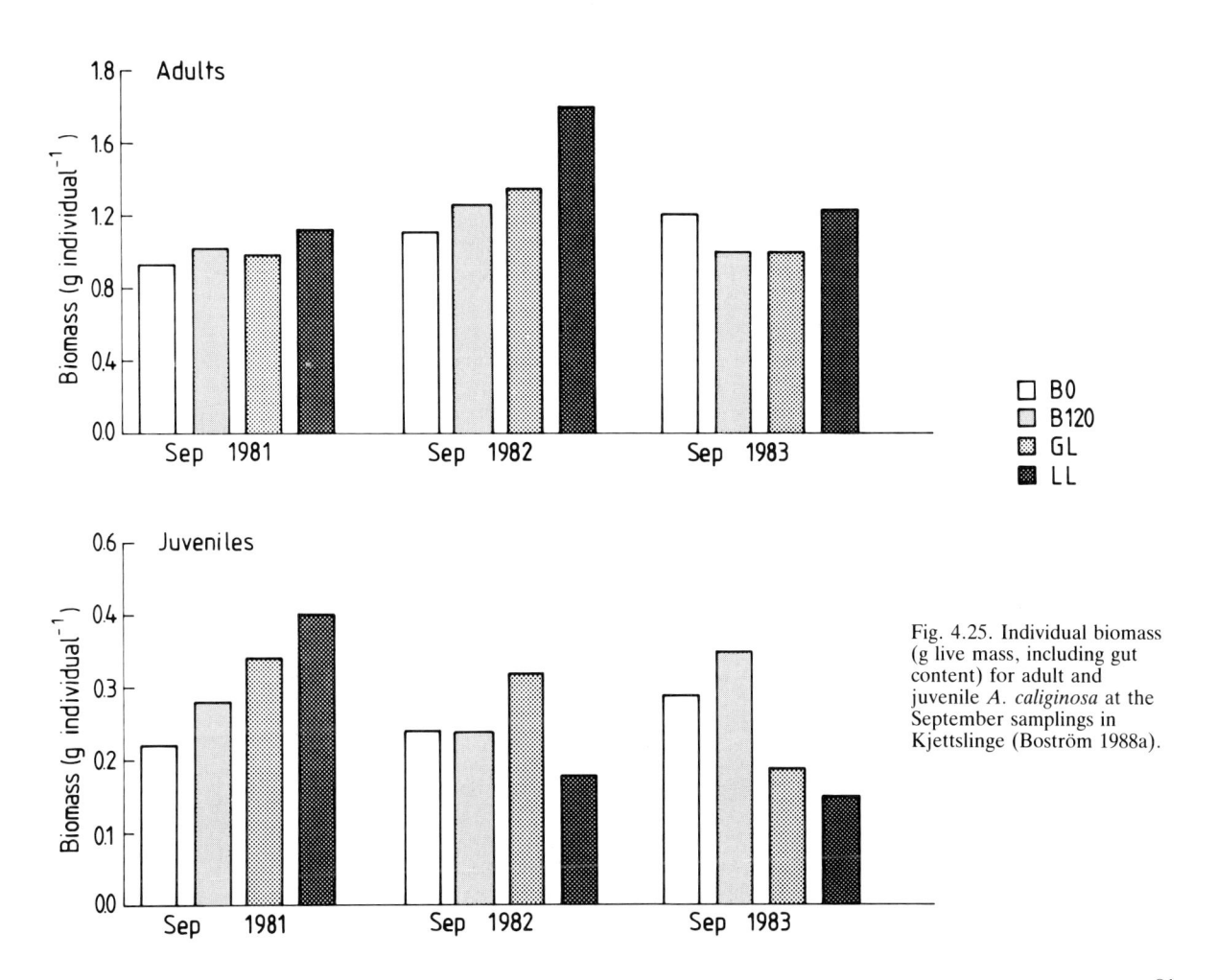

Fig. 4.25. Individual biomass (g live mass, including gut content) for adult and juvenile *A. caliginosa* at the September samplings in Kjettslinge (Boström 1988a).

Consequently, certain species can expand their distribution to greater depths in ploughed soil. However, the depth distribution for some of the smaller-sized animals did not differ between ploughed and unploughed soil, e.g., *Tullbergia krausbaueri* s.l., Pauropoda and in some cases prostigmatid mites. These small animals could move into deeper regions even through the more narrow pore system of the unploughed soil.

In general, Collembola and Acari are favoured by nitrogen fertilization because it results in an increased production of crop residues, their major source of energy (Curry 1986). However, the beneficial effects of fertilization, in terms of the higher production of crop residues, were probably counteracted by other effects of nitrogen fertilizers. For example, the soil was drier in B120 than in B0 owing to the higher rates of plant growth and transpiration (Alvenäs et al. 1986). In addition, high fertilizer concentrations may have had toxic effects on soil animals.

The differences in soil moisture between B0 and B120 may explain the lower enchytraeid abundance in B120 than in B0 towards the end of the experiment. Low soil moisture may also explain the comparatively low enchytraeid densities in GL, but this crop received the greatest amount of fertilizer, so fertilizer toxicity cannot be ruled out as contributing factor.

The summers of 1982 and 1983 had long periods with dry, warm weather. This coincided with low enchytraeid abundance in GL, and in LL in 1982. The low abundances were probably due to the shallow depth distribution of enchytraeids in the leys, which made them more exposed to drought than the deep-dwelling enchytraeids in the barley crops were. The concentration of enchytraeids in the uppermost part of the profile in GL may have been due to the root distribution. In GL 420 g dry mass m^{-2}, i.e., 77% of the total root biomass was in the top 0–10 cm (Hansson 1987). In LL 65% of the root biomass was in the upper 10 cm and 15% of the roots were below 27 cm. In barley, root mass was more evenly distributed throughout the top soil (0–27 cm). This means that organic matter input to the soil from roots was very superficial in GL, somewhat deeper in LL and uniform throughout the top soil in barley. GL with its shallow root-system takes up water from the surface layers which become very dry during summer droughts. LL, with its deeper root system and long taproots, penetrating down to the subsoil, takes up water at greater depths and some moisture remains in the surface layer. This difference could be clearly seen when taking soil cores.

Soil and soil surface macroarthropods
See the corresponding section under "Seasonal dynamics of the components".

Earthworms
Few projects have studied the influence of crops on earthworm populations over a succession of years. Samplings have mainly been made in different years in fields with varying soil types and agricultural history. Irrespective of the actual crop, the total numbers and proportion of different species of earthworms differ greatly between soil types (Guild 1948).

The total biomass of earthworms at Kjettslinge did not show any clear tendency to increase or to decrease in any of the cropping systems during the five-year study but it varied considerably between years (Fig. 4.24). The mean biomass was highest in lucerne, where the fluctuations also were most pronounced.

In barley the mean earthworm biomass was 21 g m^{-2} (fresh weight) which is similar to what Graff (1953b) reported from arable land in Germany. In Scotland the biomass in 5-yr-old arable fields varied from 7 to 14 g m^{-2} (Evans and Guild 1948a). In GL and LL mean biomass were 38 and 102 g m^{-2} (fresh weight), respectively (Tab. 4.23). This is well in line with the 7–70 g m^{-2} reported from meadows in Sweden (Nordström and Rundgren 1973), and the 50–72 g m^{-2} reported from leys in Scotland (Evans and Guild 1948a).

No significant differences in biomass or numbers were found between GL and the barley cropping systems (Fig. 4.24, Tab. 4.20), although more earthworms are usually found in leys than in annually cultivated soils (Evans and Guild 1948a, Heath 1962). This probably occurred because the dominant species, *A. caliginosa*, grows poorly on meadow fescue roots compared with lucerne or barley roots (Boström and Lofs-Holmin 1986). The major part of the younger organic matter in GL probably consisted of roots. It has been found that the food supply of the adult earthworms also influences the number of cocoons produced (Evans and Guild 1948a, Boström 1987).

The biomass of adult *A. caliginosa* was usually higher than the biomass of juveniles, while juveniles were usually more numerous than adults. Guild (1948) and Reinecke and Ljungström (1969) also usually found more juvenile than adult *A. caliginosa* (Fig. 4.24, Tab. 4.23).

The abundance of *L. terrestris* did not increase in the leys, although biomass often is higher in leys than in annually cultivated soils (Evans and Guild 1948a). The slow population increase probably was due to the low cocoon production rate and the long time of development of this species (Satchell 1967, Lofs-Holmin 1983). The number of species did not differ between cropping systems, although Graff (1953b) reported higher numbers of species in grassland than in arable land.

The mean individual biomass of adults was lower in the barley cropping systems than in the leys, and until autumn 1984 the adults weighed more in LL than in GL (Fig. 4.25). Plant materials of varying origins and stages of decomposition promoted earthworm growth differently (Boström and Lofs-Holmin 1986, Boström 1987). The individual mass of adult worms decreases during senescence (Phillipson and Bolton 1977). Hence, the individual mass varies with the type and amount of food and also with the age of the adult.

Until autumn 1982 the juveniles tended to be smaller in barley than in the leys, but in autumn 1983 juveniles weighed more in the barley. The reasons for low individual biomass are more complicated for juveniles than for adults. The mean individual mass of juveniles can be low due to a low growth rate or large numbers of newly emerged juveniles. In the first case the population may decrease while the second case may lead to a population increase. To decide which is the most probable situation, earthworms divided into different mass classes should be studied.

Summary

– Peak leaf area index ranged from 1 in B0 to 7 in GL.
– B120 had the highest peak above-ground plant biomass (900 g m^{-2}). In both leys peak above-ground biomass was around 600 g m^{-2}.
– Maximum above-ground growth rate was 7 g m^{-2} d^{-1} in B0 and 20 g m^{-2} d^{-1} in B120.
– The shoot/root peak biomass ratio was lowest in B0 (0.15) and highest in GL (1.0).
– Mean root depth was 30 cm in B0 and B120, 20 cm in GL and 40 cm in LL.
– Mean biomass of above-ground arthropods in the experimental field was 180 mg m^{-2}.
– Over 70% of bacterial biomass and fungal lengths were found in the top soil.
– No significant differences were found between or within sowing rows for bacterial biomass or fungal lengths.
– Mean nematode biomass in the experimental field was around 300 mg m^{-2}, and bacterial feeders were most numerous.
– Soil microarthropod biomass, i.e., springtails and mites, was around 60 mg m^{-2} and enchytraeid biomass was 500 mg m^{-2}.
– Mean earthworm biomass was 28 g m^{-2} (fresh weight).
– *Aporrectodea caliginosa* constituted more than 80% of the earthworm biomass. Other common species were *A. longa* and *Lumbricus terrestris*.
– The dynamics of biomass and its distribution between different parts of the plant in relation to phenological development differed in magnitude but not in temporal distribution between B0 and B120.
– Above-ground growth dynamics in the leys were similar, largely due to similar management measures.
– Root biomass in barley reached its maximum at anthesis (flowering).
– Root biomass increased during autumn in both leys.
– The amount of litter on the ground was negligible in B0 and B120.
– The amount of litter on the ground in the leys ranged between 25 and 75 g m^{-2} in both leys, showing no clear seasonal trends.
– Significant amounts of roots died in GL before the second harvest.
– Total fungal length, FDA-active fungi and bacterial numbers varied only slightly during the growing season, but their dynamics were positively correlated with soil moisture dynamics.
– In an irrigation experiment total hyphal length increased significantly compared with an unirrigated plot, but the increase in bacterial biomass was less pronounced.
– Root- and bacterial-feeding nematodes decreased in barley after harvest and soil cultivation. Other feeding groups were rather stable in B120 and GL.
– Collembola and Acari decreased in B120 during the autumn after ploughing.
– The dynamics of enchytraeids were positively correlated with soil moisture.
– Earthworm biomass tended to increase during the growing season.
– The larger root biomass and the presence of a litter layer (at least during parts of the growing season) in the leys, compared with the barley cropping systems, constituted the main structural differences between the crops.
– Microbial abundance and biomass, including Protozoa, did not differ significantly between the cropping systems.
– Plant- and bacterial-feeding nematodes were most abundant in LL, whereas fungal and omnivorous/predaceous nematodes were most abundant in GL.
– Total mesofaunal biomass was roughly similar for all cropping systems.
– Both abundance and biomass of soil fauna were highest in LL, followed by GL, B120 and B0.
– Nitrogen fertilization in barley decreased enchytraeid abundance.
– Collembola and Acari were distributed uniformly throughout the top soil in B0 and B120, while they were concentrated in the surface layers in the leys.
– In all cropping systems earthworms were the dominant faunal group, in terms of biomass, followed by enchytraeids.

5. Organic carbon and nitrogen flows

Olof Andrén, Torbjörn Lindberg, Ullalena Boström, Marianne Clarholm, Ann-Charlotte Hansson, Gerd Johansson, Jan Lagerlöf, Keith Paustian, Jan Persson, Roger Pettersson, Johan Schnürer, Björn Sohlenius and Maria Wivstad

5. Organic carbon and nitrogen flows. O. Andrén, T. Lindberg, U. Boström, M. Clarholm, A.-C. Hansson, G. Johansson, J. Lagerlöf, K. Paustian, J. Persson, R. Pettersson, J. Schnürer, B. Sohlenius and M. Wivstad

Introduction	87
Primary production	87
Photosynthesis	88
Plant biomass production	89
Isotope carbon budget	91
Plant uptake of nitrogen	94
Symbiotic nitrogen fixation	94
Uptake of fertilizer and soil nitrogen	97
Decomposition and nitrogen mineralization	98
Mass loss and abiotic control	99
Heterogeneous resource	100
A decomposition model including resynthesis	102
Nitrogen mineralization and immobilization	103
The decomposer organisms	107
Microorganisms, including Protozoa	109
Nematodes, including root consumers	111
Microarthropods and enchytraeids	113
Macroarthropods, including plant-feeding arthropods	115
Earthworms	115
A synthesis of organism activities and decomposition processes	118
Summary	125

Introduction

Carbon and nitrogen flows follow a main pathway from the plant, via decomposer organisms, to the soil organic matter. This chapter is structured according to this pathway, starting with primary production and plant nutrient uptake, and ending with litter decomposition and nutrient mineralization (Fig. 5.1). In organic forms, carbon and nitrogen are intimately linked. There are, however, important differences in their movements through the ecosystem.

Carbon is fixed in the plant and usually only a small fraction of the living plant is consumed by herbivores. When the plant or parts of it die, decomposer organisms, i.e., microorganisms and saprovorous animals, attack the litter. Part of the litter is assimilated by the organisms, forming new biomass or is respired (i.e., mineralized). The non-respired decomposition products are incorporated into the soil organic matter (SOM), where they may remain for periods ranging from hours to several centuries before being mineralized. Most transfers involve respiration, i.e., carbon dioxide returns to the atmosphere and becomes available for photosynthesis. Atmospheric CO_2 is then eventually fixed by plants and the cycle is completed.

Nitrogen is taken up mainly by plant roots and is incorporated into biomass. However, microorganisms compete with plants for inorganic nitrogen that has been added as fertilizer or mineralized from organic matter. In an agroecosystem the harvest is a major nitrogen sink, and the system has to be replenished with nitrogen from outside sources, such as manure, fertilizers or biologically-fixed N. The soil organic matter usually contains over 95% of the nitrogen in the system. The fluxes to and from this pool represent a large, sometimes the largest, flux of nitrogen in the agroecosystem (See Chapter 7).

Comparable agroecosystem studies on carbon and nitrogen flows are rare, but similar investigations have been made in Poland (Ryszkowski 1984) and the USA (Elliott et al. 1984, Hendrix et al. 1986). This area of research is receiving increasing attention, and agroecosystem projects similar to "Ecology of Arable Land" have recently been initiated in, e.g., the Netherlands (Brussard et al. 1988), the USA (G. P. Robertson, pers. comm.), and West Germany (F. Beese, pers. comm.).

Primary production

Net primary production can be defined as the total amount of organic matter assimilated minus that lost to respiration (Roberts et al. 1985). Production may either be estimated directly (e.g., by measurement of CO_2-exchange) or indirectly by a harvest method, i.e., cutting and soil coring. Primary production within the present project has, as in most ecosystem studies, been estimated using the latter approach (Hansson and Steen 1984, Hansson and Andrén 1986, Pettersson et al. 1986, Hansson et al. 1987, Hansson and Andrén 1987, Andrén et al. 1987, Hansson and Pettersson 1989, Pettersson 1989). The harvest method gives valuable information about the temporal and spatial distribution of biomass. However, to accurately estimate net primary production based on changes in biomass, losses through death, grazing, root sloughing and exudation must be considered. A drawback of the harvest method is that short-term responses of a crop to rapid changes in external factors cannot be observed (Long 1986).

In addition to field studies of primary production, plants were grown in pots under $^{14}CO_2$ atmosphere, with a gas-tight seal at the soil surface. Labelling made it possible to calculate a complete carbon budget for a young plant, including rhizodeposition, i.e., epidermal tissues sloughed from the roots during growth and soluble compounds released from roots by exudation or after cell autolysis. These substances are not included in field estimates of root production, since they are either decomposed between samplings or lost during the washing procedure. The laboratory estimates of the partitioning of labelled carbon into biomass and rhizodeposition were used to correct the field data for the lack of data on rhizodeposition (see Chapter 7).

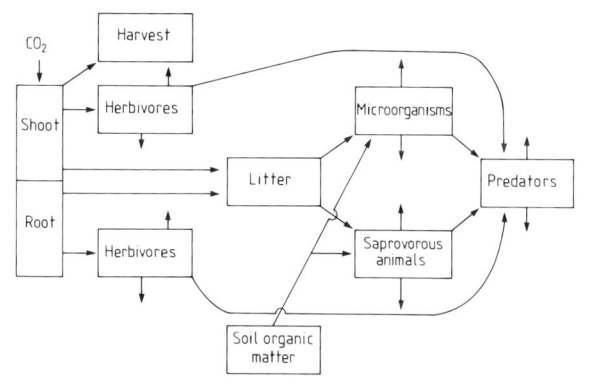

Fig. 5.1. The main pathways for carbon and nitrogen flows in an agroecosystem. Arrows pointing upward indicate respiration and those pointing downward indicate carbon and nitrogen contributions to soil organic matter (SOM) as well as nitrogen mineralization. For simplicity, in this figure microbivores are considered as predators. Nitrogen fertilizer input and nitrogen losses are not shown.

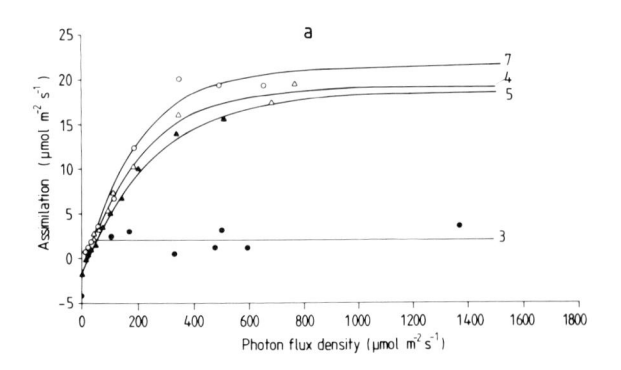

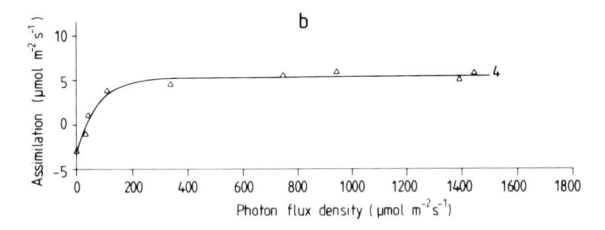

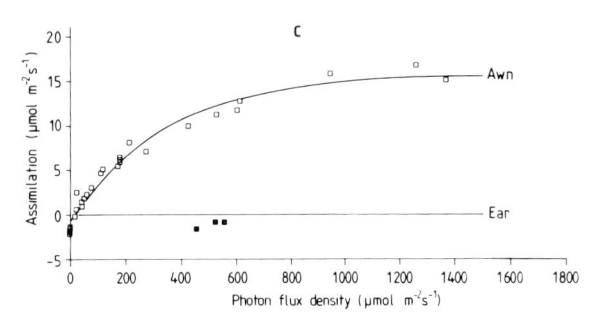

Fig. 5.2. Photosynthesis light-response curves, measured in the field on 4 July (a), 8 July (b) and 18 July 1985 (c). Leaf numbers are counted from bottom to top, i.e., the flag leaf is number 8.

Tab. 5.1. Net carbon assimilation (μmol m^{-2} s^{-1}) of leaves, ears and awns in the fertilized barley crop in 1985. All values are obtained by fitting the field data (photosynthesis – light response) to a simple photosynthesis model (Peat 1970).

Organ	4/7	8/7	16/7	18/7
Awn	–	–	4.75	15.75
Ear	–	–	0.00	0.00
Flag leaf	–	17.99	ND	30.00
Penultimate leaf	21.42	17.21	8.71	24.52

Photosynthesis

Gasometric measurements of net photosynthesis in the field provide an estimate of total carbon fixed. In principle, the amount of carbon translocated below ground can be calculated as the difference between total net canopy assimilation and the amount found in the aboveground crop. If the production of macroscopic roots is estimated with soil coring, the remainder of belowground production, i.e., root respiration and rhizodeposition can be estimated. While photosynthesis studies were not part of the initial research plan, subsequent interest in improving field estimates of plant carbon inputs led to their inclusion. However, only orienting measurements during the last field season of the project were made.

Field measurements of CO_2-exchange by individual assimilating plant parts (i.e., leaves, ears and awns) were performed in B120 with the ADC LCA system, i.e., a portable leaf chamber with CO_2 analyzer (Analytical Development Co. Ltd., Hoddesdon, Herts., U.K.), described by Long (1986). The assimilation rate (A) of a particular plant organ was measured over a range of photon flux densities (Q) by using a series of neutral density filters, and dark respiration was obtained by covering the leaf chamber with black polyethene.

The fixation of carbon by crop communities can be monitored at different levels: 1) the individual assimilating organ, 2) the whole plant, and 3) the crop. A plant community consists of a complex hierarchy of photosynthetically active units of different ages and types, which have different potentials for carbon assimilation and varying degrees of interaction, such as self-shading. Besides light intensity, the main factors controlling the assimilation rate of leaves in the field are temperature, leaf age and water availability.

As leaves age the photon flux density (Q), at which the assimilation rate is light-saturated (A_{max}), progressively decreases (e.g., Biscoe and Gallagher 1978). A representative set of photosynthesis-light response curves for leaves measured on 4 July 1985 is shown in Fig. 5.2a. Leaf number 7, which at that date was the top leaf in the canopy, together with leaf numbers 5 and 4 all had an A_{max} of about 20 μmol m^{-2} s^{-1}, and the apparent carbon fixation efficiencies (the elevation of the initial slope of the curve) were similar. Above about 500 μmol m^{-2} s^{-1}, further increases in Q did not increase A, indicating complete saturation of the photosynthetic apparatus (A_{max}). Below 500 μmol m^{-2} s^{-1} there was a proportional decrease in A as Q decreased. The oldest leaf which was still green at that time (i.e., leaf 3), had a A_{max} slightly above the compensation point (5 μmol m^{-2} s^{-1}). Four days later A_{max} had also decreased in leaf 4 (Fig. 5.2b). Typical rates of A_{max} of awns were about 15 μmol m^{-2} s^{-1}, and ears with the awns removed did not exceed the compensation point (Fig. 5.2c and Tab. 5.1), which is in accordance with the results presented by Legg et al. (1979).

Water stress may substantially affect crop productivity. The soil water potential decreases as the soil dries out, and as a consequence the plant water potential also decreases. This affects several physiological processes in plants which may influence the productivity of the crop. The water stress occurring during July 1985 (see Chapter 2) not only induced premature senescence of the leaves (see Chapter 4), but at the end of the stress period (16 July) A_{max} had decreased by about 60% (Tab. 5.1). The dry period was then followed by rain, and on 18 July the A_{max} recorded for the flag leaf and the penultimate leaf (i.e., leaves 8 and 7) were the highest recorded during the growing season (Tab. 5.1).

These results include both environmental effects (light regime and water stress) and changes in photosynthetic capacity due to internal factors such as age. The bulk of the results from the photosynthesis measurements at Kjettslinge are still under compilation, but results so far are similar to those reported by others, for barley and cereals in general. The decease in A_{max} with age is well documented and so is the effect of water stress on leaf area and photosynthesis (e.g., Biscoe and Gallagher 1978). The maximum rate of photosynthesis that we found (30 µmol m^{-2} s^{-1}) is normal for leaves of C_3 plants (Lawlor 1987), and the A_{max} at about 500 µmol m^{-2} s^{-1} was also in the normal range (Beadle et al. 1985).

A tentative conclusion drawn from the photosynthesis measurements is that the low productivity of the barley crop during the dry year of 1985 compared with the previous years was not only due to the low leaf area index, but also partly due to the reduced photosynthesis rate during the dry period.

Plant biomass production

The net change in biomass during a given time interval depends on the balance between production of new biomass and loss of existing biomass. Thus, production (P) can be calculated as the change in biomass (B) plus consumption and death:

$$P = \Delta B + \text{Consumption} + \text{Death}.$$

Losses by death are equal to the change in the amount of dead material (D) plus the amount disappearing as a result of decomposition:

$$\text{Death} = \Delta D + \text{Decomposition}.$$

Consequently, when estimating production using the harvest approach, fluctuations in both living and dead parts should be monitored and decomposition rates estimated. Wiegert and Evans (1964) estimated aboveground production according to these principles, and Milner and Hughes (1968) presented them in one of the first IBP handbooks. However, published production estimates have usually been calculated either as the difference between annual maximum and spring minimum biomass (the maximum method) or as the sum of significant *increases* in biomass or in both living and dead parts (the summation method). Decomposition has often been neglected (Kelly et al. 1974, Sims and Coupland 1979, Risser et al. 1981). Since production, death and decomposition can be simultaneous, the changes in biomass and dead plant mass between samplings are net results of these processes. This means that biomass may be stable or may even decrease in spite of high production if the death rate exceeds the production rate. On the other hand, if only *increases* in biomass and dead parts are summed, the same production may be calculated twice if the plant material is first measured as biomass and then as standing dead. Consequently, such production calculations may give either over- or underestimations, depending on how the processes interact. To avoid counting production twice, a group of methods, called "balancing transfers" or "decision matrices", have been developed that take into account the relationships between fluctuations in living and dead parts (Fairley and Alexander 1985).

Although some fluctuations in biomass are due to translocations between above- and below-ground parts of the crop, only a few authors have considered this source of error when calculating production. Balsberg (1982) suggested that the level of error in production estimates is increased by ignoring translocation from senescent organs into living ones and from below-ground rhizomes to developing above-ground organs.

When calculating production, appropriate statistical tests should be used to evaluate changes. Any method considering all changes between sampling dates is subject to error, proportional to the number of intervals considered and inversely proportional to the precision of the data. However, even when production is calculated as the sum of exclusively significant increases in biomass, it may still be an overestimate owing to random variation, since only the positive parts of the variation are considered when summing increases (Cochran 1977, Singh et al. 1984). Thus a production estimate not including losses through death and decomposition, which should theoretically yield underestimations, may instead result in overestimations for purely statistical reasons.

Persson (1978a) realized the statistical consequences of summing increases and therefore used a correction term to compensate for overestimations introduced by random variation. However, a more straightforward solution to the problem is to calculate production from significant changes, i.e., both increases and decreases.

Both biological and statistical aspects must be considered when calculating production; i.e.,

a) an adequate number of replicates of samples of adequate size should be taken;
b) sampling frequency should be adapted to phenology;
c) both living and dead parts should be sampled and then treated separately;
d) decomposition rates of dead parts should be estimated;
e) adequate statistical tests should be applied; and
f) both positive and negative changes should be included in the calculations.

To fulfil the criteria above is not easy and compromises are unavoidable. For example, in our project as in many others, practical and economical factors forced us to choose between a few sampling occasions with optimal numbers of replicates and sample sizes, or frequent samplings with fewer replicates and smaller sample sizes. Since rather little is known about when, e.g., peak root biomass occurs, especially in perennial crops, and few sampling occasions means an increased risk of missing the real biomass maximum, the latter alternative was preferred. In spite of this, an acceptable precision (SD/mean); 30, 24 and 50% for barley, GL and LL respectively, was achieved for the root biomass estimates. The above-ground biomass estimates naturally had higher precision, around 10%.

When analyzing different methods for calculating production, a conceptual model of the carbon flows in an annual arable crop was devised (Fig. 5.3). The model was based on the fractions of plant material that were possible to sample with the existing field equipment. In the model, carbon is incorporated into the plant biomass by photosynthesis (F_n). During growth, parts of the plant die and become dead plant material (flow F_A to DP). Part of this dead material is shed (F_B) and forms litter on the ground (AL). At harvest the main part of the standing crop is exported from the field. In an annual crop, the crop residues are incorporated into the soil during ploughing in autumn (F_C) and form the soil litter (SL) fraction, i.e., by definition this material consists of dead organic material supplied to the soil during previous years. Since the above-ground litter production in annual crops is very low, the transfer of litter by earthworms from AL to SL was not considered.

The other main pathway for carbon to the soil is translocation (F_1) from shoots to roots (LR). The amount of dead roots (DR) increases due to root death (F_2) and decreases due to decomposition (F_3). By definition the DR compartment only contains roots that have died during the current year, since older dead roots belong to the SL fraction. As a result of the methods for separating soil and soil organic material, two other fractions were defined. Unseparated material, only freed from soil, was named macro-organic material (MOM). When living roots were removed the remaining material was called dead organic material (DOM), consisting of dead roots from the current year and SL.

From the model, equations were formulated for calculating production from significant increases and decreases in the amount of living and dead plant fractions, including losses due to death and decomposition but not due to rhizodeposition (Tab. 5.2). To use these equations, not only LR and DOM must be sampled during the growing season, but SL must also be sampled in early spring of the current year, before sowing or start of growth. In addition, the decomposition rates (k) of SL and DR must be estimated in separate experiments. Using the decomposition rates, the amount of SL and DR (i.e., DR = DOM–SL) at each sampling occasion can be calculated.

In a perennial ley there are some additional difficulties: a) There is no natural zero starting point in standing crop of roots, shoots and litter, and b) the contribution to DOM from above-ground litterfall may not be insignificant. Provided these points are considered, the same model can be applied to a perennial ley. The amount of living roots at the first spring sampling can be used as the zero starting point, and litter contributions can be added to DOM from calculations based on the disappearance of AL.

During the first years of the project, root production in barley was estimated as the maximum amount of LR.

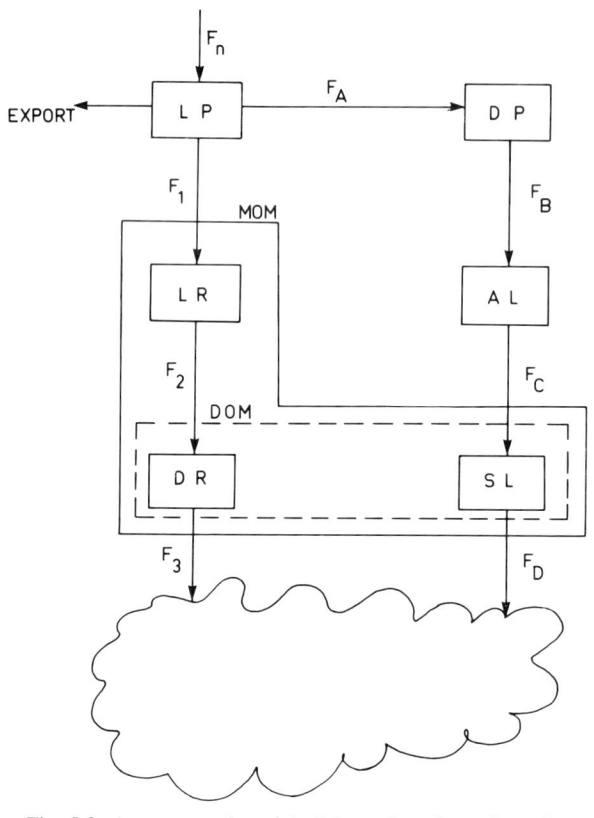

Fig. 5.3. A conceptual model of the carbon flows through an annual arable crop, used for calculations of flows not directly measurable. (LP = living above-ground plant parts, DP = dead above-ground plant parts, AL = above-ground litter, DOM = dead organic material in soil (DR + SL), SL = soil litter, LR = living roots, DR = dead roots, which died during the current year, MOM = soil macro-organic material (LR + DR + SL) (Hansson and Steen 1984).

Tab. 5.2. Equations for calculation of root production in an annual crop. (Hansson and Steen 1984).

$$SL_t = SL_{(t-1)} k_{SL}$$
$$DR_t = DOM_t - SL_t$$
$$F_3 = \Sigma\, k_{DR}\, DR_{t-1}$$
$$F_2 = \Sigma\, (DR_t - DR_{t-1}) + F_3$$
$$F_1 = \Sigma\, (LR_t - LR_{t-1}) + F_2$$

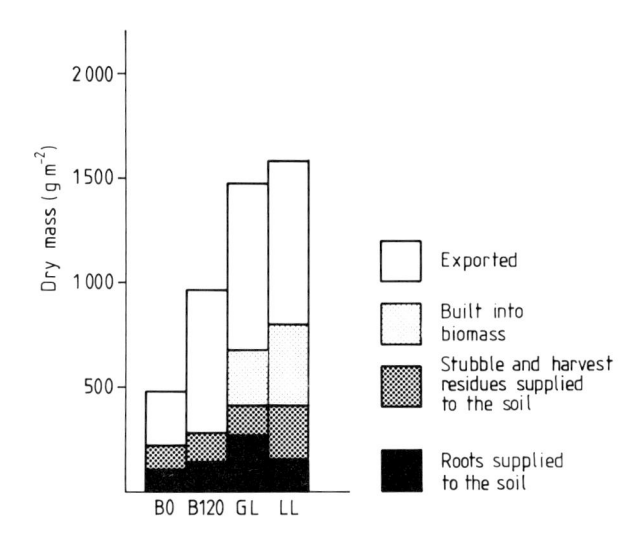

Fig. 5.4. Annual primary production as ash-free dry mass m^{-2} yr^{-1}, divided into exported amounts, biomass increase and the supply to soil by mortality.

Since it was extremely difficult to separate DR from SL, losses due to root death were not included in the production estimates (Hansson and Steen 1984, Hansson et al. 1987). However, the conceptual model and the equations were developed in parallel with the field work and during 1985, LR and DOM were sampled simultaneously with decomposition studies of SL and DR.

When calculating root production in the meadow fescue ley, dynamics in DOM and decomposition rate of SL were used to calculate DR. Production was then calculated as the sum of the maximum amounts of LR and DR. Consequently, the production estimate included losses due to death without requiring the tedious task of separating dead roots from other debris. Since the increase in DR occurred before the increase in LR, the risk of including the same production twice in the calculations was assumed to be very small (Hansson and Andrén 1986).

Above-ground production was calculated by summing significant increases in LP, increases in DP were included if they occurred simultaneously with an increase in LP, and increases in AL were included if they occurred simultaneously with an increase in both DP and LP (Pettersson et al. 1986). Decreases during the growing season were small, and generally not included in the calculations.

Net primary production (NPP), estimated by the harvest method, was 1.0 kg m^{-2} yr^{-1} in the fertilized barley. This production was almost twice that in unfertilized barley, but only about 60 and 66% of the NPP in lucerne and meadow fescue, respectively (Fig. 5.4). The proportion allocated below ground in the annual crop was influenced by nitrogen fertilization, i.e., a higher proportion (20%) was allocated below ground in B0 than in B120 (16%). In both leys about 30% of NPP was allocated below ground. About 50% of NPP was exported as harvest in B0, GL and LL, whereas as much as 70% was exported in B120. In barley both grain and straw were exported. Stubble, harvest residues and roots averaged 230 g m^{-2} in B0 and 300 g m^{-2} in B120. In the perennial crops, 18–25% of NPP was incorporated into non-harvested perennial parts, i.e., stubble and below-ground biomass, and about 400 g m^{-2} was annually delivered to the soil through root death and litterfall.

The response of crops to nitrogen deficiency could be expected to have a significant impact on agroecosystem nitrogen cycling. However, in B0 and B120 the amount of organic material entering the soil was not greatly affected even though nitrogen limited total production in unfertilized barley (Fig. 5.5). This was a result of the increased below-ground allocation in B0 and the removal of almost all above-ground production at harvest. However, the above conclusion depends on whether other components of the below-ground production, i.e., rhizodeposition which is not measured by the soil coring method, responded similarly to changes in N availability. This assumption is at least partly corroborated by the laboratory experiment with ^{14}C-labelling discussed in the following section. Thus the level of biotic activity and the maintenance of soil organic matter is less affected in a nitrogen deficient system than would be implied from the radical decrease in above-ground production.

The differences between the leys and barley in their yearly inputs of organic matter were smaller than expected (Fig. 5.5). However, when the leys were ploughed, the inputs were three to four times the annual input from barley (Andrén et al. 1987). The differences in inputs of organic matter between the two leys and barley are thus dependent on which years of the crop rotation are compared.

Isotope carbon budget

In plant carbon budgets, the total amount of carbon fixed by the plant, the amount translocated below-ground and the amount of CO_2 evolved through root respiration and root decomposition all constitute major fluxes. It is presently not possible to directly measure the total carbon input from roots to soil organisms; the major difficulty being the separation of CO_2 from root respiration from that evolved from decomposing root-derived material. However, by measuring the root-derived carbon remaining in the soil after removal of roots, it is possible to calculate the original amount delivered to the soil, provided the "humification quotient" (final/initial mass) for root-derived material is estimated independently. Thus all these components in the carbon budget can be separated if the plant is grown in $^{14}CO_2$-atmosphere, rhizosphere respiration is collected and determined, and the humification quotient of root-derived material can be estimated.

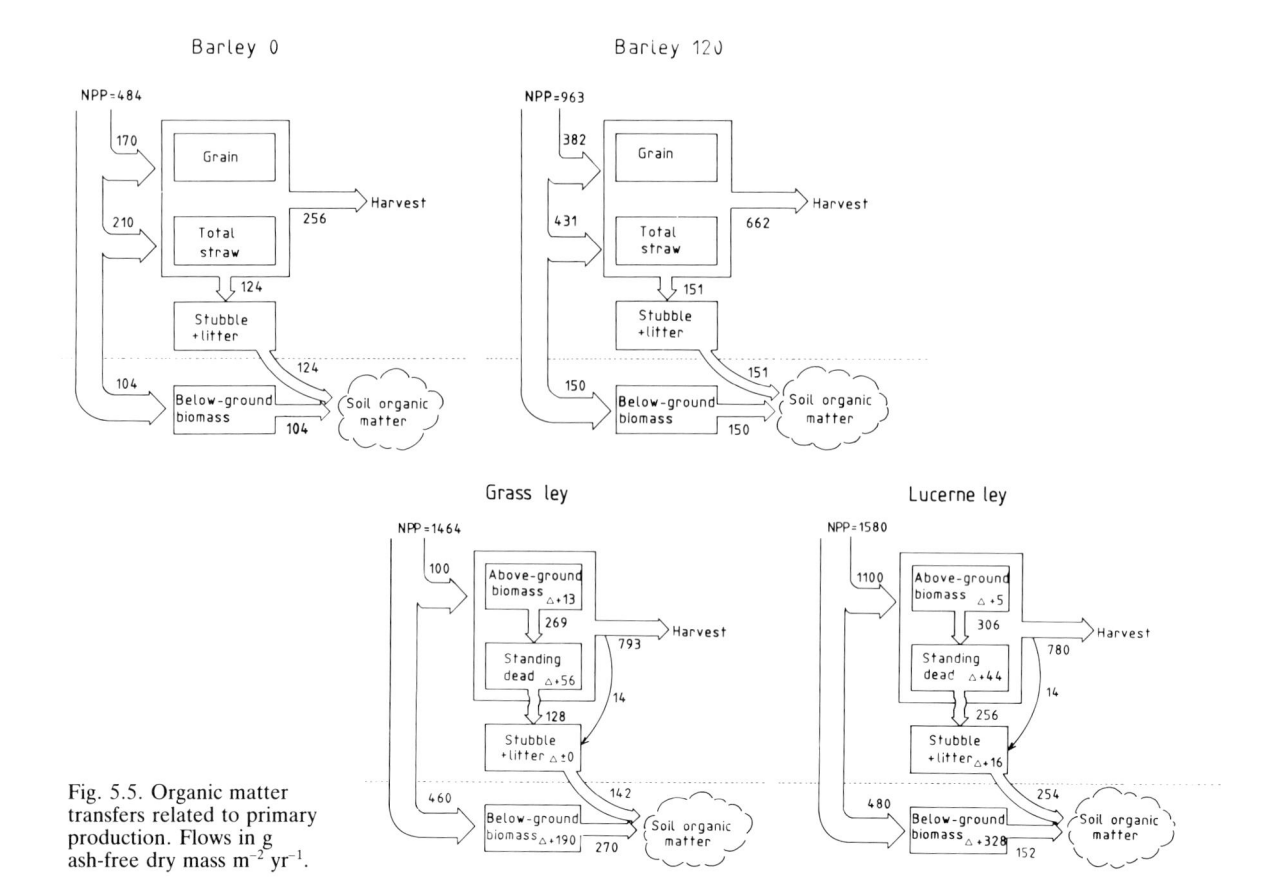

Fig. 5.5. Organic matter transfers related to primary production. Flows in g ash-free dry mass m^{-2} yr^{-1}.

These requirements were met using an assimilation chamber, where air temperature, light, moisture and $^{14}CO_2$-supply were controlled (Fig. 5.6). The chamber was 1.2 m long, 0.6 m wide and 1.0 m high and built of plexiglass panels, which were glued and screwed together. The plants were grown in PVC pots (diameter 0.15 m, height 0.23 m) with an air-tight lid, made of butyl rubber, at the soil surface. Each pot was watered individually according to values shown by a soil moisture meter. Since water must be removed from the air in the chamber, a water-chilled condenser was used. The chamber atmosphere was forced through the condenser by a hygrostat-controlled fan, which also ensured circulation of the air in the chamber.

Carbon dioxide had to be supplied in pace with its assimilation by the plants. The $^{14}CO_2$-supply was controlled by means of a gas analyser (Uras 2). A membrane pump continuously sampled air from the chamber for analysis. When the CO_2 content in the chamber fell to a certain level, a solution of labelled sodium carbonate was automatically pumped into a beaker containing phosphoric acid. At the same time a mixer in the beaker and a fan placed above it were activated. CO_2-free air was supplied to the soil compartment in the pot, and the exhaust air was led to a sodium hydroxide trap for collecting CO_2 from the soil.

When plants are grown with ^{14}C-labelled CO_2 in the atmosphere, assimilated ^{14}C will be found in all plant material above- and below-ground. Carbon translocated below-ground is deposited in root biomass, lost to the soil as rhizodeposition, or respired. Dead roots and deposited material will subsequently be decomposed by soil microorganisms and soil fauna. If the soil is isolated from the atmosphere it is possible to determine the amount of $^{14}CO_2$ originating from root respiration and decomposition of roots and root-derived material. Thus it is possible to distinguish between CO_2 evolved from root respiration plus decomposition of root material (^{14}C-labelled) and that evolved from decomposition of older (non-labelled) organic matter in the soil.

Barley was grown in the assimilation chamber for seven weeks in soil taken from the field experiment. The cultivation was conducted both with fertilizer nitrogen equivalent to 12 g N m^{-2} (B120) and without a nitrogen supply (B0). Thus it was possible to estimate the effects of nitrogen on root mortality and the quotient roots/rhizodeposition.

The distribution of labelled C in plant parts, rhizosphere respiration and soil is given in Fig. 5.7. In B0, 30% of the fixed carbon was translocated below-ground and the fraction in B120 was 25%. About 40% of the below-ground carbon was recovered in the harvested

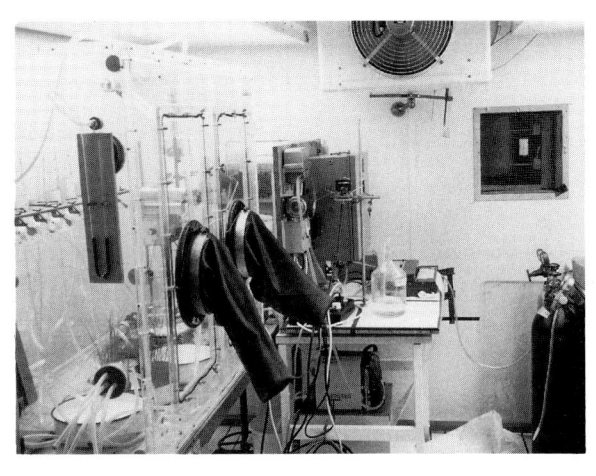

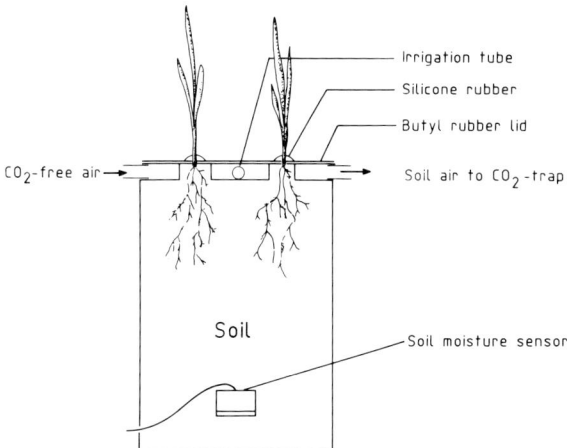

Fig. 5.6. The complete ^{14}C assimilation chamber used for carbon budget experiments and for labelling plant material for decomposition experiments. The drawing shows one of the pots used in the chamber.

root biomass of B120 and somewhat less in B0. At harvest, the root/shoot biomass ratios were 0.144 for B0 and 0.140 for B120. These ratios were lower than those found after the same growing period in the field, which were between 0.25 and 0.30 for both crops (Hansson et al., in press). The difference could be due to a higher phenological development rate in the laboratory experiment, caused by higher temperature and artificial light conditions.

Meadow fescue plants were labelled in the assimilation chamber during six weeks. At harvest, the plants from two pots were divided into shoots, crowns and roots. Crowns were defined from morphological characteristics as the first 5–10 mm above the soil surface and 1–3 mm below the soil surface. The contents of labelled carbon were determined (Fig. 5.7). The proportion of fixed carbon translocated to the roots was 44–49%, of which 37–43% was released into the soil. Thus a higher proportion of labelled carbon was translocated to the roots of meadow fescue compared with barley. Root

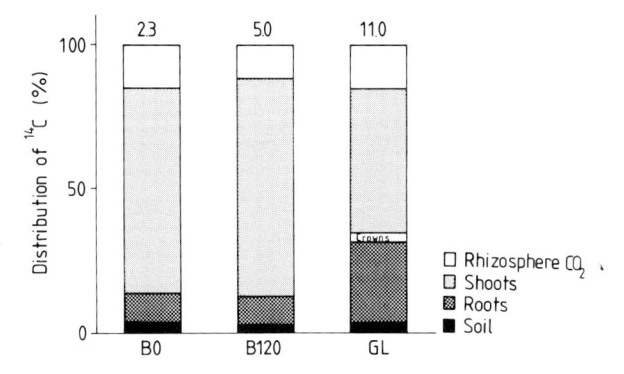

Fig. 5.7. The distribution of assimilated ^{14}C in B0, B120 and GL after seven (B0, B120) or six weeks (GL) in the assimilation chamber. The absolute amounts of assimilated ^{14}C (mean of two pots, g pot^{-1}) are also indicated.

and microbial respiration constituted the major part of the root-derived carbon, and only 9% of the carbon translocated to the roots was recovered in the soil outside roots. However, rhizodeposition constituted a smaller part of the carbon transferred below ground in meadow fescue, compared with barley (Fig. 5.7).

Several experiments in growth chambers have indicated that considerable amounts of carbon may enter the soil as rhizodeposition; e.g., Johnen and Sauerbeck (1977) grew wheat to maturity with the shoots in an atmosphere containing $^{14}CO_2$ and the roots in a gas-tight container so that rhizosphere $^{14}CO_2$ production could be measured. At harvest, three times the amount of ^{14}C that was recovered in root biomass had been supplied to the soil as rhizodeposition.

Root respiration has been measured by, e.g., Sauerbeck et al. (1976), who determined root respiration from plants cultivated in a sterile nutrient solution. However, the absence of microorganisms probably affected root development and respiration. Instead of separately determining root respiration, we estimated the amount of root material mineralized into CO_2 during the growing period.

To determine how much fresh root material the labelled C remaining in the soil at harvest originally represented, i.e., the amount of root material mineralized during the chamber period, a soil sample was taken directly after cultivating meadow fescue, and all roots visible under a microscope were removed. The remaining soil was incubated until the decomposition of root-derived material was very slow, i.e., when the root-derived material remaining in the soil was stabilized. Other ^{14}C-labelled materials (roots and shoots of meadow fescue and glucose) were incubated in soil from the field under the same experimental conditions. The total amount of ^{14}C-labelled root material originally supplied to the soil was calculated from comparisons of the stabilization of the different ^{14}C-labelled materials.

It was found that 11%, 17% and 47% of the rhizosphere respiration came from decomposition of root-derived C, if roots, shoots and glucose, respectively were used as model substances for the rhizodeposited material. The true value for the stabilization of rhizodeposited carbon is probably intermediate between those for glucose and roots, and therefore an estimate of 20–40% of rhizosphere respiration from microbial decomposition of rhizodeposited material may be a good approximation. This would mean that ca. 10% of the carbon taken up by the plant was respired by the roots.

The carbon distribution patterns found in our growth chamber experiments are well in accordance with those reported in the literature (see Hansson et al. (in press) for review). For example, in a field experiment with wheat in which a pulse label of $^{14}CO_2$ was supplied at eight different growth stages, 326 g m^{-2} was translocated below ground in a crop with an above-ground production of 1100 g m^{-2} (Keith et al. 1986). Half the amount of carbon translocated below ground was lost as respired $^{14}CO_2$, whereas 25% was found in root biomass and 25% remained in the soil.

The respired $^{14}CO_2$ either originated from root respiration or from microbial degradation of root-derived material. Our barley experiments in the growth chamber indicate that the distribution of recovered carbon between above- and below-ground parts was about the same as that found by Keith et al. (1986), whereas a higher proportion was found in the roots. Consequently, both from the literature and our field and laboratory experiments we can conclude that an amount of carbon as large or larger than that found as root biomass in cereals at harvest entered the soil as rhizodeposition during growth (Hansson et al., in press).

Plant uptake of nitrogen

In arable soils, the root system has two or three major sources of nitrogen. The first source is the nitrogen mineralized from litter and soil organic matter. The second source is nitrogen added through fertilizers and atmospheric deposition. In legumes, e.g., lucerne, a third source is available – atmospheric nitrogen fixed through bacterial symbiosis. In this section, nitrogen uptake from these three sources is presented and discussed.

Symbiotic nitrogen fixation

Investigations of symbiotic nitrogen fixation in the lucerne ley were conducted during 1981 to 1983. The objectives were: a) to estimate the seasonal distribution and total input of biologically fixed nitrogen, b) to estimate the relative amounts of nitrogen input to the lucerne crop by biological nitrogen fixation and by uptake from the soil solution, c) to define the main influencing environmental factors and to evaluate nitrogen fixation as a function of these factors.

Three different methods were used for measuring nitrogen fixation in the lucerne ley, i.e., the total-N difference method, the acetylene reduction method and the ^{15}N method.

The total-N difference method compares total nitrogen yield in a nitrogen fixing crop and a non-nitrogen fixing reference crop. The method is simple and cheap and does not require advanced equipment, but only rough estimates of N_2 fixation can be obtained because the method is based on the assumption that the fixing and non-fixing crops take up equal amounts of soil nitrogen (Rennie et al. 1978).

The acetylene reduction (AR) technique is based on the ability of the nitrogenase to reduce acetylene to ethylene instead of reducing atmospheric nitrogen (Turner and Gibson 1980). When using the AR method three moles of acetylene, C_2H_2, are reduced to ethylene, C_2H_4, instead of reducing one mole of atmospheric nitrogen, i.e.,

$$N_2 + 6\,H^+ + 6\,e^- \rightarrow 2\,NH_3,$$
$$3\,C_2H_2 + 6\,H^+ + 6\,e^- \rightarrow 3\,C_2H_4.$$

However, not all of the electron flow to nitrogenase is used for dinitrogen reduction. Some is lost through hydrogen evolution, and since the reduction of hydrogen is inhibited by acetylene (Schubert and Evans 1976) the true conversion factor $C_2H_2:N_2$ is often higher than 3.

An in situ AR assay was used when measuring nitrogen fixation in the field at Kjettslinge (Fig. 5.8). Plastic cylinders were pushed into the soil in spring and were left stationary throughout the season. Gas-tight plastic

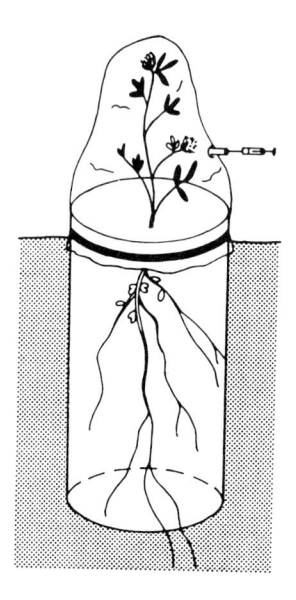

Fig. 5.8. The field cuvette used for estimations of nitrogen fixation by lucerne with the acetylene reduction method.

bags were temporarily attached to the cylinders, thus enclosing the plant-soil system. Acetylene and an internal standard, propane, were injected into the plastic bags. Gas samples were taken twice during the 1–1.5 h incubation and were analyzed within 24 h by gas chromatography (Balandreau and Dommergues 1973, Mårtensson and Ljunggren 1984a, Wivstad et al. 1987).

The AR method is very sensitive and rapid, but the uncertainty in the conversion factor and the existence of diel and seasonal variations in N_2 fixation makes extrapolation to quantitative estimations over a growing season questionable (Bergersen 1970). Consequently, complementary methods are necessary for seasonal estimates of N_2 fixation.

The ^{15}N method used is based on isotope dilution, where ^{15}N-labelled fertilizer is added to the soil. The nitrogen fixing crop dilutes the ^{15}N taken up from the soil-fertilizer pool with atmospheric $^{14}N_2$. The non-fixing reference crop does not have that N source. This method for estimating N_2 fixation is an attractive complement to the AR technique, because it integrates N_2 fixation over time. The use of a suitable reference crop is important, since it greatly influences the quantitative estimates of N_2 fixation, e.g., soil N availability and uptake pattern of soil N may differ between the fixing crop and the reference crop (Rennie 1982, Witty 1983). The ^{15}N excess in the soil decreases with time as unlabelled N is mineralized. If the fixing and non-fixing crops take up soil N at different times during the season, the estimates of N_2 fixed becomes unreliable. Further, it is necessary that the root systems of the fixing and non-fixing crop are similar, since both root systems must exploit a soil volume with similar ^{15}N-excess.

The N_2 fixation of the lucerne ley was evaluated using A-values, which are measures of soil N availability in relation to fertilizer N uptake by the crop and which are dependent on the rate of fertilizer N addition (Fried and Broeshart 1975). Using A-values, it is possible to use different fertilizer rates for the fixing and non-fixing crops. A meadow fescue ley was used as non-fixing reference in the experiments. During 1981, uninoculated lucerne was also used as a reference crop. The experimental plots of lucerne and the reference crop were supplied with ^{15}N labelled fertilizer in aqueous solution. Crop biomass, total N content and N_2 fixation were estimated at harvests and at the end of the growing season (Wivstad et al. 1987).

During 1981, when the lucerne ley was established, the three methods were compared (Mårtensson and Ljunggren 1984a). The N_2 fixation ranged between 79 and 104 kg N ha^{-1} yr^{-1}, depending on the method used (Tab. 5.3). When using the AR method, a conversion ratio of 4.41 was used between produced ethylene and reduced nitrogen, as estimated by Mårtensson and Ljunggren (1984b). The highest value, obtained with the AR method includes fixed N in the whole plant-soil system. The two other methods are based on the amount of fixed nitrogen found in the above-ground plant biomass and should give lower values than the AR method. The uncertainties introduced when short-term AR measurements are extrapolated to a whole season may partly obscure the real differences.

Quantitative estimates of N_2 fixed and translocated into above-ground plant biomass, using the ^{15}N technique, are shown in Tab. 5.4. The annual amounts were 84, 242 and 319 kg N ha^{-1} in the three successive years, respectively (Mårtensson and Ljunggren 1984a, Wivstad et al. 1987). The proportion of N in the plant that was fixed from the atmosphere was 70% and 80% for 1982 and 1983, respectively. The low estimate for 1981 is explained by the late sowing in June and the slow establishment of the ley. Ley harvests in 1982–1983 were normal for a good year in this area, and the estimates of fixed N_2 in the above-ground plant parts are considerably higher than normal fertilizer N additions (200 kg N ha^{-1} for a grass ley harvested for hay). Significant amounts of fixed N could also be found in the roots and crowns. For example, nitrogen going into below-ground production in 1983 amounted to 140 kg ha^{-1} (Pettersson et al. 1986). Assuming that 80% of the root N was derived from the atmospere, a total N fixation for 1983 of more than 400 kg ha^{-1} can be calculated.

The establishment and development of a ley are important factors influencing N_2 fixation. The percentage of plant nitrogen derived from N_2 fixation increased with the age of the ley, making up 40, 70 and 80% during the three successive years (Tab. 5.4). Fixation capacity is highly dependent on the carbohydrate reserves of the plant (Vance et al. 1979), which may explain some of the variation within and between years. For example, fixation before the first harvest in 1983 was twice that in 1982, and the fraction of total N derived from fixation was significantly higher (Tab. 5.4). This period accounted for most of the total difference in fixation between the two years. Total plant biomass was about 50% greater in the autumn of 1982 than in 1981 (Pettersson et al. 1986), and thus there was probably a larger carbohydrate reserve in the overwintering plant biomass, allowing higher fixation rates during spring of 1983.

Tab. 5.3. Nitrogen fixation of the lucerne ley, estimated by an in situ acetylene reduction (AR) method, a ^{15}N dilution (A-value) method, and a total-N difference method. Mean and standard error, n = 10 (AR), n = 4 (^{15}N and difference method) and probability value (P) for the hypothesis that the method was not different from the AR method (Adapted from Mårtensson and Ljunggren, 1984a).

Method	Reference crop	N fixation (g N m^{-2})	P
AR		10.4 (1.3)	–
A-value	Uninoculated lucerne	10.1 (1.0)	NS
	Meadow fescue	8.4 (1.3)	NS
Total-N difference	Uninoculated lucerne	7.9 (1.3)	0.05

Tab. 5.4. Amount and proportion of fixed nitrogen in the herbage of the lucerne ley estimated by the ^{15}N method. Mean values ± standard error, n = 4 in 1981 and 1982, n = 12 in 1983 (Adapted from Mårtensson and Ljunggren 1984a, Wivstad et al. 1987).

Date of sampling	Total N g m-2	Fixed N g m-2	Fixed N/ total N (%)
1981 Aug 9	12.3±1.3	4.6±1.0	37
Nov 4	8.4±0.3	3.8±0.3	45
Total	20.7±1.6	8.4±1.3	40
1982 Jun 18	10.5±1.3	6.4±1.7	61
Aug 12	15.1±0.6	11.7±1.6	77
Nov 2	9.1±0.7	6.1±0.6	67
Total	34.7±1.4	24.2±2.4	70
1983 Jun 20	17.2±0.6	12.3±0.8	72
Aug 24	15.5±0.9	13.5±1.0	87
Oct 18	7.3±0.2	6.1±0.9	84
Total	40.0±1.1	31.9±1.6	80

Harvest patterns as well as phenology greatly influenced the seasonal dynamics of N_2 fixation (Fig. 5.9). The nitrogenase activity of the lucerne was followed with in situ AR assay in three different harvest patterns (Wivstad et al. 1987). The periodic harvesting of the lucerne ley resulted in a cyclic pattern of N_2 fixation. A decline in nitrogenase activity was observed after shoot removal (Fig. 5.9a, b) and the N_2 fixation rate remained low during at least two weeks after harvest. The maximum rate of N_2 fixation occurred during budding and early flowering (Fig. 5.9) and was followed by a rapid decline as flowering proceeded (Fig 5.9b, c).

A number of environmental factors influence N_2 fixation. Soil moisture markedly affects nitrogen fixation in lucerne, although lucerne is a drought-resistant crop (Sprent 1971, Carter and Sheaffer 1983). The effect of water stress on N_2 fixation of lucerne was investigated using the in situ AR assay in a separate experiment in 1983 (Wivstad et al. 1987). July and August were dry in both 1982 and 1983, and in early August 1983, when the soil matrix potential was close to or below wilting point in the top soil, irrigation with 30 mm water almost doubled the nitrogenase activity.

The average air temperature during the growing season (April-October) was 10.8°C in 1982 and 11.7°C in 1983 (Alvenäs et al. 1986), and from this difference it is not possible to make any assumptions concerning the influence of temperature on N_2 fixation. However, in 1982 there was a cold period in early June, with temperatures below zero during some nights. This could partly explain the low proportion of fixed N_2 in the first harvest of 1982.

Diel variation of N_2 fixation may be important (Abdel Wahab 1980) but in the lucerne ley at Kjettslinge no such variation was found (Mårtensson and Ljunggren 1984a). The suggested explanation for the lack of diel variation was that the nodules within the soil were protected against short-term fluctuations in temperature (Chapter 3).

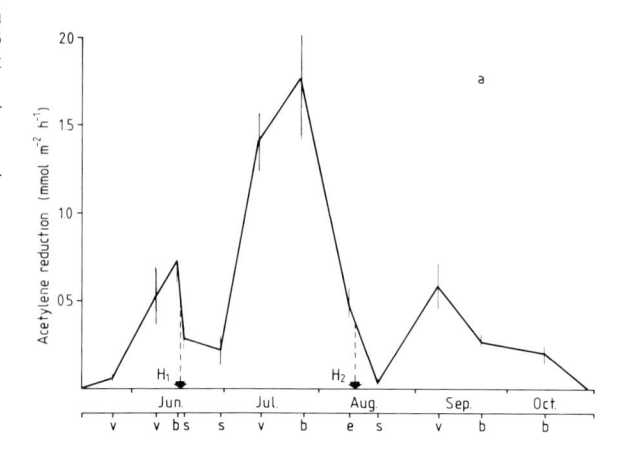

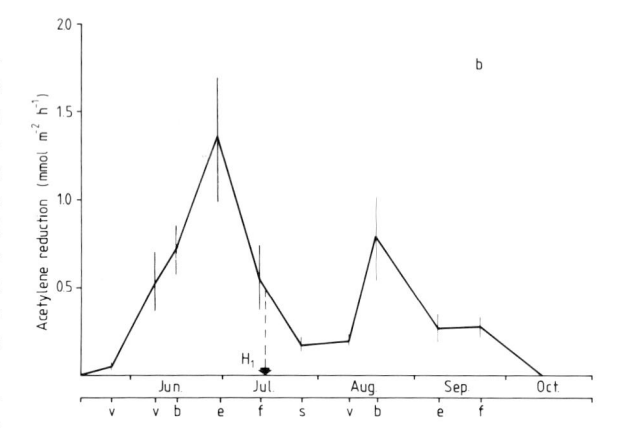

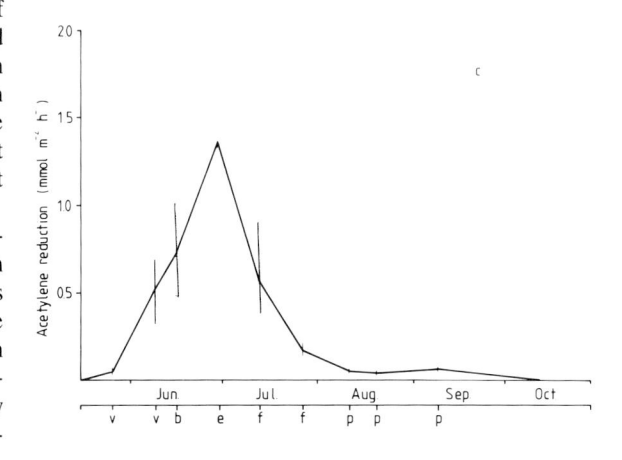

Fig. 5.9. Seasonal dynamics of N_2 fixation in LL in 1982, subjected to different harvest practices. Two harvests (a), one late harvest (b), no harvest (c). H1, H2 = harvests. The estimations were made by in situ acetylene reduction assay. Means ± standard error are given; n = 8 until 15 June, n = 4 after 17 June. Phenological stages are indicated as: v = vegetative, b = budding, e = early flowering, f= full flowering, p = pod filling, s = stubble. (Adapted from Wivstad et al. 1987).

Available mineral N in the soil is inversely correlated to the rate of N_2 fixation (Groat and Vance 1981). The seasonal fluctuations in the N_2 fixation rate at Kjettslinge partially reflected the availability of soil N. The low proportion of fixed N_2 in spring may be explained by the accumulation of mineral N in the profile from decay occurring in the lucerne ley during the spring and preceding autumn. The high content of mineral N in the soil profile, 70 kg N ha^{-1}, in the summer of 1981 (Bergström 1986) in addition to the initial time lag due to establishment, could partially account for the low proportion of fixed N_2 in the seeding year. In the established lucerne ley the soil profile was almost devoid of mineral N down to 1 m.

In an evaluation of the nitrogen economy and energy balance in Swedish agriculture, Jansson and Simán (1978) calculated the over-all input of symbiotically fixed nitrogen to be 27 kg ha^{-1}. About 30% of Swedish arable land is covered with ley crops in various combinations and amounts of legumes and grasses. From these calculations the average nitrogen input to the leys through symbiotic N_2 fixation would amount to ca. 80 kg ha^{-1} yr^{-1}.

The cultivation of legume crops has a number of advantages. They provide a protein-rich fodder of high value in animal production. The nitrogen input from these crops can provide an alternative to nitrogen fertilizer, which when used excessively, may cause leaching problems. Legumes are a favourable preceding crop in a crop rotation because of their nitrogen input into the soil and the beneficial effects of their deep root system on soil structure. However, when a lucerne ley is ploughed, high amounts of nitrogen can be leached (Bergström 1987a), so proper management is crucial to reduce potential environmental impacts of lucerne leys.

Uptake of fertilizer and soil nitrogen

Nitrogen is taken up by the plant roots either from the mineral N pool or from symbiotic fixation. The nitrogen is allocated above and below ground; one part is removed by harvest and the rest is bound in biomass or returned to the soil (Fig. 5.10).

Since nitrogen fertilization mainly promoted aboveground production, proportionally more of the plant nitrogen was removed by harvest in B120 (70%) than in B0 (55%). The amount of nitrogen incorporated into the root systems constituted 18 and 26% of the total nitrogen uptake in B120 and B0, respectively. Approximately 50–60% of the N uptake in the leys was harvested, and about 20% was used for root growth (Fig. 5.11).

By using ^{15}N-labelled Ca(NO$_3$)$_2$-fertilizer in a field lysimeter experiment, Lindberg et al. (1989) estimated

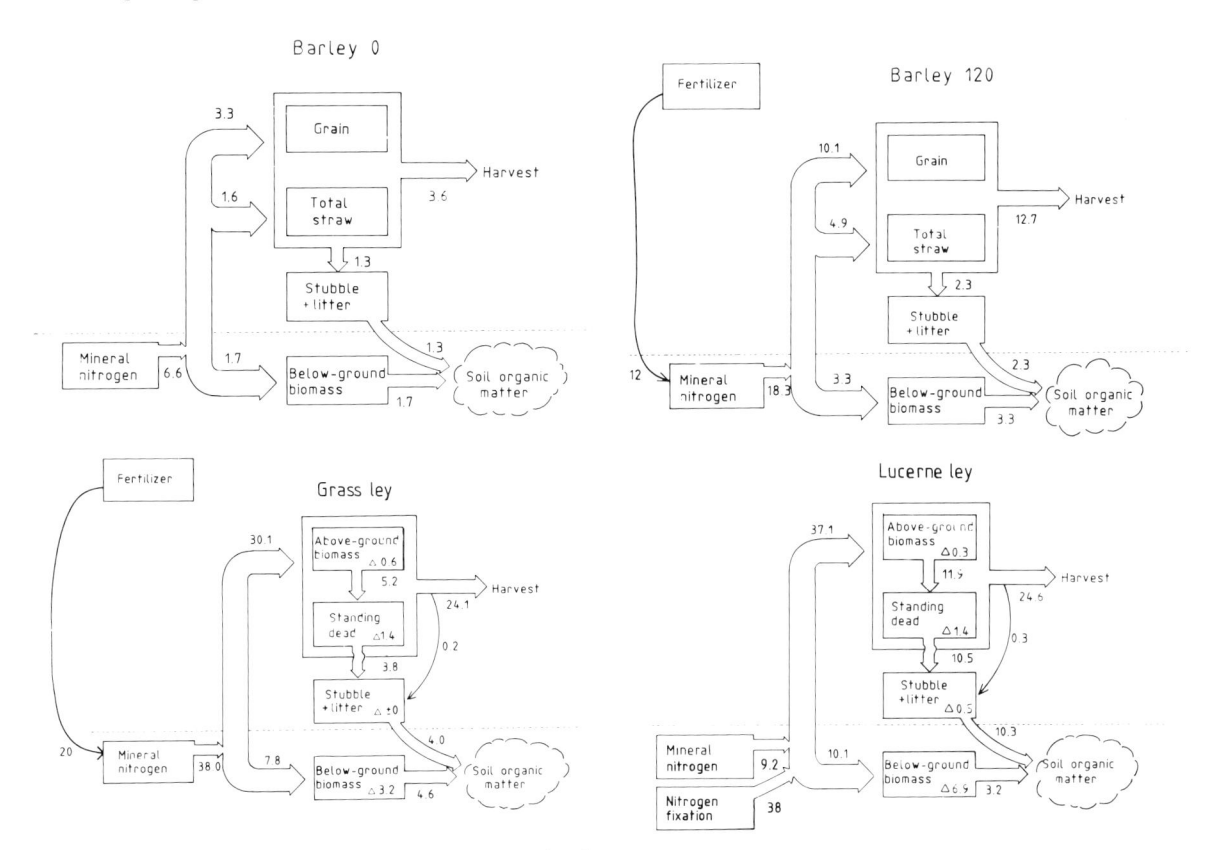

Fig. 5.10. Nitrogen flows in the crop subsystem (g N m^{-2} yr^{-1}).

fertilizer recovered in the harvested crop to be 59% of the fertilizer addition in B120 (Fig. 5.12c), which was similar to that estimated by the difference method (N uptake in the fertilized treatment − N uptake in the unfertilized treatment) the same year (Tab. 5.5). Calculated by the difference method, recovery varied between 59 and 92% with a mean of 75%, comparable with the 71–83% reported by Dowdell (1982) in spring barley. Fertilizer recoveries calculated from isotope methods are often lower than those calculated by difference methods (Hauk 1971). Several tracer studies have yielded values of 40–50% (Jansson 1963, Myers and Paul 1971, Dowdell 1982, Nørlund et al. 1985). The lower values often found with tracer methods could result from an apparent increase in soil N uptake due to mineralization-immobilization turnover between labelled and unlabelled nitrogen (Hauk and Bremner 1976, Jansson and Persson 1982). Conversely, the higher values obtained with the difference method could be due to larger root systems of fertilized crops, which results in a greater soil volume accessible for N uptake (Sørensen 1982). Other factors limiting growth, e.g., drought, may affect the growth of fertilized/unfertilized crops differently, resulting in misleading differences in nitrogen uptake.

In ^{15}N-lysimeters with meadow fescue ley, the overall export of fertilizer ^{15}N from the two fertilizer applications was 65% (13.0 g N m^{-2}) as determined by the isotope method. The recoveries in the harvest from the two fertilizations are presented in Fig. 5.12. Hansson and Pettersson (1989) in a parallel microplot experiment with ^{15}N found a similar (63%) fertilizer export, and concluded that virtually all fertilizer N was found in, or had passed through, the meadow fescue plants (Tab. 5.6).

Tab. 5.5. Fertilizer nitrogen recovery in barley (B120) harvest (straw + grain) as measured by the isotope (^{15}N) method and by the difference method. (N D = not determined).

Year	Isotope method[1] (g m^{-2})	(%)	Difference method[2] (g m^{-2})	(%)
1981	N D	N D	7.7	64
1982	7.1	59	7.1	59
1983	N D	N D	11.0	92
1984	N D	N D	7.4	61
1985	N D	N D	10.3	86

[1]Lindberg et al. (1989).
[2]R. Pettersson pers. comm., Hansson et al. 1987.

Decomposition and nitrogen mineralization

The organic compounds produced through primary production are decomposed and mineralized in a series of complex processes. Decomposition is not simply carbon loss through respiration and the liberation of inorganic nutrients, even if this simple view can be sufficient for some purposes. Decomposition includes secondary production of microbial and animal biomass and organic metabolites, which in turn become resources for decomposition (Swift et al. 1979). The heterogeneous mixture of products resulting from this continuous breakdown and resynthesis is referred to as humus or stabilized soil organic matter.

In this section the processes involved in decomposition are viewed in the context of conceptual models of gradually increasing complexity. Abiotic controls on decomposition and mineralization are considered first. Then the activity of decomposer organisms are explicitly introduced. Finally, the possible interactions of soil organisms in decomposition processes are discussed.

The decomposition and mineralization experiments included estimates of decomposition rates made with

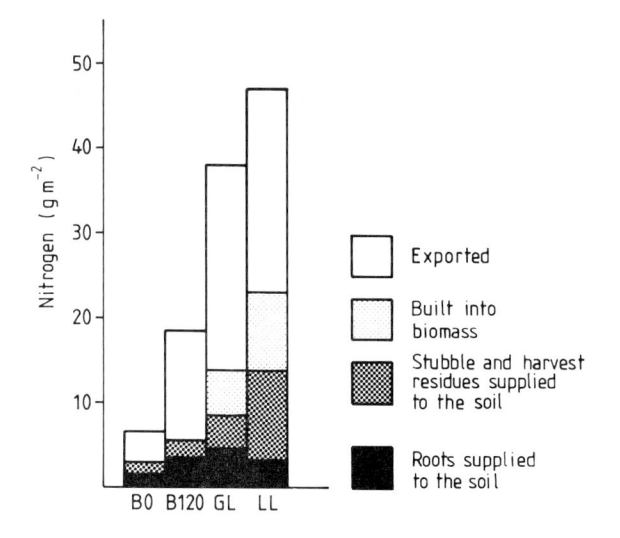

Fig. 5.11. Nitrogen taken up by the crops, expressed as g N m^{-2} yr^{-1}, divided into exported amounts, biomass increase and the supply to soil by death.

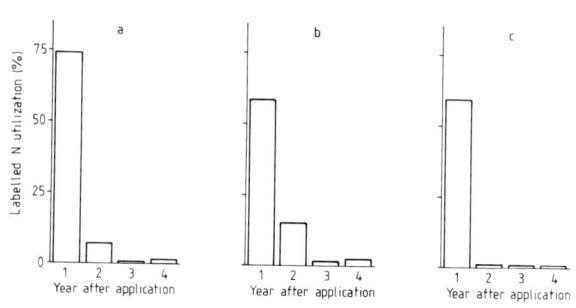

Fig. 5.12. The export of labelled nitrogen from grass ley (a, b) and barley (c) during four years after application. Ca(^{15}NO$_3$)$_2$ was added to field lysimeters during year 1 as follows:
a) First fertilization (12 g N m^{-2} in May) to the grass ley labelled.
b) Second fertilization (8 g N m^{-2} in June) to the grass ley labelled.
c) A single fertilization (12 g N m^{-2} in May) to the barley labelled.
Bars are means of two replicate lysimeters. Data from Lindberg et al. (1989).

Tab. 5.6. Recovery of Ca(^{15}NO$_3$)$_2$ (19.3 g N m^{-2}) in GL. Data from Hansson and Pettersson (1989). (N D = not determined).

	Year 1 (g N m^{-2})	(%)	Year 2 (g N m^{-2})	(%)
Exported as harvest	12.2	63	2.15	11
Above-ground plant parts not harvested	1.89	10	0.83	4
Live roots	3.67	19	N D	N D
Dead organic matter	2.06	11	N D	N D
Soil	0.84	4	N D	N D

both litter-bag techniques and ^{14}C-labelled plant material. Nitrogen mineralization was studied using litter-bags, ^{15}N-labelled material, and laboratory incubations of soil cores. Decomposer organisms were studied in field experiments using litter-bags and soil samplings in the four cropping systems as well as in controlled laboratory experiments.

Mass loss and abiotic control

The aim of the abiotic approach is usually to describe and predict mass loss of a resource or several resources within a given range of climatic conditions. One of the simplest models is to consider a homogeneous resource decomposing under constant environmental conditions according to first-order kinetics. The model assumes that a constant fraction (k) of the resource is lost per time unit, i.e., mass loss follows a negative exponential curve.

This model with the differential equation,

$$\frac{dM}{dt} = -k \cdot M,$$

constitutes the foundation for decomposition modelling. The equation is usually shown in the integral form,

$$M = M_{(0)} \cdot e^{-k \cdot t},$$

and is called "the exponential decay equation". $M_{(0)}$ denotes the mass of the resource when time (t) equals zero. A basic property of this model is that the absolute mass loss rate, e.g., g litter-bag^{-1} d^{-1}, decreases with time. This is true for a homogeneous resource in a constant environment. It is thus not necessary to assume a heterogeneous resource or varying climatic conditions to explain a lower absolute mass loss rate towards the end of an experiment.

However, a constant abiotic environment rarely occurs outside the laboratory. In central Sweden there is frost in the soil for ca. 100 d during the winter, and in summer the top soil can reach 25°C (Chapter 3). Soil moisture during a year can vary from water-logged conditions to drought beyond the wilting point. Since temperature and moisture are the major control factors for decomposition, their dynamics usually have to be included or compensated for in the model. An exception may be long-term investigations, with one or two samplings per year, since long-term climatic variation is considerably less than seasonal variation within a given year.

A simple way to compensate for temperature fluctuations is to transform time according to measured temperature and express the mass of the resource as a function of the transformed time. A first step would be to remove all days with a temperature < 0°C, thus removing periods of frost. A second step would be to use the cumulative temperature sum instead of time, assuming a linear relationship between temperature and decomposition rate. However, transformations using the Q_{10} power function is more in line with results obtained from controlled experiments investigating the relationship between temperature and biological activity. The Q_{10} value is the increase in activity for a 10°C rise in temperature. The function can be written as:

$$E_T = Q_{10}^{(T-T_b)/10},$$

where E_T = the temperature factor, T = temperature, and T_b = base temperature, i.e., the temperature corresponding to $E_T = 1$. For decomposition modelling, values of Q_{10} are normally within the range of 2 – 5.

Soil moisture may be expressed in several ways, of which volumetric water content (Θ), and soil water tension (Ψ) are common (Chapter 3). The relationship between soil moisture and decomposer activity in the soil may vary considerably, but assuming boundary conditions such that there is no activity below a certain moisture level and no moisture limitations above a certain level is reasonable. Between these limits, a logarithmic relationship between soil water tension and activity of microorganisms is often found (Sommers et al. 1980). Under water-logged conditions, decomposition rates can decrease due to oxygen stress, but this is seldom the case in well-drained agricultural soils.

In our investigations, we used a moisture factor of the type:

$$E_\Psi = \frac{\ln(\Psi_{min}/\Psi)}{\ln(\Psi_{min}/\Psi_{max})},$$

where Ψ is the soil water potential and Ψ_{max} and Ψ_{min} are boundary values for maximum and zero activity.

This approach, i.e., assuming a homogeneous resource and compensating for fluctuations in temperature and moisture, was applied to the results from a two-year litter-bag field experiment with barley straw. The litter-bags were buried 10–15 cm deep under B0, and 25 litter-bags were sampled on 14 occasions. A one-compartment simulation model, assuming first-order kinetics, was fitted to the ash-free mass data, with measured temperature and soil moisture as driving variables. The rate constant k, Q_{10}, Ψ_{max} and Ψ_{min} were optimized for best fit and the resulting values were 0.00575 d^{-1}, 1.78, -0.006 MPa and -0.35 MPa, respectively (Andrén and Paustian 1987). This model gave a near-perfect fit, $R^2 = 0.997$, and is shown together with the driving variables in Fig. 5.13.

Heterogeneous resource

Plant litter is not chemically or structurally homogeneous. It consists of a great number of components, each with a distinct decomposition rate in isolation (Minderman 1968). However, a simple model consisting of two components decomposing in parallel at different rates can often give a good fit to mass or carbon loss data (Hénin et al. 1959, Jenkinson 1977).

The corresponding differential equations are:

$$\frac{dM_R}{dt} = -k_R M_R, \quad \frac{dM_L}{dt} = -k_L M_L; \quad M_R + M_L = M,$$

with the index R denoting a refractory (slowly decomposing) fraction and L denoting a labile (rapidly decomposing) fraction.

This model, usually called "double exponential", was applied to the results from a field experiment using ^{14}C-labelled plant material, grown in the assimilation chamber described under "Isotope carbon budget". B0, B120, GL and LL were grown for two months in steel cylinders (diameter 25 cm, height 30 cm) with soil from the experimental field. After harvest and mixing of the soil, the cylinders were dug into the soil in the experimental field. Soil cores were taken from each cylinder to obtain initial ^{14}C-amounts. In addition to the cylinders originating from the growth chamber, cylinders containing labelled B120 straw mixed with soil were also put in the field. The soil in the cylinders was sampled every spring and autumn and analyzed for ^{14}C, and the experiment was conducted for 4.5 yr. The double exponential decay equation was fitted to the results, and Fig. 5.14 shows the model fits for the different plant litters.

It was not necessary to apply corrections for temperature or moisture dynamics, since these tend to level out when samplings are made at long (6-month) intervals. After the initial year, the mass loss rate was determined by the rate constant for the refractory fraction, k_2 (Fig. 5.14). The different resources all lost around 50% of the ^{14}C during the first year, and thereafter they decomposed at a rate of 0.06 to 0.15 yr^{-1}. The model did not fit the data for lucerne roots very well, probably because this root system consisted of comparatively few and thick roots, which increased sampling errors.

To further investigate the decomposition rates of labelled organic material, the soil samples were incubated in the laboratory, at 30% of water holding capacity for 145 d at 25°C. The soil was placed in a plastic tube together with a smaller plastic tube containing CO_2-absorbing Na_2CO_3, which was changed at regular intervals and analyzed for ^{14}C.

In Fig. 5.15 the ^{14}C loss during the 145 d is given as a percentage of the total amount of labelled carbon in the sample. Almost 50% of the loss was recorded during the first 12 d, which is a typical effect of sample preparation, drying and rewetting (not shown). The decomposition rates were considerably higher than those found in the field even when the initial 12-d flush was excluded. All labelled material followed the same pattern, i.e., the decomposed fraction initially decreased with time after placement in the field, but after 1.5 years a more or less constant rate was observed. These results support the two-component model shown in Fig. 5.14, indicating that there were no significant changes in resource quality during the later stages of the field incubation.

There were also some differences in ^{14}C loss rates between the litter types. For instance, the barley roots did not decompose as rapidly as the other litter types. The leys showed an intermediary decomposition rate, and barley straw showed the highest rates, around 20% of the sample's ^{14}C content was lost during laboratory incubation. This ranking order of decomposition rates in the different litter types is fairly consistent with the k_2 constants estimated in the field (Fig. 5.14).

There are alternative ways to model a heterogeneous material. Since water-soluble components rapidly decrease after litter placement in the field (Tab. 5.7), it may be appropriate to subtract the initial water-solu-

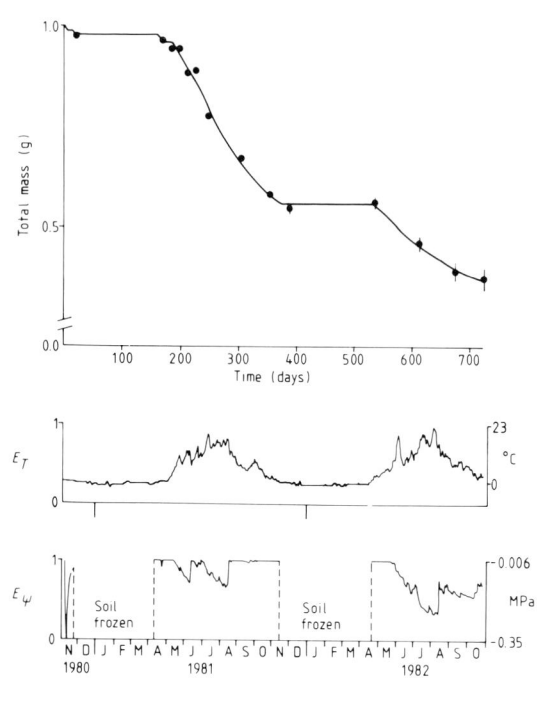

Fig. 5.13. Remaining mass (● with standard error bars) of barley straw, buried 10–15 cm deep, on 4 November 1980. Predicted mass (—) using a one-compartment first-order simulation model with multiplicative factors for temperature (E_T) and soil moisture (E_Ψ). (Andrén and Paustian 1987).

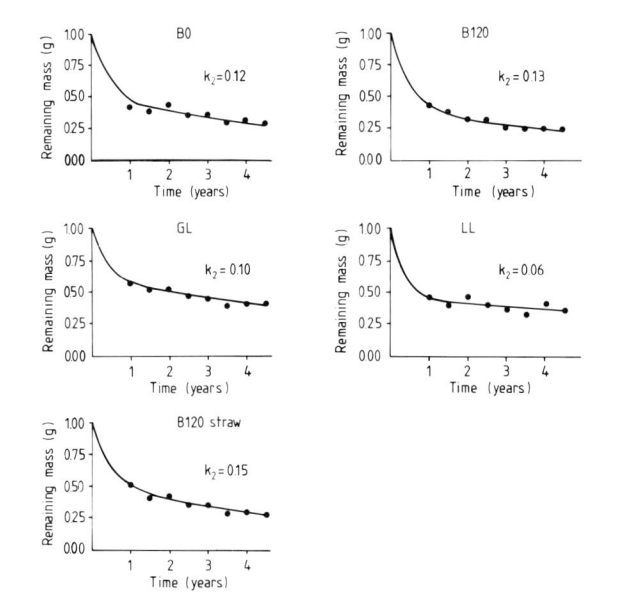

Fig. 5.14. Remaining ^{14}C (●) in root and B120 straw litter, incubated in the field for 4.5 yr. Predicted mass (—) using a double exponential model (J. Persson, unpubl.).

bles, determined by chemical analysis of the litter, and fit a one-component model to the remaining fraction. When this approach was used and all days with frost in the soil were removed from the data set, good fits were obtained for all above- and below-ground parts, incubated in litter-bags under their parent crop (Andrén 1987). The barley straw was incubated 10–15 cm deep, mimicking ploughing under of stubble, whereas the ley material was put on the soil surface, mimicking wilting during winter. All root litter-bags were buried in the soil. Predicted mass losses for the first year were 63% (B0), 55% (B120), 73% (GL) and 67% (LL) for the above-ground parts (Fig. 5.16). Corresponding mass losses of roots, all placed 15 cm deep, were 40%, 28%, 54% and 91%, respectively. The different decomposition rates were not only due to differences in the resource. This was obvious from the results of incubating B0 straw in all cropping systems where the general pattern of decomposition rates was: B0 > B120 > GL < LL (Andrén 1987). Differences in soil moisture conditions under the different crops probably accounted for most of the difference (Chapter 3), but other factors may have contributed, e.g., temperature or soil physical differences between leys and barley plots. Since the mean root biomass was higher in the leys, and GL showed a high concentration of plant roots about 15 cm deep (Hansson and Andrén 1987), influences from plant roots through organic C input, soil aggregation, biomass activation or aggregate disruption could also have contributed to the differences in decomposition rates (Helal and Sauerbeck 1987).

The two techniques for measuring decomposition can to some extent be compared using the results presented here. The litter-bags (1 mm mesh size) represented a barrier to earthworms and macroarthropods and may also have contained unnaturally high concentrations of litter. Measurements of ^{14}C-loss from labelled litter mixed with the soil were hampered by variability in recovery rates, possible effects of the container on moisture and earthworm access and few replicates due to the high cost of labelled litter. However, if the results are similar when using the two methods, the credibility of the observations increases. The simplest approach may be to compare the predicted fraction remaining after one year, since the methods had different incubation times and models. The mean remaining fraction for the four root types in litter-bags was 47%, which was well in accordance with the 46% remaining after one year in the ^{14}C experiment. The corresponding figures for B120

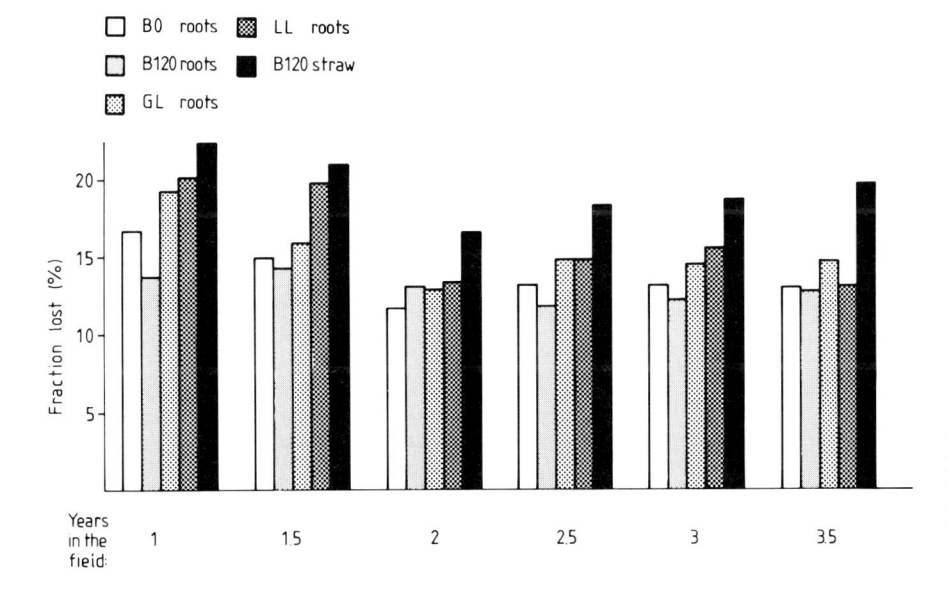

Fig. 5.15. Fraction ^{14}C lost during 145 d of laboratory incubation of soil samples taken from the field at different occasions.

Tab. 5.7. The initial concentration of water-soluble material in various litter types, and the fraction of the initial water-soluble lost at the first sampling after being placed in the field (Andrén 1987).

Crop	Shoots Initial (%)	Lost (% of init.)	Roots Initial (%)	Lost (% of init.)
B0	4.3	42	4.6	43
B120	4.8	43	5.8	35
GL	28.3	88	34.6	92
LL	33.8	93	43.8	92

straw was 45% and 49% for litter-bags and ^{14}C, respectively. If the different root types are compared separately, however, the correlation between the methods was not particularly good. The decomposition rates for the root types, in increasing order, was B120 < B0 < GL < LL for litter-bag roots and GL < LL < B120 < B0 for ^{14}C-labelled roots. These discrepancies may have many causes, but one explanation may be that the roots placed in the litter-bags were taken from the field in autumn, while in the ^{14}C-experiment roots were grown in a cultivation chamber for two months before the pots were taken to the field. Roots of different ages can have different contents of easily decomposable components (Steen and Larsson 1986). For example, LL roots collected in autumn for the litter-bags contained high amounts (44%) of water-soluble storage carbohydrates (Andrén 1987), that were probably not found in the younger LL roots grown in the cultivation chamber. Also differences in sample preparation may have influenced the results, since the roots for litter-bags were washed from the soil and dried before being put into the bags. The roots in the ^{14}C-experiment were left in situ in the soil before being placed in the field.

A decomposition model including resynthesis

To model nitrogen immobilization or, more generally, absolute increases of components during the course of decomposition, models including resynthesis must be used. Resynthesis can be defined as the production of new compounds during decomposition of a resource. Decomposition models encompassing synthesis of components have recently been used for modelling, e.g., humus formation or nitrogen immobilization. A widely used class of models (Paul and van Veen 1978) consists of refractory and labile fractions of fresh litter, and one or more secondary pools consisting of decomposition products (Fig. 5.17). This class of models has been used for litter decomposition in prairies (Hunt 1977), soil organic matter in long-term agricultural experiments (Parton et al. 1983) and barley decomposition and nitrogen dynamics in the field (Andrén and Paustian 1987). Andrén and Paustian (1987) fitted such a model simultaneously to measured amounts of total mass, water-solubles and nitrogen (Fig. 5.18). The model fit the observations well, including the increase in water-solubles after 150 d and the initial loss and subsequent increase of N. The increases in water-solubles and nitrogen were due to an accumulation of decomposition products in the "active" pool, which had an intermediate residence time. This pool is conceptually defined in the model as microbial biomass and metabolites, and the best fit was obtained when 35% of this pool consisted of water-soluble compounds. Towards the end of the experiment, the model output suggested that the "active" pool constituted 10% of the litter-bag contents. The total microbial biomass as measured in the field was about 2% of the total mass (Wessén and Berg 1986), which implies that the main part of this pool consisted of metabolites and constituents from lysed cells. The best-fit value for nitrogen concentration in the "active" fraction (3%) is reasonable for material of microbial origin (McGill et al. 1981), and the N concentration assumed for the "stabilized" fraction (5%) is typical for humified material (Allison 1973). Thus the dynamics of total mass, water-solubles and N could be explained by the

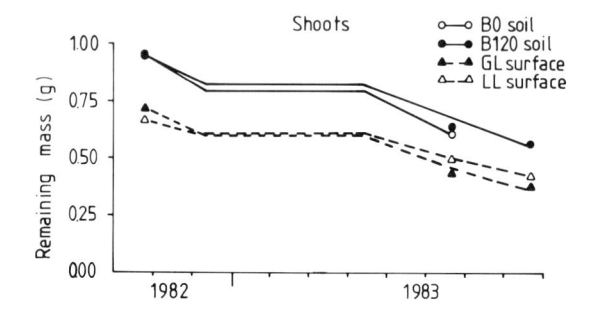

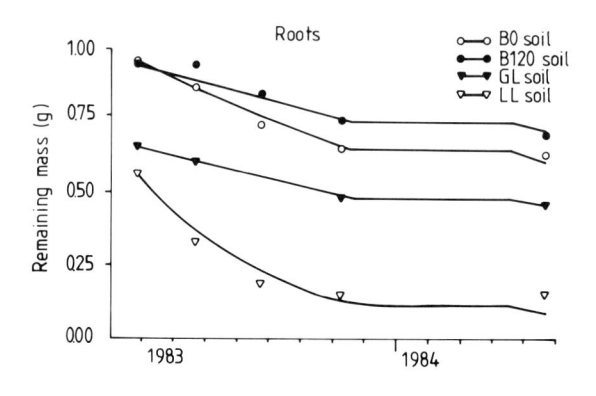

Fig. 5.16. Remaining ash-free mass of shoots, incubated 10–15 cm deep (B0, B120) and on the surface (GL, LL) in plots cultivated with the same crop as the incubated shoots. Below is shown the ash-free mass of roots, incubated 10–15 cm deep in plots cultivated with the same crop as the incubated roots. The initial masses were reduced by the analyzed amount of water-solubles, and a one-component exponential model was fitted to the results (Andrén 1987).

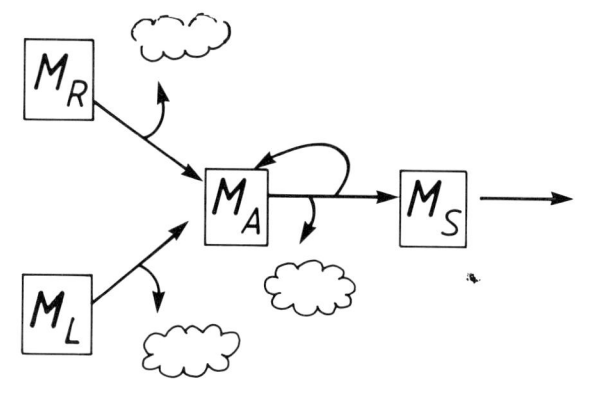

Fig. 5.17. A four-compartment decomposition model, assuming first-order kinetics, biosynthesis and organism turnover. M_R = refractory part of resource, M_L = labile part of resource, M_A = active pool, M_S = stabilized pool. (Adapted from Andrén and Paustian 1987).

simulation model, without explicitly including measured biomass of microorganisms or fauna. However, this model is a step towards including biotic components in decomposition modelling, and an interesting future task would be to directly couple organism activity and biomass data to decomposition process rates.

Nitrogen mineralization and immobilization

In this section mineralization/immobilization is described from the standpoint of the resource (i.e., organic nitrogen), while Chapter 6 focuses on the dynamics of the product (i. e., inorganic nitrogen). Estimates of net annual mineralization rates are presented in Chapter 7. The results presented in this section include estimates of nitrogen mineralization from litter and crop residues, mineralization of residual fertilizer nitrogen (determined from ^{15}N-lysimeter studies) and analyses of nitrogen mineralization from soil organic matter (SOM).

The major differences in nitrogen dynamics and mass loss between different types of litter occurred during the early phases of decomposition, and were related to the initial amount of nitrogen, which in turn was closely correlated to the initial amount of water-solubles. The early phase of rapid change in N mass lasted from 4 to 9 months, depending on the initial N mass and date of placement in the field. During winter, when soil temperatures were below 0°C, the chemical composition of litter did not change. Thus, litter with an initial mass of approximately 6 mg N g^{-1} organic matter, or more, rapidly lost nitrogen (Fig. 5.19), probably due to loss of soluble N through leaching as well as biological processes (Wessén and Berg 1986). Litter with an initial content of 3–5 mg N g^{-1} immobilized nitrogen after a short initial phase of mineralization. As a consequence the wide range of nitrogen contents found in the different litters was gradually reduced with time (Fig. 5.20). Later contributions to mineralization thus became similar for all types of litter.

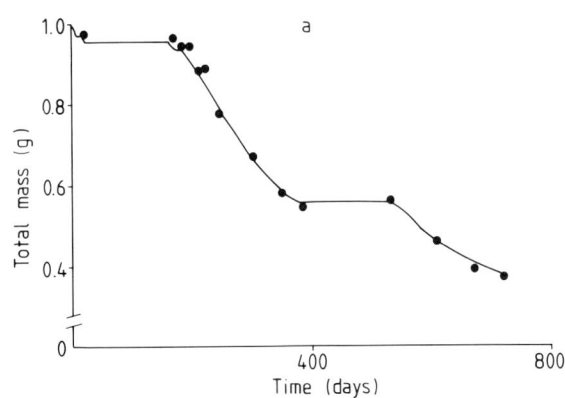

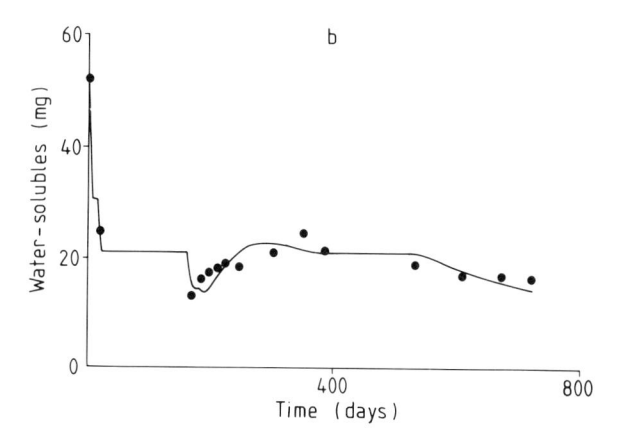

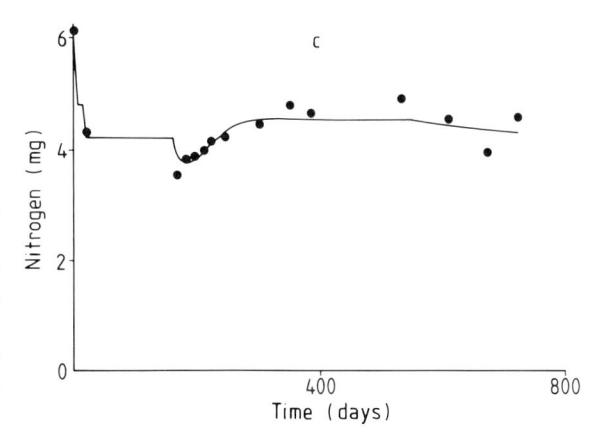

Fig. 5.18. Simulated (—) and measured (●) data from a two-year decomposition experiment with barley straw. The four-compartment decomposition model shown in Fig. 5.17 was simultaneously fitted to total mass a), water-soluble mass b) and nitrogen mass c). (Andrén and Paustian 1987).

Net changes in N mass in barley litter were relatively small, from 2 mg N immobilized to 4 mg N mineralized. Ley litter initially mineralized more nitrogen, 14–21 mg N, while later changes, as for barley, were much smaller.

Barley straw with an initial N mass of 3.4 mg g^{-1} immobilized 0.9–4.9 mg N during 8 months, depending on the starting date, cropping system and placement in the field (Fig. 5.21). Differences seemed to reflect the availability of mineral nitrogen in the top soil, including the effect of the surface-applied fertilizer-N, as well as differences in soil moisture (Andrén 1987). Straw placed 15 cm below the soil surface in the leys, where root density was high, immobilized less N than straw placed at the surface, although mass loss was higher. The explanation may be competition for available nitrogen between plant roots and decomposers.

The results on barley straw decomposition and nitrogen dynamics presented by Andrén and Paustian (1987) indicated an initial mineralization during the first autumn/winter followed by a phase of immobilization during the first growing season. The immobilization phase coincided with the start of ingrowth of fungal mycelium into the decomposing barley straw (Wessén and Berg 1986). They estimated the ingrowth to be 2.8 mg mycelium, which corresponded to an addition of approximately 0.07 mg N. The bacterial biomass increased with 0.35 mg dry mass, which corresponded to approximately 0.02 mg N. Although it was not apparent from the estimates of bacterial biomass, a high bacterial production in the straw during the spring was indicated by transient high numbers of protozoa and bacterivorous nematodes. However, standing crops of protozoa and microfauna contributed much less than the standing

Fig. 5.19. Mineralization/immobilization of nitrogen (mg g^{-1} initial ash-free litter) in a range of litters as a consequence of initial N mass. The fresh litters were incubated 4 to 9 months in the Kjettslinge soil. The change in N mass was calculated as final N mass subtracted by the initial N mass. Data are compiled from Andrén and Paustian (1987) and Andrén (1987). The dotted lines were fitted by linear regression. (Triangles: B0, B120, GL and LL above-ground litter, $R^2 = 0.94$. Squares: B0, B120, GL and LL root litter, $R^2 = 0.86$.)

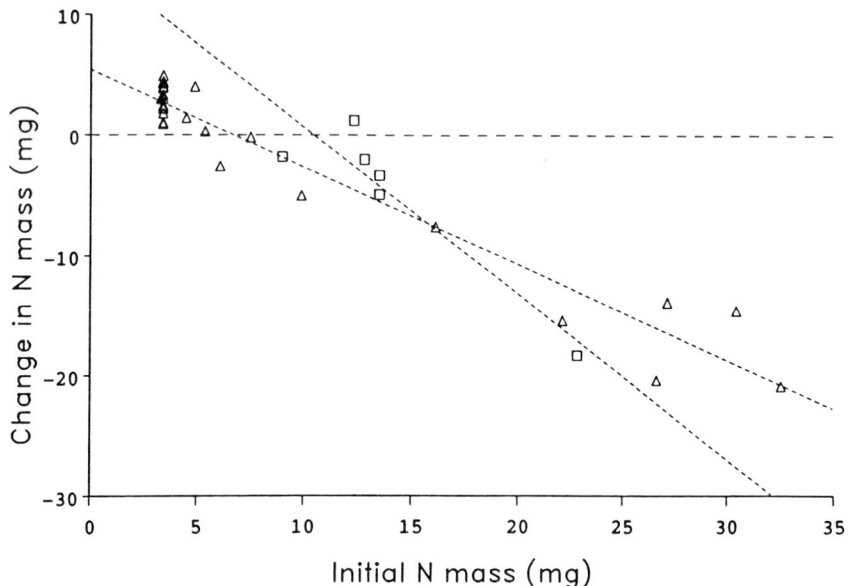

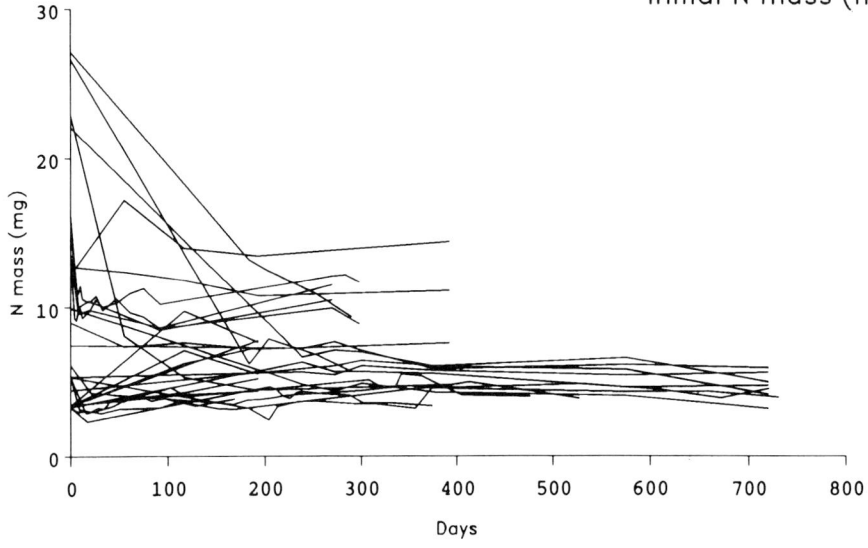

Fig. 5.20. Nitrogen mass (remaining from 1 g initial ash-free litter) in above-ground and root litter during decomposition in the Kjettslinge soil. Data are compiled from Andrén and Paustian (1987) and Andrén (1987).

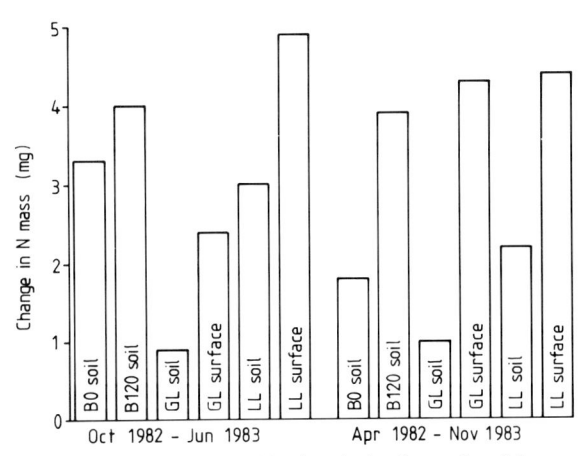

Fig. 5.21. Nitrogen immobilization during 8 months of decomposition of B0 straw placed in the four cropping systems. The straw had an initial nitrogen concentration of 3.4 mg g^{-1} ash-free straw. Data from Andrén (1987).

crops of bacteria and fungi to the increase in litter-bag nitrogen.

The increase in microbial biomass explained only a small part of the total of 1.1 mg N immobilized (Andrén and Paustian 1987), and a major part of the increase may have been present as microbial residues. The four-compartment model presented earlier in this Chapter includes an "active" component. i.e., microbial biomass plus metabolites. Optimization of the model parameter values resulted in an estimated size of the active pool that was five times as high as the measured microbial biomass.

Selected results on early changes in N mass in decomposing litter were extrapolated to an areal basis of g N m^{-2} (Tab. 5.8). In barley, the yearly input of decomposable material could be fairly accurately estimated, since all roots die in the autumn, while for leys the values are more approximate.

During the first 8–10 months after the addition of the fresh litter to the soil, the change in N mass varied between an immobilization of 0.6 g N m^{-2} to a net mineralization of 1.3 g N m^{-2} for the different above- and below-ground litters. The values are comparable to the decreases in N in the DOM fraction, which ranged from 1–2 for B0 and 3–5 g N m^{-2} for the other crops (Hansson and Steen 1984, Hansson 1987). Since the DOM fraction only constitutes about 3% of the total soil organic N (Hansson et al. 1987), a major portion of the yearly net mineralization must have originated from older soil organic N. Budget calculations indicated a total annual net mineralization of 8.0 and 9.2 g N m^{-2} in B0 and B120, and 21.4 and 14.7 g N m^{-2} in GL and LL, respectively (Chapter 7).

The contribution of fresh litter to total net mineralization can be indirectly estimated by examining the amount of litter N recovered in the harvest. During the spring of 1981, ^{15}N-labelled straw was incorporated in the soil of a microplot experiment at Kjettslinge simulating B120. Year 1 (autumn 1981) 5.9% of the straw N was recovered in the harvest and during year 2 another 2.9% was recovered (J. Persson, unpubl.). These values are similar to those found for wheat straw N (Myers and Paul 1971). In a greenhouse experiment with ^{15}N-labelled barley grown in Kjettslinge soil, Schnürer and Rosswall (1987) found that 7% of added barley root N was recovered in the harvest. In the greenhouse experiment, the selection of rather coarse roots probably resulted in material more resistant to decomposition than barley roots in general. Using these values for N availability and the annual amounts of litter input (Tab. 5.8), it can be calculated that only 0.04–0.12 g N m^{-2} from B120 litter would appear in the next harvest. This can be compared with the mean N export in the B120 crop during 1981–1985, which was 13.7 g N m^{-2} (Pettersson 1988).

Ploughing the leys on 11 August 1984 after 5 growing seasons, resulted in a pulse incorporation of approxi-

Tab. 5.8. Net change in nitrogen mass in litter during the first 8–10 months. Positive values indicates a net immobilization and negative values a net mineralization.

Cropping system	Resource	Initial amounts[a] (kg ash-free dry mass m^{-2})	Duration of decomposition study (start – end)	Net change in resource N[c] (mg N g^{-1} initial ash-free dry mass)[d]	(g N m^{-2})[e]
B0	Stubble	0.10	Oct 1982–Jun 1983	+ 3.3	+ 0.33
	Roots	0.10	Apr 1983–Nov 1983	− 2.0	− 0.20[f]
B120	Stubble	0.14	Oct 1982–Jun 1983	+ 4.0	+ 0.56
	Roots	0.14	Apr 1983–Nov 1983	+ 1.2	+ 0.17[f]
GL	Stubble	0.09	Oct 1982–Jun 1983	−14.6	− 1.3
	Stubble	0.21[b]	Aug 1984–Jun 1985	− 7.2	− 1.5
	Roots	0.27	Apr 1983–May 1984	− 1.4	− 0.38[g]
	Roots	0.55[b]	Aug 1984–Jun 1985	− 1.7	− 0.94[g]
LL	Stubble	0.10	Oct 1982–Jun 1983	−20.9	− 2.1
	Roots	0.90[b]	Apr 1983–Nov 1983	−18.3	−16.5

[a]Andrén et al. 1987; [b]At leybreak; [c]Final N mass subtracted by initial N mass; [d]Based on litter-bag decomposition studies (Andrén unpubl.); [e]Initial amounts multiplied with the change in N-mass; [f]Assuming that the seasonal root biomass maximum was available for decomposition after harvest; [g]Assuming that the total dead root supply decomposed as indicated.

mately 15 g N m^{-2} in stubble, litter, and roots in the grass ley, and 29 g N m^{-2} in the lucerne ley (Pettersson et al. 1986). Litter-bags containing grass straw and roots in ploughed plots and the non-ploughed control showed very small differences in mineralization patterns, with an initial autumn mineralization and an immobilization during spring. Extrapolated on an areal basis the stubble and root residues that were ploughed under contributed to a net mineralization of from 0.9 to 16.5 g N m^{-2}. Lucerne roots showed the highest contribution (Tab. 5.8).

The export as harvest of added fertilizer-N varied between 59–65% in the fertilized cropping systems during 1982, as measured by ^{15}N (see Uptake of fertilizer and soil nitrogen). The uptake of residual labelled N by the crop during the succeeding years may be used to estimate the remineralization of fertilizer nitrogen. However, in the perennial grass ley, a substantial part of the residual labelled N was contained in the roots, and it is probable that some of this N was translocated to the above-ground parts during subsequent years. Theoretically, some fertilizer NO$_3$–N in the barley might have escaped incorporation into SOM, and been taken up by the crop during the following growing season.

In a field lysimeter experiment (Lindberg et al. 1989), and a field pot experiment (Persson, unpubl.) at Kjettslinge, the export of residual labelled fertilizer N in harvested barley was 1–3% of that added, and in the harvested grass ley it was 1–13% (depending if the first or second fertilization is considered) the second year after application. The third year's export was 0.3–1.4% in both crops (Figs 5.12, 5.22). Ploughing the grass ley mobilized some of the labelled N, and the recovery of N in the barley planted after the meadow fescue the fourth year increased to 2–3% of the applied labelled N (Fig. 5.12). The yearly contribution of residual labelled N in the harvest never exceeded 2.1 g N m^{-2} for GL and 0.2 g N m^{-2} for B120.

N-mineralization potentials (Stanford and Smith 1972) have been widely used to determine the effects of various agricultural practices on soil fertility (Carter and Rennie 1982, El-Haris et al. 1983). The mineralization rate, determined during long-term aerobic laboratory incubations and corrected for field temperature and moisture, has also been used to calculate N-mineralization in the field (Smith et al. 1977, Campbell et al. 1984). According to this usage, the N-mineralization potential (N$_0$) is a measure of the fraction of soil nitrogen that supplies the major portion of plant-available nitrogen during the growing season. If this is the case, seasonal changes in mineralizable N may reflect the availability of soil nitrogen to plants.

The amounts of N mineralized during 13-wk incubations of Kjettslinge soil in 1984, were equivalent to 18.8–48.3 g N m^{-2} (Bonde and Rosswall 1987). The cropping systems were ranked, in the order of increasing mineralization rates, B0 < B120 < LL < GL. The cropping systems showed a steady decline of 10.3–16.5 g N m^{-2} from June to August and a subsequent increase of 3.2–8.1 g N m^{-2} during autumn. Seasonal differences were as large as the differences between cropping systems. Nitrogen mineralization dynamics and the relative differences between cropping systems coincided reasonably well with other estimates of annual net mineralization (Chapter 6). The autumn increase in mineralizable soil N occurred during a period with net mineralization and N-loss from the systems (Bergström 1986, 1987a, b).

The unfertilized barley, which had the lowest amounts of mineralizable N at all samplings, showed the largest post-harvest increase (8.1 g N m^{-2}). Bonde and Rosswall (1987) suggested that the increase was partly due to a buildup of the microbial biomass and immobilization of N caused by the addition of harvest residues with a high C/N-ratio.

Bonde and Rosswall (1987) used three related models to describe the kinetics of N-mineralization during the 13-wk incubations of Kjettslinge soil: (1) first-order, (2) two-component (sum of two first-order models), and (3) a simplified special case of the two-component model.

The first-order model (Stanford and Smith 1972) is represented in product-appearance form as follows:

$$N_m = N_0(1 - e^{-k_0 t}).$$

N$_m$ is the amount of N mineralized at time t, N$_0$ is the amount of mineralizable nitrogen present at time 0, and k$_0$ is the rate constant.

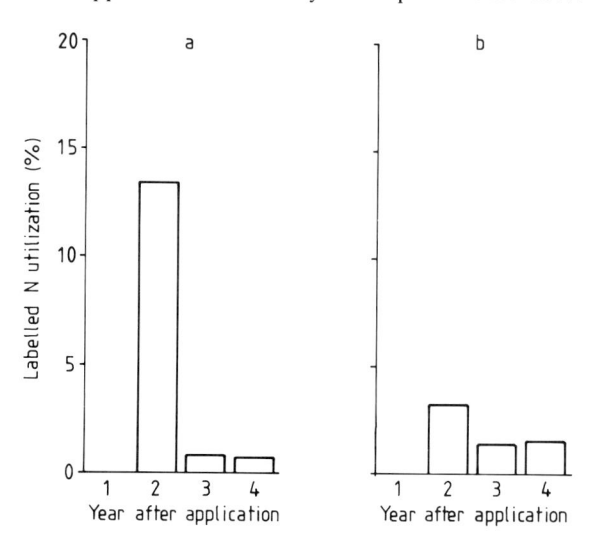

Fig. 5.22. The export of residual labelled nitrogen from grass ley (a) and barley (b) during four years. ^{15}N labelled plant material was produced in an incubation chamber (Fig. 5.6), whereafter frames with the labelled stubble, roots and soil were transplanted into the experimental field and replanted. Bars are means of two replicate frames and 3 consecutive years (year = 2), two consecutive years (year = 3) or one year (year = 4). Data from J. Persson (unpubl.).

Because different fractions of the organic soil nitrogen may be differentially susceptible to mineralization, the first-order model is sometimes modified to a two-component model:

$$N_m = N_a(1 - e^{-ht}) + N_r(1 - e^{kt}).$$

N_a and N_r are the amounts of organic N present in available and resistant fractions, respectively, and h and k are the first-order rate constants for the two fractions. Note that these models are similar to those used for mass loss in sections "Mass loss and abiotic control", and "Heterogeneous resource", above. They differ in that the models presented here are formulated for product-appearance, i.e., production of mineralized nitrogen whereas those presented earlier are concerned with substrate disappearance.

The simplified special case of the two-component model is described by:

$$N_m = N_a(1 - e^{ht}) + kN_r t.$$

In all cases, the special case of the two-component model offered the best description of the curves of accumulated mineral-N during the 13-wk incubation as determined by least-square regression. In the special case model, the mineralization of the resistant fraction (N_r) was approximated by a constant mineralization rate, C_t, where $C = kN_r$. This was considered to be a good approximation when the incubation period is short compared with the half-life of the resistant fraction (N_r) of soil organic N.

Including the constant term in the third version of the model means that a finite quantity for potentially mineralizable nitrogen, N_0, is not estimated. To obtain N_0 values comparable with those in the literature, the first-order model was reanalyzed (Paustian and Bonde 1987). The N mineralized during the first two weeks of incubation was excluded from the determination of the mineralization rate constant k_0 (Stanford and Smith 1972). The first-order model and the 13-wk incubations showed similar patterns of N mineralization in the different crops, but the model showed a continuous decline in N_0 from June to August, and an increase in k_0, in autumn. N_0 values varied from 187–131 µg N g^{-1} soil (73–51 g N m^{-2}) in B120. The first-order model was not appropriate for the August samplings in LL and B0, since cumulative mineralized N in the incubation increased linearly.

Some authors have noted an apparent inverse relationship between k_0 and N_0 (El-Haris et al. 1983, Bonde and Rosswall 1987). Bonde and Rosswall (1987) suggested that the relationship may reflect variation in the "quality" of the potentially mineralizable N, for example due to seasonal changes in the size of the microbial biomass compared with the size of the active non-biomass SOM. However, analysis of results from several published studies showed that the inverse relationship may also be due to variation in analytical procedures (e.g., incubation time) and statistical artifacts associated with regression analysis using the product-form of the first-order model (Paustian and Bonde 1987).

When Kjettslinge soil from a field lysimeter experiment was subjected to long-term (44-wk) incubations (Lindberg et al. 1989) the curves of accumulated mineral-N were best described by the true two-component model. The sum of N_a and N_r was slightly lower than N_0, as estimated by the 13-wk incubations, and amounted to 61 g N m^{-2}. The long-term incubation was performed on one occasion on soil collected in May. Contrary to the 13-wk incubation, there was no difference between B120 and GL. Labelled N added three years earlier as Ca(^{15}NO$_3$)$_2$ had mineralized almost twice as fast as the bulk of the soil N.

The decomposer organisms

Mathematical models are only approximations of reality. Many aspects of soil biology are not accounted for when using mathematical models. In soil, the great diversity of the microbial and faunal community makes it difficult to explicitly include them in models of decomposition. An alternative approach towards quantifying their role in decomposition processes is to define food webs, into which species are aggregated on the basis of functional characteristics. This approach was used to estimate C and N flows through decomposer organisms, together with several more detailed studies of specific organism groups.

Organism abundance, biomass dynamics and taxonomical composition are reported in Chapter 4, with a summary of sampling methods (Tabs 4.3, 4.4). These data were used to estimate carbon and nitrogen flows through soil organisms, using energetics calculations (Petrusewicz and Macfadyen 1970, Heal and MacLean 1975, Persson et al. 1980). No significance tests were performed on biomass or C and N fluxes, since these are calculated from abundances and individual biomass estimates, temperature correction factors and energetic quotients, of which the precision is unknown. The significance tests of abundance dynamics reported in Chapter 4 may, however, be used for tentative evaluation of the significance of differences reported here.

Microorganisms are the main actors in decomposition processes, and are responsible for more than 90% of heterotrophic soil respiration (Chapter 7). At the same time, microorganisms can be viewed as a component of the soil organic matter. Soil microbial biomass as a source and sink for C and N was studied on a low-resolution non-taxonomic level. Microbial biomass measurements were not used directly to estimate C and N flows. However, they are reported to enable comparisons with other soil organism data.

Soil respiration as CO$_2$ efflux was measured in soil cores taken in B120, GL and LL as part of the denitrification measurement programme (Chapter 6). The measurements of soil respiration, together with the independent estimates of root and faunal respiration, were used to estimate microbial respiration (Chapter 7).

Carbon flows for soil fauna were derived from calculations of respiratory activity, based on empirical relationships between O_2 consumption rates and body mass. A power function, relating mass and oxygen consumption, was used (Petrusewicz and Macfadyen 1970). This function can be written as:

$$R = aW^b,$$

where R is oxygen consumption (mm^3 ind^{-1} h^{-1}) and W is individual biomass (g fresh mass) at 20°C. The parameters a and b differ between animal groups and are given in Tab. 5.9. Respiration rates were modified for daily temperature fluctuations using the Q_{10} relationship, and the algorithm of Ågren and Axelsson (1980) and daily maximum, minimum and mean soil temperatures 5 cm and 15 cm deep (Alvenäs et al. 1986). Q_{10} values and base temperatures (T_0) used for the different animal groups are given in Tab. 5.9.

Respiration was calculated for subgroups within each faunal category and for surface (0–10 cm) and subsurface (10–27 cm) layers of the plough layer. Nematodes were grouped into four feeding categories: plant feeders, bacterial feeders, fungal feeders and predators/omnivores. Enchytraeids were divided into three size classes and soil arthropods were subdivided into taxonomic groups (9 for microarthropods, 14 for macroarthropods) and further subdivided into feeding categories based on data in the literature (Persson and Lohm 1977, Persson et al. 1980). From calculated annual respiration, carbon flows were estimated using the equation C = R + P + F relating consumption (C), respiration (R), production (P) and defecation (F) (Petrusewicz and Macfadyen 1970), according to energetics quotients (R/C, P/C, F/C) for herbivores, saprovores, microbivores and predators (Heal and MacLean 1975). Budget calculations and a more detailed description of methodology are given in Sohlenius et al. (1988) for nematodes, Lagerlöf et al. (1989) for enchytraeids, Lagerlöf (1987) and Lagerlöf and Andrén (1988, in press) for soil arthropods and Curry (1986) for herbage arthropods. See Tab. 5.10 for more information.

For protozoa, estimating C flows based on biomass estimates and calculated respiratory activity was judged to be inappropriate, because an unknown but potentially large fraction of the population in soil can occur as encysted forms. Instead, consumption by protozoa was related to rhizodeposition (Clarholm 1985b) by assuming that rhizodeposited C was utilized by bacteria with a 50% yield efficiency for bacterial production. The resulting production of bacteria was considered to be entirely consumed by protozoa and bacterial-feeding nematodes. Protozoa respiration and production were then calculated from consumption, using energetic quotients (Tab. 5.10). Protozoa as grazers on the bacterial population were also studied in laboratory experiments (Clarholm 1985a).

Earthworms were sampled both by the formalin method and by hand-sorting of soil monoliths (Boström 1988a). *Aporrectodea caliginosa* constituted about 90% of the biomass and calculations were made only for this species.

Production and mortality of earthworms were estimated from life-tables (Boström 1988a). Respiration was both estimated using the power function described above, and by applying energetic quotients for saprovores (Tab. 5.10) to production data. C and N flow calculations included spring and autumn samplings of earthworms in 1982–1983.

Nitrogen metabolism of soil fauna was based on calculated C flows and the N concentration of the food. A C/N ratio of 5 was assumed for microbial material consumed by soil fauna (McGill et al. 1981) and for soil fauna consumed by predators (Persson and Lohm 1977). For saprovore consumption, C:N ratios of 20 (barley) and 17 (leys) for soil litter and of 20 (for B120 and LL) and 26 (for B0 and GL) for root consumption were used, based on mean nitrogen concentrations (Hansson 1987, Pettersson 1987). The amount of nitrogen directly mineralized by most soil fauna was calculated assuming that the nitrogen concentration in defecated material was the same as in that consumed (Pers-

Tab. 5.9. Parameters used for calculation of oxygen consumption as mm^3 ind^{-1} h^{-1} (R) from biomass as g (W) according to R = aW^b. Q_{10} values and base temperature (T_0) used for temperature correction of respiratory activity, calculated as $Q_{10}^{(T-T_0)/10}$, are also indicated. The respiratory quotient for conversion from oxygen consumption to CO_2 evolution was assumed to be 0.8 for all organisms. Values adapted from Persson and Lohm (1977) and Persson et al. (1980).

Organism	a	b	Q_{10}	T_0
Nematodes	29.25	0.72	3	20
Microarthropods				
Collembola	63	0.73	2.5	18
Acari				
Gamasina, Prostigmata, Astigmata	102	0.87	3	10
Uropodina	5	0.67	3	10
Cryptostigmata	7.2	0.69	3	10
Symphyla, Protura	63	0.73	2.5	18
Enchytraeids	33.60	0.67	2	20
Soil macroarthropods				
Myriapoda	18	0.73	2.5	10
Homoptera	207	0.66	4	18
Diptera	390	0.80	3	25
larvae	210	0.87	4	10
Coleoptera	113	0.74	2.5	10
larvae	281	0.77	2.5	20
Thysanoptera	124	0.80	4	8
Heteroptera	390	0.80	3	20
Lepidoptera	63	0.73	4	18
Hymenoptera	290	0.77	4	15
Araneae	128	0.92	2.5	15
Earthworms				
A. caliginosa	78	0.91	2	19

son 1983). To avoid excessive replication, the nitrogen flows are reported collectively for all soil fauna in the synthesis at the end of this chapter.

All faunal results presented in this section are based on the sampling depth used for each group, i.e., 0–20 cm for nematodes and enchytraeids, 0–12 cm for microarthropods and 0–10 cm for soil macroarthropods. In the synthesis section recalculated values for the entire top soil (0–27 cm) are used for comparisons between groups. The population data used in the budget calculations in this chapter are based on field samplings in September 1982 and 1983.

Microorganisms, including Protozoa
Normally 1–4% of the total soil C is found in the soil microbial biomass (Jenkinson et al. 1981), while microbial biomass N can constitute 2–6% of total soil N (Brookes et al. 1985). In a long term field experiment in Sweden, the microbial biomass C ranged from 2.0 to 3.4% of total soil C, and microbial biomass N, from 3.9 to 6.7% of total soil N (Schnürer et al. 1985). The lowest proportions were found in soil from a bare fallow and the highest in soil that had been cropped and had received straw and inorganic nitrogen. In the soil at Kjettslinge, 1.5% of total soil C was found in microbial biomass, while microbial biomass N constituted 2.6% of total soil N (Schnürer and Rosswall 1987).

Although the size of the microbial biomass may be rather constant (Chapter 4), it is the fraction of the soil organic material having the shortest turnover time, 1–3 yr (Jenkinson and Powlson 1976, Paul and Voroney 1980, Schnürer et al. 1985). The relatively rapid turnover combined with a pool size of 9.5–36 g N m^{-2} (Voroney 1983) makes the microbial biomass an important regulator of plant available N. In the soil at Kjettslinge, the total soil microbial biomass amounted to, on average, 26 g N m^{-2}, as determined by the chloroform fumigation method (Schnürer 1985). Assuming a turnover time of 1.1 yr (as indicated by Tab. 4.15) there would be a flow of 23 g N m^{-2} through the microbial biomass each year.

The microbial biomass as a source of plant available N, as well as a potential competitor with the plant for mineral N, was investigated in greenhouse (Schnürer and Rosswall 1987) and field experiments using ^{15}N. In both experiments, the rate of NO$_3$-N application was the same as in the B120 treatment, i.e., 12 g N m^{-2}. Under greenhouse conditions, the microbial biomass immobilized 12.5% of the added fertilizer N in barley after one growth cycle. In the field, the maximum immobilization (i.e., recovery from fumigated biomass) occurred during the first 7 wk after fertilization, and amounted to 1.4–1.7% of the added fertilizer N. After the initial peak, the amount of the added N found in the microbial biomass decreased to less than one percent during the three growing seasons following application. In the greenhouse experiment, the mineralization rates of different organic N sources were compared by measuring the uptake of ^{15}N by the barley. The relative amounts of ^{15}N mineralized and taken up by barley during one growth cycle were 10% for microbial biomass N, 29% for laboratory-grown fungal-N and 7% for barley root-N, compared with 38% for the Ca(NO$_3$)$_2$-N addition alone.

The microbial biomass did not act as a large net sink of nitrate fertilizer-N in the field, although the greenhouse experiment demonstrated a large potential. Fertilizer immobilization (12.5%) was higher and above ground plant recovery (38%) lower in the greenhouse than in the field (0.9 and 59%, respectively). A possible explanation for the difference was that the conditions in the greenhouse, such as favourable temperature, moisture and dense root growth in the pots, promoted microbial activity and N immobilization. It should also be noted that since microorganisms prefer ammonium to nitrate as a nitrogen source (Woldendorp 1981), immobilization might be higher if ammonium fertilizer were added (Bristow et al. 1987).

For all microorganisms, the balance between immobilization and mineralization of nutrient elements is governed by their nutrient status and metabolic activity. In fungi, most nitrogen is present in cytoplasmic constituents; hyphal wall N contents generally range between 0.5–3%, while N content of cytoplasm may be as high as

Tab. 5.10. Energetic quotients used for calculations of consumption (C), production (P) and defecation (F) from respiration (R).

Organism	R/C	P/C	F/C	Reference
Protozoa	0.40	0.30	0.30	Fenchel (1982)
Nematodes				
Herbivores	0.16	0.04	0.80	Sohlenius et al. (1988)
Microbivores	0.18	0.12	0.70	Heal and MacLean (1975)
Omnivores/predators	0.40	0.20	0.40	Sohlenius et al. (1988)
Soil arthropods and enchytraeids				
Saprovores	0.12	0.08	0.80	Heal and MacLean (1975)
Microbivores	0.18	0.12	0.70	Heal and MacLean (1975)
Predators	0.56	0.24	0.20	Heal and MacLean (1975)
Herbage arthropods				
Herbivores	0.27	0.18	0.55	Curry (1986)
Detriti/microbivores	0.21	0.14	0.65	Curry (1986)
Predators/parasites	0.49	0.21	0.30	Curry (1986)

10% (Paustian and Schnürer 1987a). The ability of fungi to translocate nitrogen-rich cytoplasm provides a mechanism for conserving nitrogen within the biomass and allowing greater exploitation of nitrogen-poor substrates. Furthermore, the disparity between wall and cytoplasm N contents has implications for the relative nitrogen requirements for growth in relation to the decomposability of the substrate. That is, if substrate availability is low and most of the non-respired assimilate is directed towards cell wall synthesis, with high rates of translocation into the new growth, N requirements relative to C incorporated into biomass will be low. Conversely, if net assimilation is higher, with more C allocated to cytoplasm synthesis, then relative nitrogen requirements will be higher. This relationship can be expressed as a change in the critical N/C ratio (i.e., the transition point between net immobilization and net mineralization of N) as a function of substrate availability (Paustian and Schnürer 1987a).

The main source of plant-available N, except for inorganic fertilizer, is the soil organic matter. When organic material is decomposed, N may be released as NH_4^+ to the soil solution. However, the plants supply of N is constrained to the volume of soil occupied by active roots. Barber (1977) estimated that wheat roots rarely occupy more than 1% of the soil volume, thus most of the soil under annual crops is not in direct contact with active roots. Since NH_4^+ is relatively immobile in most soils, mineralized N needs to be nitrified before it can be effectively transported, by diffusion or mass flow, to the roots. Therefore, local mineralization close to a growing root can be important for the N supply to plants. In this context, microbivorous grazing may act as an important regulator of plant nutrient availability (Coleman 1985, Ingham et al. 1985). Clarholm (1985b) has suggested that roots, via carbon input, induce a chain of events in the surrounding soil, involving grazing of bacteria by protozoa, which leads to local mineralization of soil organic nitrogen around the root (Fig 5.23). Bacteria in the soil are concentrated on the surfaces of organic matter, where most exist in a very low metabolic state due to restricted carbon availability (Stotzky and Norman 1964). Because of its chemical recalcitrance, soil organic matter does not provide sufficient energy for the growth of most bacteria. However, together with root-derived carbon, soil organic matter could serve as a source of nitrogen for nitrogen-limited bacteria in the rhizosphere (Woldendorp 1981). When fresh root-derived carbon is added as a pulse to any volume of soil where root tips are growing (Rovira 1969), bacteria are released from their usual shortage of carbon and become temporarily nitrogen-limited (Woldendorp 1981). Utilizing the energy source provided by roots, bacteria will start growing. Through enzymatic activities, they release enough nitrogen from soil organic matter to meet their needs for growth. Nitrogen in organic matter is transferred to biomass, but nitrogen is not yet released for plant uptake. The growth of bacteria attracts protozoa (and nematodes) which graze on the bacteria. When protozoa consume bacteria, which have approximately the same C/N-ratio as their own, about one-third of the bacterial biomass nitrogen will be assimilated into protozoan biomass, one-third will be excreted, mainly as bacterial cell wall residues, and one third will be released as NH_4^+ (Fenchel 1982). The ammonium will be released very close to the root, in an area where bacterial growth has ceased. It is therefore likely that some of the ammonium will be taken up by the root.

The importance of bacterial grazers in nitrogen mineralization was demonstrated in a microcosm experiment where plants were grown for six weeks in unfertilized soil (Clarholm 1985a). Bacteria, protozoa, glucose and nitrogen were added alone and in various combinations. Plants (shoot + root) grown in soil containing bacteria and protozoa had, on an average, 47% higher dry mass than plants grown with bacteria only. With protozoa present, nitrogen uptake by plants increased considerably in all treatments (Tab. 5.11). The importance of protozoa as mineralizers of N was most evident after repeated carbon additions, which mimicked an enlarged rhizosphere. Since fungi were carefully excluded in the experiment, the experiment showed that bacteria alone can mineralize nitrogen from the soil organic matter if supplied with a suitable C source. The presence of protozoa made more nitrogen available for plant uptake.

To investigate the importance of bacterial-protozoan interactions in the field, a series of daily observations

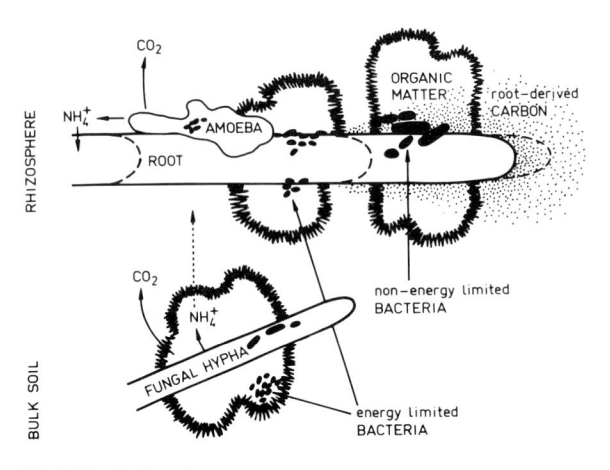

Fig 5.23. Model of possible interactions in the rhizosphere and in the bulk soil. A root is growing in the soil from left to right. Under the influence of root-derived carbon (dots) bacteria on organic matter are temporarily lifted from their normal energy limitation and start to mineralize nitrogen from the organic matter, which will be immediately immobilized in an increased bacterial biomass. At any one place, the pulse of carbon is soon depleted and the bacteria will be consumed by naked amoebae that are attracted to the site. When digesting the bacteria, the protozoa release part of the bacterial N as ammonium on the root surface, where it can be taken up by the root. (Clarholm 1985b).

were made in root-associated soil in B120 (Clarholm, in press). The development of bacteria and protozoa was followed after rainfall, which ended a dry period (Fig 5.24). Two days after the first rainfall there was an increase in bacterial biomass followed by an increase in naked amobae and a concomitant decline in bacteria. A second rainfall induced another peak of bacterial production. The bacterial and protozoan populations thus fluctuated inversely. The two periods of bacterial biomass decrease were also periods of intense nitrogen uptake by the plants (Fig. 5.24). The amount of nitrogen released from protozoa by bacterial grazing was calculated to be 7–8% of the total plant N uptake (Clarholm, in press). The experiment was performed one month after fertilization of B120, when most of the fertilizer was still available for plant uptake. The synchronization of the decline in bacterial biomass and periods of plant N uptake suggested that nitrogen mineralized as ammonium on root surfaces was preferred over nitrate fertilizer for plant uptake. A similar preference for ammonium was reported by Cox and Reisenauer (1973), who reported a 135% increase in dry mass and a 138% increase in N content for wheat seedlings when grown with 200 µg NO_3^- + 10 µg NH_4^+ as compared with seedlings grown with NO_3^- only.

Protozoan respiration, consumption and production in the four crops at Kjettslinge are shown in Fig. 5.25. The major difference in activity was between leys and barley, irrespective of nitrogen fertilization. This was due to higher carbon input in the leys, and the resultant higher bacterial production.

Protozoa activity was calculated from estimated bacterial production, and thus the values in Fig. 5.25 should not be regarded as evidence for differences in microbial activity between the cropping systems. However, the calculated activities of other soil fauna are based on independent estimates of biomass and soil temperature. These activities were calculated independently of estimates of root carbon input or microbial activity and can therefore be used as independent checks of, e.g., treatment differences in microbial activity.

Nematodes, including root consumers

The mean biomass of individual nematodes was calculated from samples of 400–500 nematodes from each cropping system according to the formula by Andrássy (1956) and converted to dry mass by assuming a dry matter content of 25% (Yeates 1979).

The variation in individual mean biomass between cropping systems (Tab. 5.12) indicated that all feeding groups in B120 and plant feeders and omnivores/predators in LL had lower mean weights than in B0 and in GL. Mean individual biomass for all nematodes was lower in B120 than in the other cropping systems. An effect of this was that the variation in total biomass (Fig. 5.26) did not correspond to abundance. Similar amounts of biomass were found in B0 and B120, and in GL and LL, respectively. Abundance increased in the sequence B0 < B120 < GL < LL (see Chapter 4).

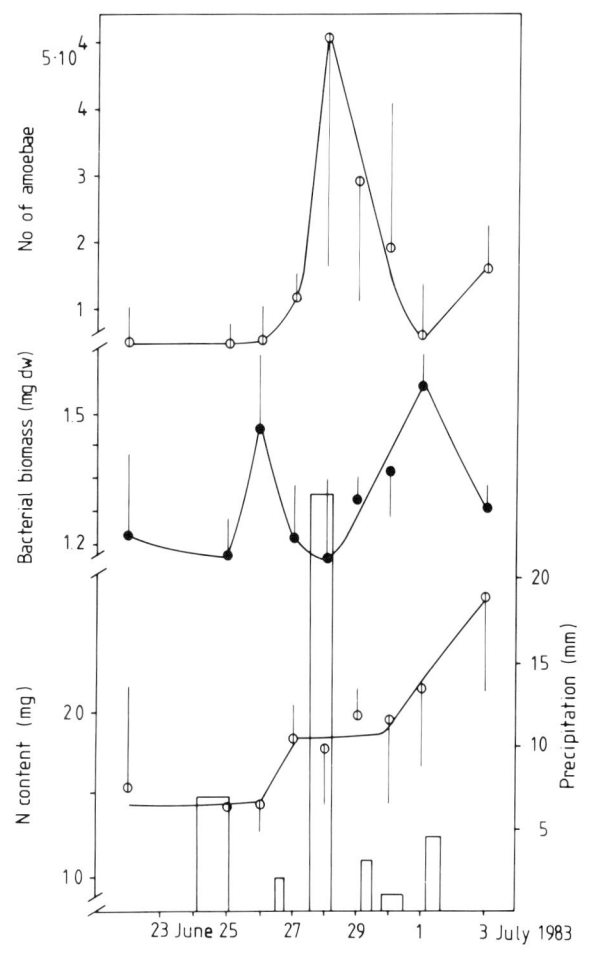

Fig 5.24. Numbers of naked amoebae, bacterial biomass and nitrogen content per plant in a short-term series of observations after a rainfall ending a dry period. The precipitation is given as bars. Mean and standard deviation. n = 4. Clarholm (in press).

Tab. 5.11. Nitrogen in plants after 6 wk of growth in a laboratory experiment (mg container^{-1}, n = 4). Bacteria only or bacteria and protozoa were added at the start of the experiment. Carbon was added as glucose at 12 occasions, 1.2 mg C each time. Nitrogen was added as NH_4NO_3, 12 × 0.3 mg N. Each container held 35 g of soil, containing 0.68 mg inorganic N after autoclaving. Nitrogen in the three seeds added to each container (3.24 mg) were subtracted from the amounts found in the plants (Clarholm 1985b).

Addition	With bacteria	Bacteria + Protozoa	Increase (%)
None	1.61	2.55	58
C	1.78	3.04	71
C + N	1.67	3.74	124

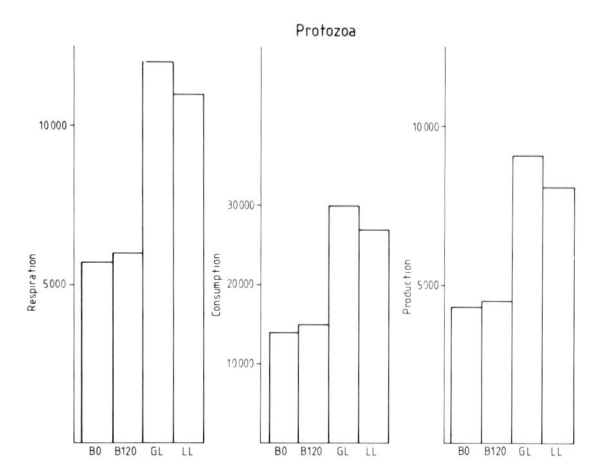

Fig. 5.25. Protozoa respiration, consumption and production (mg C m^{-2} yr^{-1}), calculated assuming that Protozoa consumed all bacterial production in the rhizosphere not consumed by other fauna, and by applying energetic quotients to calculated consumption. (Data partly from Paustian et al. 1990).

Tab. 5.12. Mean individual biomass (μg fresh mass) of various nematode feeding groups in the different cropping systems.

	B0	B120	GL	LL
Plant feeders	0.128	0.080	0.146	0.115
Fungal feeders	0.075	0.068	0.073	0.096
Bacterial feeders	0.108	0.098	0.115	0.132
Omnivores/predators	0.763	0.523	0.719	0.514
Nematoda, total	0.152	0.118	0.159	0.141

The biomass of both plant and bacterial feeders increased in the sequence B0 < B120 < GL < LL (Fig. 5.26). These two groups were positively correlated, both in numbers and biomass (r = 0.94 and 0.95, respectively; p < 0.05). Fungal feeders and omnivores/predators had the highest biomass in GL and the lowest in B120, whereas B0 and LL were intermediate. Although the patterns of variation in these two groups had some similarities, there was no significant correlation in biomass values (r = 0.78, p > 0.05).

Annual respiration was estimated assuming that the September values were representative for the other months. The mean annual respiration of all nematodes and various feeding groups largely corresponded to total biomass (Fig. 5.26). Thus lower total mean annual values were found in B0 and B120 than in GL and LL.

If the fluctuations in abundance over the year were large, the estimations of annual nematode respiration based only on September samplings from three successive years would be unrepresentative. To check this, the annual respiration based on estimates from a monthly sampling programme was used for comparison. The material for these calculations was sampled during 1982 in B120 and GL (Fig. 3 in Boström and Sohlenius 1986). It was found that the annual values based on the monthly sampling programme of 1982 were somewhat below those of the September samplings (Fig. 5.26 and Tab. 5.13). In both the September samplings and the monthly samplings, lower values of nematode respiration were found in barley than in meadow fescue.

The relative contributions of various feeding groups to biomass, respiration, consumption and production are indicated in Fig. 5.26. The relative contribution by omnivores/predators to total biomass was the highest (36–40%) in B0 and GL and second highest in B120 (32%). Bacterial feeders contributed most to the biomass in B120 (38%) and root feeders contributed the most in LL (36%). Fungal feeders were of relatively little importance in B120 (10%), and root feeders were least important (14%) in B0. The relative and absolute importance of root feeders increased in the sequence B0 < B120 < GL < LL.

Since large animals have a relatively low metabolic rate, the contribution to respiration by omnivores/predators was slightly lower than their contribution to bio-

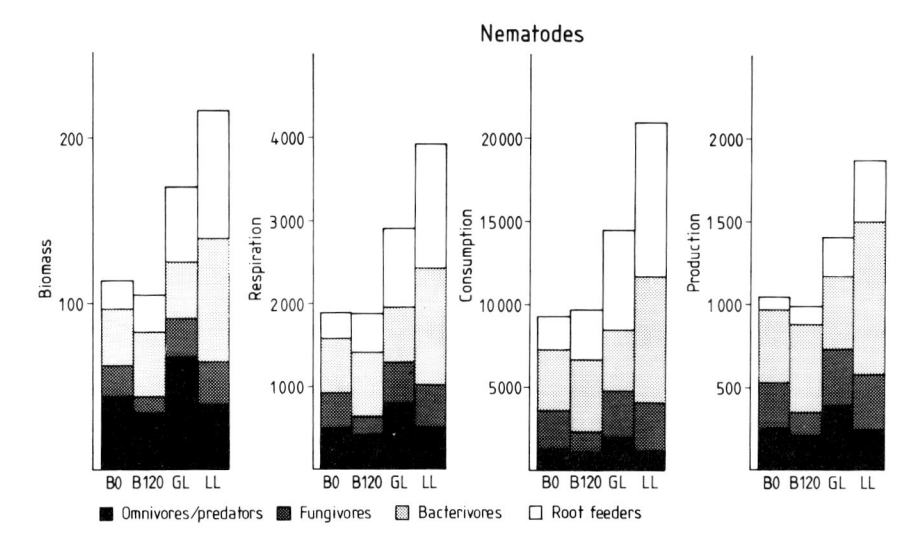

Fig. 5.26. Mean values for September 1982 and 1983 of nematode biomass (mg C m^{-2}), respiration, consumption and production (mg C m^{-2} yr^{-1}). Sampling depth: 0–20 cm. (Data from Sohlenius et al. 1988).

Tab. 5.13. Mean annual nematode biomass (mg dry mass m^{-2}) and respiration (g C m^{-2} yr^{-1}) from monthly samplings during 1982 in B120 and GL.

Feeding group	B120 Biomass	B120 Respiration	GL Biomass	GL Respiration
Plant feeders	31	0.4	58	0.5
Fungal feeders	21	0.3	45	0.5
Bacterial feeders	72	0.8	84	0.7
Omnivores/predators	61	0.4	115	0.7
Nematoda, total	184	1.9	301	2.5

mass. For example, only 22% of the nematode respiration in B120 was due to this group. Bacterial feeders contributed the largest proportion of respiration in B0 and B120 (35–41%), whereas root feeders were most important in GL and LL (33–38%). Fungal feeders contributed least to respiration (11–22%) in all cropping systems, except in B0, were root feeders contributed least (16%) to total nematode respiration.

Microarthropods and enchytraeids
Annual mean biomass, respiration, consumption and production results for Collembola in the four cropping systems, down to 12 cm are shown in Fig. 5.27. All Collembola were considered to be microbivores. (See Tab. 5.10 for energetic quotients.) Collembola biomass ranged from 5 to 28 mg C m^{-2}, and the most striking feature of Fig. 5.27 is the high biomass of *Folsomia fimetaria* in lucerne, resulting in high Collembola biomass and activity in this crop. Annual Collembola respiration during 1982–1983 ranged from 92 mg C m^{-2} yr^{-1} in B0 to 430 mg C m^{-2} yr^{-1} in LL, and matched biomass dynamics closely but not perfectly, since there were clear differences in soil temperature between the cropping systems (Chapter 3). The respiration rates corresponded to a consumption of 0.51–2.4 g C m^{-2} yr^{-1}, and Collembola annual production ranged from 61 mg C m^{-2} in B0 to 280 mg in LL.

Acari mean annual biomass and respiration were dominated by predators, whereas consumption was more evenly divided between predators and microbivores (Fig. 5.28). Mesostigmata were considered to consist entirely of predators, and Prostigmata were assumed to consist of 25% predators and 75% microbivores. The other Acari families were considered to be entirely microbivorous. In addition to having the highest Collembola biomass, lucerne had the highest Acari biomass (30 mg C m^{-2}), but the other three cropping systems were fairly similar (13–16 mg C m^{-2}). However, the Collembola biomass peaked in 1982 and 1984 and Acari, in 1983 and 1984 (Chapter 4). This may be explained by the fact that the peaks in Acari biomass consisted mainly of Mesostigmata, preying on Collembola. Thus the 1982 peak in Collembola may have been the driving force behind the 1983 Acari peak, and the Collembola decrease in 1983 may have been due to the Acari predation.

Annual Acari respiration and biomass values responded similarly to the treatments. Estimated total Acari respiration ranged from about 87 to 200 mg C m^{-2} yr^{-1}, and consisted almost entirely of predator respiration (Fig. 5.28). Acari respiration corresponded to a total consumption of 220–420 mg C m^{-2} yr^{-1}. Consumption by predators was 120–320 mg C m^{-2} yr^{-1}, while consumption by microbivores was 100–120 mg C m^{-2} yr^{-1}. Production ranged from 39 to 89 mg C m^{-2} yr^{-1}, and mainly consisted of predator production.

The Mesostigmata species found in this study feed largely on nematodes, which are the main food source for small species, such as *Alliphis siculus*, *Hypoaspis aculeifer*, *Arctoceius* spp., *Rhodacarus* spp., *Rhodacarellus* ssp. and *Dendrolaelaps strenzkei*, and for sluggish

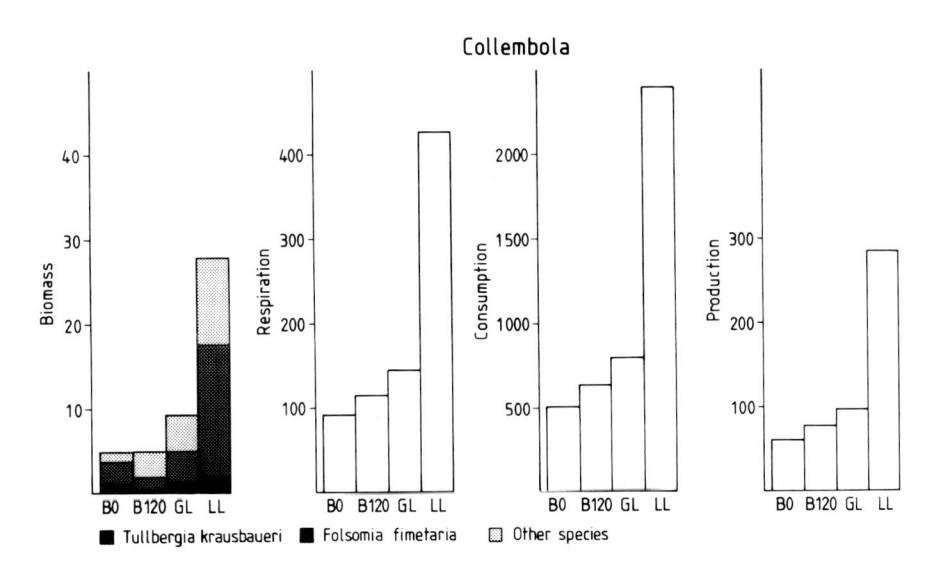

Fig. 5.27. Mean values for September 1982 and 1983 of Collembola biomass (mg C m^{-2}), respiration, consumption and production (mg C m^{-2} yr^{-1}). Sampling depth: 0–12 cm. (Data from Lagerlöf and Andrén in press).

ones like *Nenteria breviungulata*. The large and active species, like *Amblygamasus stramenis*, *Paragamasus lapponicus* and *Veigaia* spp., mostly feed on Collembola. Among the Prostigmata, Rhagidiidae and Eupodidae are the most common families with predatory species (Krantz 1975). *Tyrophagus* (Astigmata) has also been reported preying on nematodes (Walter et al. 1986). In addition, the abundant enchytraeid fauna in the Kjettslinge soil (see below) may also be an important food source for predatory mites.

The consumption of living plant material by soil mites at Kjettslinge was considered to be insignificant and was not accounted for in the respiration calculations. However, *Siteroptes graminum*, which feeds on aboveground parts of grasses and can cause damage of economic importance, was occasionally found. Herbage infestation by this species is initiated by mites in the soil, where they most likely are microbivores. Other related species can also exhibit similar behaviour (Emmanuel et al. 1985).

In Fig. 5.29 estimates of annual enchytraeid biomass, respiration, consumption and production, down to a 20 cm depth under the four cropping systems are shown. The values are calculated from September samplings during 1982–1983 and temperature data as described above. Enchytraeid feeding habits were assumed to be 50% saprovorous and 50% microbivorous (Persson et al. 1980). The lowest biomass estimate was in GL and the highest in B0; about 81 and 320 mg C m^{-2}, respectively.

Enchytraeid respiration ranged from 650 to 2800 mg C m^{-2} yr^{-1} (Fig. 5.29), which was considerably higher than microarthropod respiration and similar in magnitude to nematode respiration. The only other faunal group with higher respiration, excluding protozoa, was earthworms. Microbial and saprovorous consumption each ranged between 2200 and 9300 mg C m^{-2} yr^{-1}. Enchytraeid production was lowest in GL and highest in B0 (430 and 1900 mg C m^{-2} yr^{-1}, respectively).

Enchytraeid respiration was low compared with that

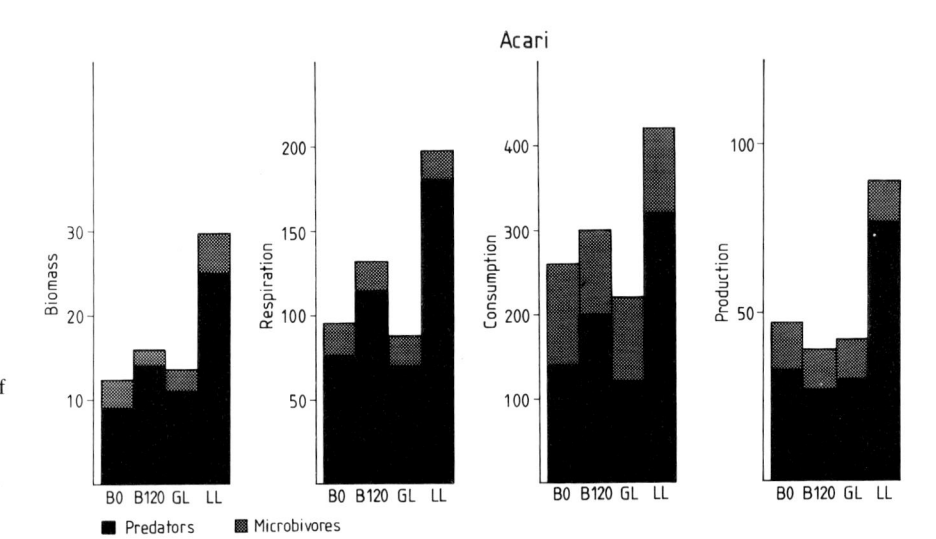

Fig. 5.28. Mean values for September 1982 and 1983 of Acari biomass (mg C m^{-2}), respiration, consumption and production (mg C m^{-2} yr^{-1}). Sampling depth: 0–12 cm. (Data from Lagerlöf and Andrén 1988).

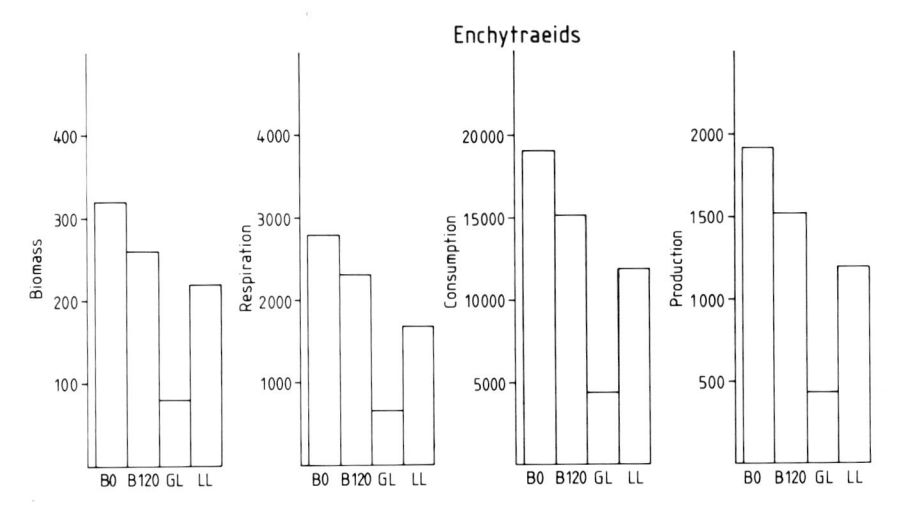

Fig. 5.29. Mean values for September 1982 and 1983 of enchytraeid biomass (mg C m^{-2}), respiration, consumption and production (mg C m^{-2} yr^{-1}). Sampling depth: 0–20 cm. (Data from Lagerlöf et al. 1989).

reported from abandoned Swedish farmland by Persson and Lohm (1977). They found respiration to be about 10 g C m^{-2} yr^{-1}, about four times the values in our investigation. Compared with Polish arable fields, the Kjettslinge values were similar or somewhat lower (Kasprzak 1982).

Macroarthropods, including plant-feeding arthropods
This section concerns both arthropods in the herbage, caught by suction sampling, and macroarthropods extracted from soil using funnel extraction. The results concerning herbage fauna are compiled from Curry (1986), who investigated the above-ground macroarthropod fauna in eight samplings during 1982–1983.

Biomass, respiration, consumption and production of the herbage arthropods are summarized in Fig. 5.30. Total biomass was highest in lucerne, mainly due to detriti/microbivores, and unfertilized barley harboured the lowest biomass with relatively few herbivores. Respiration was highest in lucerne and B120. GL was intermediate, and B0 was lowest. Herbivore respiration was greatest in B120, but the other crops had almost equal proportions of herbivore and detriti/microbivore respiration. Predator/parasite respiration ranged, in percent of total, from 11% in B120 to 30% in B0. The leys had an intermediate proportion of predator respiration, slightly below 20%. Consumption showed a similar pattern to respiration, except that there was a proportionally lower consumption by predators than herbivores, due their high assimilation/consumption quotients, i.e., their comparatively high-quality food.

The soil and surface macroarthropods were sampled extensively during the experiment, and individual specimens were measured at each sampling for length/mass regression calculations. However, the results in the following are based on one mean biomass for each of 14 separate taxonomic groups and one sampling in September 1982. More exact data, based on measured biomass, will be published later.

Biomass, respiration, consumption and production results are shown in Fig. 5.31. Clearly, the barley harboured less soil macroarthropods than the leys. Lucerne had an especially high biomass of soil macroarthropods. In all crops, Coleoptera were dominant, both in terms of biomass and carbon flows. Diptera larvae, the Collembola species *Isotoma viridis,* and spiders were also important. Divided into feeding habits detritivores/microbivores dominated, followed by predators and herbivores. The herbivore (root feeding) component was considerably smaller than that found for herbage macroarthropods. The consumption by predators, ranging from 0.88 to 2.3 g C m^{-2} yr^{-1}, indicated that a considerable part of the soil animal production was consumed. Beetles (Carabidae, Staphylinidae) as well as spiders dominated the predators. Their prey includes arthropods as well as enchytraeids and earthworms.

Earthworms
Earthworms were sampled by the formalin method in September and May from 1980 until 1984, as described in Chapter 3. Cocoons and earthworms were sampled monthly from June until October 1983 in the GL by soil coring followed by hand sorting of soil. Since soil coring usually yielded higher numbers and biomass of earthworms than formalin extraction, formalin data was corrected for this underestimation (Boström 1988a).

In GL and LL the amount of casts transported to the soil surface was estimated in 1980 and 1982 (Boström 1988a). Old casts were removed from 1 m^2, randomly chosen in each block, and during two following days all new casts were collected, dried and weighed. Casts belonging to different species were not separated. The amount of casts produced by *A. caliginosa* during varying soil conditions was estimated in the laboratory.

The influence of earthworms on litter disappearance was studied in GL during periods in 1983, 1984 and 1985. Frames of two types, those excluding earthworms through 1-mm mesh bottoms and those accessible to

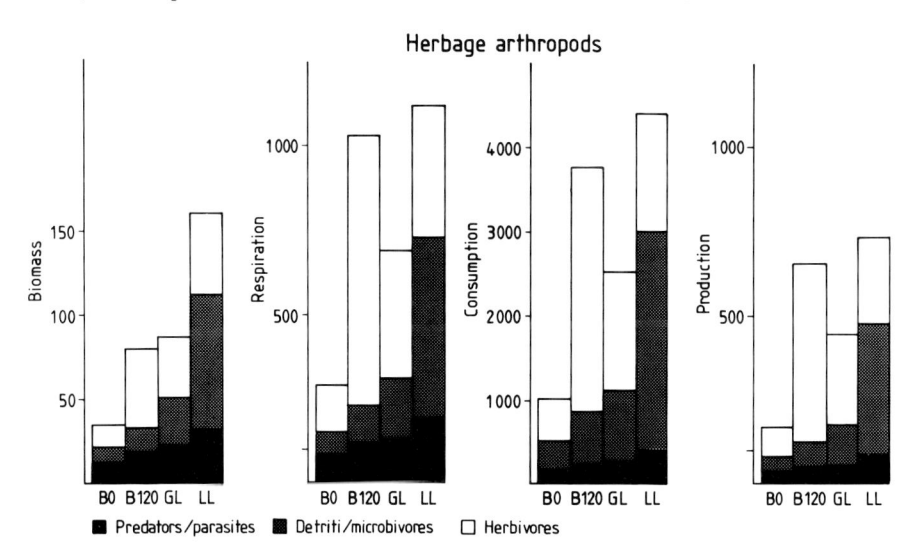

Fig. 5.30. Mean values for June to September 1982–1983 of biomass (mg C m^{-2}). respiration, consumption and production (mg C m^{-2} yr^{-1}) of arthropods in the herbage. (Data from Curry 1986).

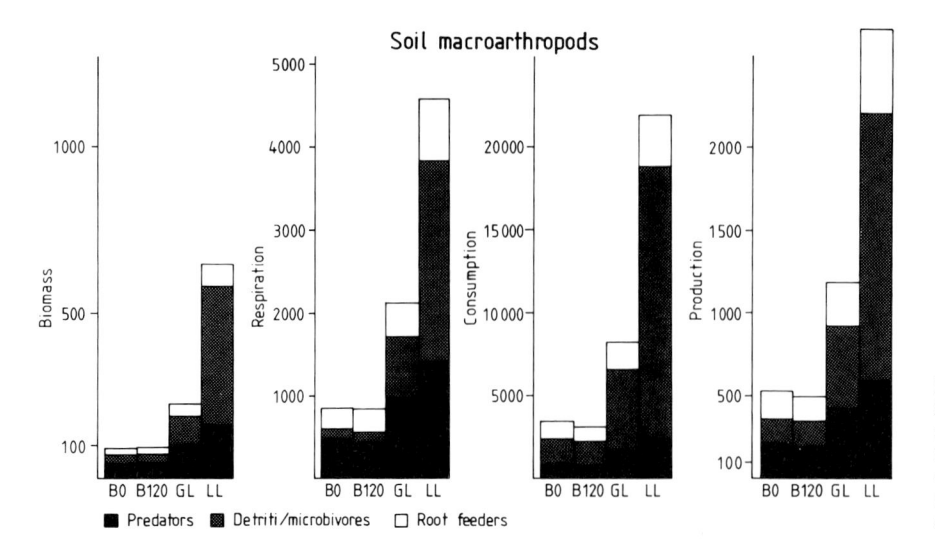

Fig. 5.31. Mean values for September 1982 of soil macroarthropod biomass (mg C m^{-2}), respiration, consumption and production (mg C m^{-2} yr^{-1}). Sampling depth: 0–10 cm.

earthworms through 6-mm mesh bottoms, were placed on the soil surface between plant rows. A known amount of litter was added to each frame and litter disappearance was measured. The two types of frames were used in combination during 1985 to estimate the amount of litter falling through the 6-mm mesh bottom. Soil with all visible litter removed was put into the fine mesh frame and the coarse mesh frame with added litter was put inside. At the end of the experiment the litter left inside the frames and the litter on the initially litter-free soil were weighed separately.

To compare growth rates in different treatments, juvenile *A. caliginosa* were cultured in the laboratory in a mixture of soil and a known amount of various plant material. The earthworms were weighed weekly until mass increase ceased (Boström and Lofs-Holmin 1986). Cocoon production rates at different temperatures were estimated (Boström 1988a).

Earthworms in the laboratory can live for several years, but the lifespan of earthworms in the field is probably much shorter. Although the earthworm populations seldom differed significantly between seasons and years, there is a continuous production and mortality of earthworm biomass. Thus, net population growth is not a reliable measurement of production. To calculate the production and mortality of *A. caliginosa* in the leys, the following assumptions were made:

(1) Individuals that were juveniles in mid May were considered to be either adults or dead in September, while earthworms hatched later were still juveniles in September.
(2) Juveniles in September were considered to still be juveniles in May of the next year.
(3) Earthworm growth or mass loss was assumed to be linear, and individuals were assumed to die halfway between the autumn and spring sampling.
(4) It was assumed that no cocoons hatched during spring before the samplings in May, as supported by soil temperature data.

Based on the number of adults and juveniles at each sampling, data were analyzed according to these assumptions and production and mortality were estimated (Boström 1988a). The estimated values of production and mortality in the leys were considerable higher than the measured biomass dynamics (Tab. 5.14). Production was higher than mortality indicating a net increase in population biomass during 1982–1983 (Fig. 4.24). Since earthworm samplings were made only once each year in barley the life-table approach could not be used for estimation of tissue production. In barley mortality and production were assumed to be equal to the mean annual biomass of *A. caliginosa*, i.e., 90% of the autumn values in 1982 and 1983.

Based on soil moisture, temperature, cocoon production rates and number of adults at monthly samplings, it was estimated that *A. caliginosa* produced 12 cocoons adult^{-1} in GL during 1983 (Boström 1988a), which is lower than the 27–35 cocoons yr^{-1} estimated by Evans and Guild (1948a) and Graff (1953a). It was estimated that 125, 145, 270 and 385 cocoons m^{-2} yr^{-1} were produced in B0, B120, GL and LL, respectively, during 1982 and 1983. The individual cocoon mass was assumed to be 0.030 g in all cropping systems i.e. similar to that in GL.

Since *A. caliginosa* constituted 90% of the earthworm biomass, most surface casts were probably deposited by this species. The total amount of casts produced during 1982–1983 was estimated to be 3.5, 3.4, 6.4 and 11.0 kg m^{-2} yr^{-1} in B0, B120, GL and LL, respectively. In September 1982 only 20 and 50% of the casts produced in B120 and the leys, respectively, were deposited on the soil surface. Earlier studies indicate that between 50 and

Tab. 5.14. Measured biomass changes and estimated tissue production and mortality of *A. caliginosa* (g live mass m^{-2} yr^{-1}, ± SE) during 1981–1983. In alternative A and B, juveniles and adults, respectively, were assumed to have the highest mortality. See text for more information.

Treatment	Measured biomass increase	Estimated production Alt. A	Alt. B	Measured biomass decrease	Estimated mortality Alt. A	Alt. B
GL	13±7	27±11	34±14	10±10	24±8	31±5
LL	46±8	56±15	62±19	19±10	32±18	36±21

90% of the casts produced by *A. caliginosa* are deposited below ground (Barley 1959, Andersen 1983).

When feeding, earthworms fragment organic matter and mix it with soil, making more particle surfaces accessible to microorganisms. Casts contain higher concentration of nutrients than the surrounding soil (Lunt and Jacobson 1944, Parle 1963, Boström 1988a) probably because earthworms select organic matter to feed on. During aging, the nitrate concentrations in casts increase (Parle 1963, Syers et al. 1979, Boström 1988a) while the ammonium concentrations decrease (Parle 1963). About 6% of the non-available nitrogen ingested by juvenile *A. caliginosa* was excreted as faeces and urine in forms available to plants (Barley and Jennings 1959). A potential for higher denitrification rates in casts than in the surrounding soil was demonstrated by Svensson et al. (1986), while others have proposed higher rates of nitrification in casts than in the surrounding soil (Parle 1963, Scheu 1987).

Due to a number of factors, the data from different studies on nitrogen content in casts vary widely. Variability in earthworm species, soil type, organic matter, climate, edaphic factors and age of casts influence the nitrogen content of casts. Surface casts in B120, GL and LL contained 0.3, 0.3 and 0.5% of Kjeldahl N, respectively. The carbon content was 3.2%.

Needham (1957) estimated that half of the nitrogen excreted from earthworms is secreted as mucoproteins and half as urine. He found that more nitrogen is excreted by *A. caliginosa* during wound healing and fasting than when they were feeding, i.e., 133.3 and 87.5 µg g^{-1} biomass d^{-1}, respectively. Since some of this nitrogen is probably recovered in casts, the excretion in non-faecal forms can be estimated to be 50% of this value or 43.8 µg N g^{-1} d^{-1}.

Assuming $Q_{10} = 2$ and $Q_{10} = 3$ for temperatures above and below +5°C, respectively, it can be estimated that 0.1 g N m^{-2} yr^{-1} was excreted as mucus and urine in barley and GL, while the excretion in LL amounted to 0.2 g N m^{-2}.

Plant litter on the soil surface is covered by earthworm casts, and species such as *Lumbricus terrestris* collect litter and transport it down into the soil. The close contact between soil and organic matter creates environments favourable for microorganisms and prolongs humidity, thereby probably increasing the rate of decomposition. Earthworms contributed to the disappearance of 15–50 g dry mass m^{-2} yr^{-1} of surface litter during 1983–1985, which corresponded to 10–36% of the litter produced annually. Only 2% of the mass loss in 1985 was due to litter falling through the coarse meshes and could be assigned to experimental error. *L. terrestris* entered only a few frames and deposited casts there, and probably only litter from those had been actively transported down into the soil. Earthworm contribution to litter removal from litter bags has been found to differ with soil type, from 5% of total mass loss in one year in a coarse sand to 21% in a sandy loam (Boye Jensen 1985). On the other hand, Curry (1969) did not find that soil fauna increased the rate of decomposition, but hypothesized that his results were influenced by the extremely high quantities of organic matter used.

Estimating the influence of soil fauna on disappear-

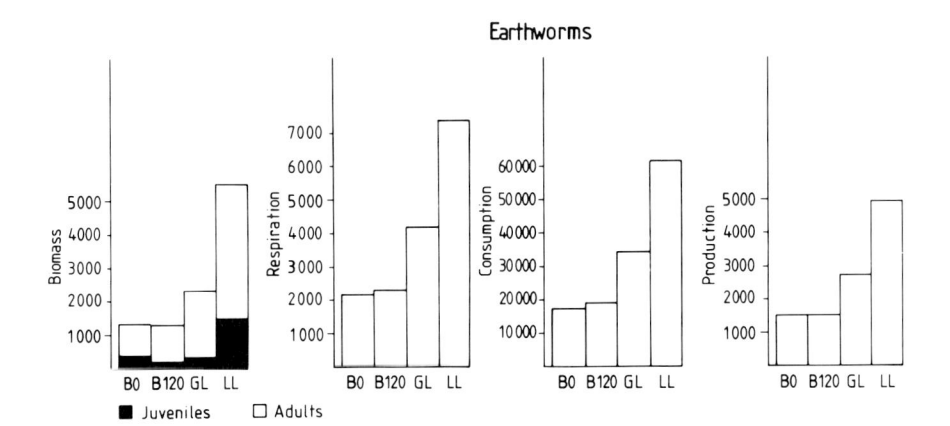

Fig. 5.32. Annual mean values of earthworm biomass (mg C m^{-2}), respiration, consumption and production (mg C m^{-2} yr^{-1}).

Tab. 5.15. Content of nitrogen and carbon (g m^{-2} yr^{-1}) in different compartments of the *A. caliginosa* population as means of 1982 and 1983. (Consumption includes ingested soil).

Compartment	B0	B120	GL	LL
Mean annual biomass				
Carbon	1.11	1.14	2.10	3.85
Nitrogen	0.25	0.25	0.46	0.85
Respiration				
Carbon	1.78	2.25	2.41	4.65
Production of tissue				
Carbon	1.11	1.14	1.96	3.79
Nitrogen	0.25	0.25	0.43	0.83
Production of cocoons				
Carbon	0.35	0.41	0.75	1.08
Nitrogen	0.05	0.06	0.12	0.17
Excretion				
Faecal				
Carbon	112	109	205	352
Nitrogen	11	10	19	55
Non faecal				
Nitrogen	0.09	0.11	0.11	0.21
Consumption				
Carbon	115	113	210	363
Nitrogen	11.4	10.4	19.7	56.2

ance of organic matter in soil gives only an indication of their influence on decomposition rate. It is uncertain whether organic matter transported from litter bags by earthworms decompose faster than litter inside the bags. However, decomposition of plant residues is usually more rapid below than above soil surface (Brown and Dickey 1970, Andrén 1987).

The dry content of cocoons was 19%, of adult *A. caliginosa* containing soil, 25.9% and for juveniles, 23.0%. Adults, juveniles and cocoons contained 54.0, 52.6 and 98.0% organic matter, respectively. The nitrogen content in earthworms and cocoons was 11.0 and 7.7%, respectively, while the content of carbon for both earthworms and cocoons was 50% of their dry mass corrected for ash.

A summary of the estimations of nitrogen and carbon flows through the *A. caliginosa* populations in the four cropping systems is shown in Tab. 5.15. Respiration was calculated by the function shown above, and defecation was calculated from cast production. Consumption was calculated from the equation C = R + P + F (Heal and MacLean 1975).

The values shown in Fig. 5.32 and in the following figures were calculated using a technique similar to that used for the other components of the soil fauna. The only difference was that earthworm respiration was calculated from production instead of biomass. The estimates in Fig. 5.32 were based on energetic quotients for saprovores (Tab. 5.10). Respiration calculated in this way was somewhat higher, and consumption was considerably lower than the values reported in Tab. 5.15. The differing results were caused by differences in the way earthworm consumption was calculated. If, only the litter carbon consumed by earthworms is considered as a potential food source, the assimilation/consumption quotient may approach that of other saprovores. If, on the other hand, all ingested carbon including humus carbon is regarded as potential food for earthworms, the assimilation quotient becomes considerably lower. The calculated ratio using the second alternative (Tab. 5.15), 0.03, is similar to the estimates reported by Bolton and Phillipson (1976) for *A. rosea*; A/C ≤ 0.03.

A synthesis of organism activities and decomposition processes

Earlier in this chapter we showed that it is possible to accurately model decomposition, mineralization and immobilization of N without considering organisms. While this does not imply that the organisms are without significance, it suggests that their activities are largely governed by the controlling factors used in the model, i.e., temperature, moisture and resource availability. Along the same line, it might be argued that since microorganisms account for more than 90% of the soil respiration (excluding root respiration), the soil fauna are of little significance. Earthworms may still be interesting, since they may influence both soil properties and the decomposability of plant litter by burrowing and mixing mineral soil and organic material.

Before arguing further it may be wise to reexamine some of the results pertaining to faunal activities. To enable comparisons, all results in the following were extrapolated to the whole top soil layer, 0–27 cm depth. The extrapolations were made separately for each treatment, assuming that the sampling data from 10 cm depth down to the actual sampling depth were representative for the deeper part of the top soil. Biomass, respiration, consumption and production for the organism groups previously presented are compiled in Figs 5.33 and 5.34.

Earthworms dominated the metazoan biomass in all systems, followed by enchytraeids, soil macroarthropods and nematodes (Fig. 5.33). Acari and Collembola showed the lowest biomass of the groups in all systems. However, other microarthropod groups not reported here, such as Symphyla and Pauropoda, were also found, but with a biomass one order of magnitude lower than Acari or Collembola (Lagerlöf and Scheller 1989). Comparing the systems, lucerne contained the highest biomass for all groups, except for enchytraeids. For nematodes, soil macroarthropods and earthworms the second highest mean biomass was found in the meadow fescue, which showed the lowest biomass of enchytraeids. The general relationship of soil fauna biomass in the four systems was B0 = B120 < GL < LL. Acari and Collembola did respond positively to LL, and to a smaller extent to fertilization of barley. The biomass of herbage fauna in the four systems increased in a pattern similar to that of most other soil animals, B0 < B120 = GL < LL. Enchytraeids, the only obviously deviating group, showed the pattern B0 > B120 > GL < LL (Figs 5.33–34).

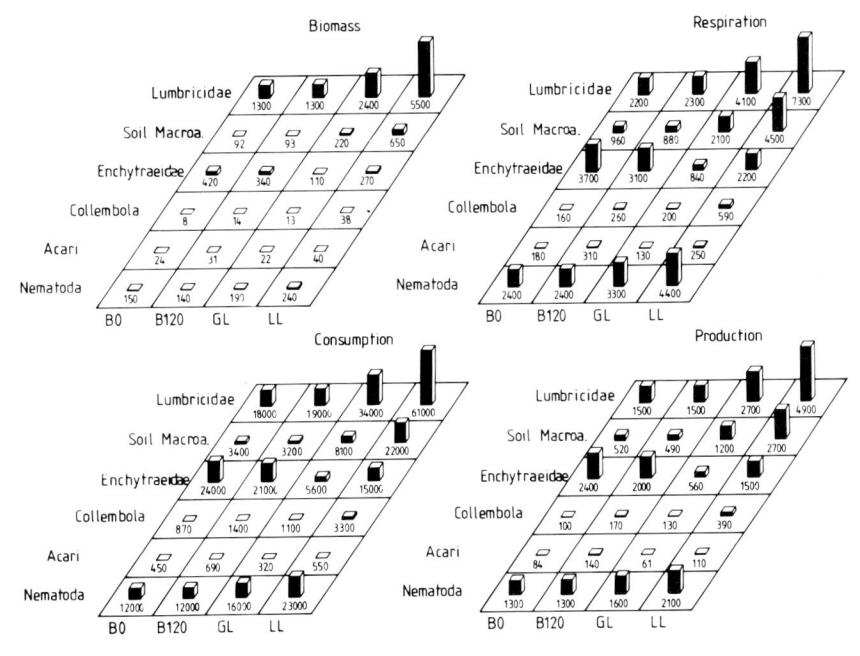

Fig. 5.33. Summary diagram of soil faunal biomass (mg C m^{-2}), respiration, consumption and production (mg C m^{-2} yr^{-1}) in the four cropping systems. Means of September samplings in 1982 and 1983. Values recalculated for the whole top soil, 0–27 cm depth. The carbon flows for all groups except earthworms were calculated by applying energetic quotients to respiration values, calculated from biomass and soil temperature. Earthworm carbon flows were calculated by applying energetic quotients to production, estimated by a life-table approach. (Data partly from Paustian et al. 1990).

Fig. 5.34. Summary diagram of heterotroph organism biomass (g C m^{-2}) and respiration (g C m^{-2} yr^{-1}) in the four cropping systems. Means of September samplings in 1982 and 1983 for all faunal groups. Values calculated for the whole top soil, 0–27 cm depth. Microorganism biomass, including protozoa is based on one sampling in September 1982. Flows through microorganisms were not calculated from biomass and temperature, but from soil respiration and carbon input to the soil. (Data partly from Paustian et al. 1990).

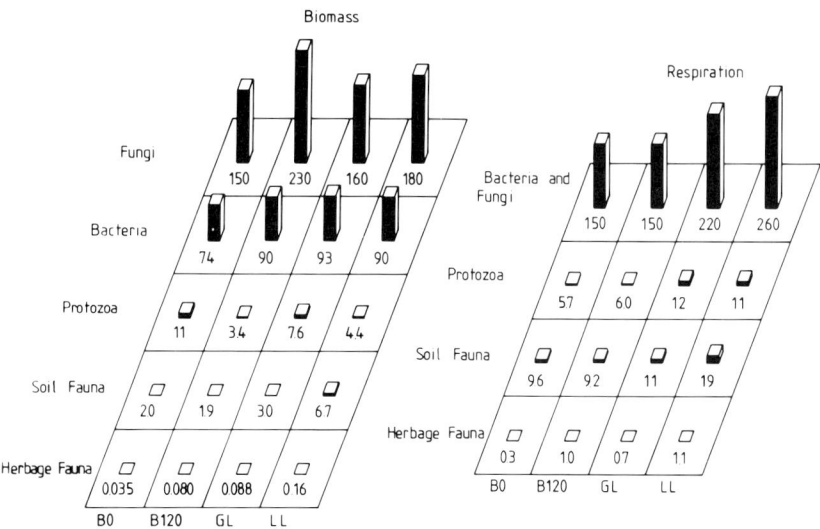

There were several differences between cropping systems that may have caused the differences in animal biomass. One factor was the amount of energy (carbon) delivered to the system. This can be ordered as B0 < B120 < GL = LL (Chapter 7), which fits rather well with the biomass distributions for most faunal groups, excluding enchytraeids. However, it is not only the amount of carbon but also the quality that determines the supply of energy, since low-quality, i.e., slowly decomposing litter, will not contribute much to organism biomass. Quality, measured as decomposability, can roughly be ordered as B120 < B0 < LL < GL for shoots, and B120 < B0 < GL < LL for roots (Andrén 1987, see also section Heterogeneous resource). Quality can also include the amount of nitrogen that is delivered from the litter to the soil. Microorganisms growing on substrate with higher nitrogen concentration maintain a higher nitrogen concentration in their biomass (Paustian and Schnürer 1987a). Microbivores, e.g., Collembola, grow more rapidly and lay more eggs when feeding on nitrogen-rich microorganisms, and this results in increased biomass (Booth and Anderson 1979).

However, the net amount of nitrogen delivered from litter to the soil is also dependent on the availability of nitrogen already in the soil (see section The decomposer organisms in this chapter), and a better measure of the nitrogen available for decomposers may be the nitrogen concentration of the litter. Nitrogen concentration of the litter at Kjettslinge, in increasing order, was B0 < B120 ≪ GL < LL for shoots and GL < B0 = B120 ≪ LL for roots (Andrén 1987, Hansson 1987, Pettersson 1987). The quality aspects of the litter inputs coincide well with the faunal biomass observations, again excluding enchytraeids.

Another factor determining faunal biomass is habitat space, which may be determined by the presence or absence of a surface litter layer (Gill 1969) or the available pore space down the soil profile (Haarløv 1955, Elliott et al. 1980). The surface litter layer, present in GL and LL, is of great importance for surface-living macroarthropods and large, predatory microarthropods. The observed differences in macroarthropod biomass coincided well with the amount of surface litter.

Plants constitute the habitat of the herbage fauna, and a greater crop biomass results in more habitat as well as food. The differences in herbage fauna biomass between B0 and B120 were partly due to differences in habitat size. If the faunal biomass is expressed as g g^{-1} herbage (shoot) biomass instead of g m^{-2} land area, the B120/B0 ratio of faunal biomass in the herbage is reduced from 2.3 to 1.1. Even though differences in herbage biomass during most of the growing season were smaller than the maximum differences used here, this example shows the importance of choosing the relevant unit for comparisons of, e.g., population densities.

The annual ploughing of barley created higher soil porosity (lower bulk density) than in the leys and distributed litter through the soil profile but may have killed or trapped parts of the soil fauna. The decrease in bulk density after ploughing in the barley as compared with the permanent leys was only temporary, and one year after ploughing the bulk density under barley was higher than that under GL (Chapter 2). Still, the positive effects of ploughing may be one of the reasons for the comparatively high biomass of enchytraeids under barley.

Soil moisture in the top soil during dry summer periods was, due to differences in plant uptake and rooting depth, lowest in GL and highest in B0 (Chapter 3). This may have been the most important factor governing the distribution of enchytraeids (Chapter 4 and Lagerlöf et al. 1989).

The respiration sums differed less between groups than biomass did (Fig. 5.33), since the faunal groups with low biomass consisted of small species with higher specific respiration rates than those of larger species. Earthworm respiration still dominated, but nematode and enchytraeid respiration were not much lower, and in B0 they had even higher sums of respiration than earthworms.

Consumption (Fig. 5.33) reflects not only the biomass of a certain group, but also food quality. Earthworms, eating low-quality food, dominated the consumption by soil fauna. The mainly predaceous Acari, consuming high-quality food such as Collembola, were insignificant in comparison.

Production (Fig. 5.33) was more similar between the groups than consumption was, again because smaller species have a more active metabolism. The relations between the groups were fairly similar to those for respiration, and enchytraeids again showed higher values than earthworms in B0.

The next step may be to compare all heterotroph organisms, including bacteria, fungi and protozoa (Fig. 5.34). The heterotroph biomass in all cropping systems was clearly dominated by microorganisms. It should, however, be pointed out that microorganism biomass is not a good measure of activity, since large and varying parts of the microbial biomass can be inactive at any given time. The high reproductive potential of microorganisms also makes "snapshot measurements", in this case in September 1982, of limited value. Thus the biomass values for bacteria, fungi and protozoa should be interpreted with care, especially concerning apparent treatment differences, which were not statistically significant (Schnürer et al. 1986a).

Abundance and biomass data are appropriate measures for comparing structures of faunal communities, but respiration values, in this case calculated from total soil respiration, root carbon input, faunal biomass and soil temperature, are better for comparisons of carbon flows and organism activity in general. The respiration of bacteria and fungi was about one order of magnitude greater than that of soil fauna or Protozoa (Fig. 5.34). The leys, with their higher carbon input and greater root activity, induced higher respiration for the majority of soil organisms than barley did. Nitrogen fertilization of barley apparently did not affect soil organism activity at this low level of taxonomic resolution, but the activity of herbage fauna was clearly reduced in B0. However, as stated earlier, this difference in herbage fauna is not apparent if the differences in plant biomass are taken into account, i.e., the differences in habitat size per unit land area can explain the differences in abundance and activity.

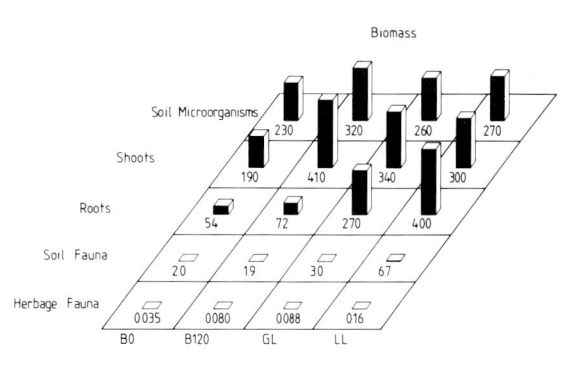

Fig. 5.35. Summary diagram of organism biomass (g C m^{-2}) in the four cropping systems. Means of September samplings 1982 and 1983 for all faunal groups. Values recalculated for the whole top soil, 0–27 cm depth. Microorganism biomass, including protozoa is based on one sampling in September 1982. (Data partly from Paustian et al. 1990).

When above- and below-ground plant biomass were included in the analyses (Fig 5.35), the following observations became evident. First, soil microorganism biomass and root biomass were in the same order of magnitude. Second, above-ground maximum plant biomass seemed to have had little influence on other biotic components. This was not unexpected, since we compare perennial crops harvested twice, with barley harvested once. Third, the root biomass correlated fairly well with general soil organism activity. Since roots supply both various organic compounds to the soil organisms and affect soil structure as well as moisture, this was not surprising.

In an overview of the agroecosystem, the interactions among the organisms, e. g., predation and competition, must be included. To examine these interactions, a carbon food web can be constructed (fig. 5.36). This food web, although suffering from the "spaghetti syndrome", is extremely simplified. The species list in Chapter 4 should be consulted to get a picture of the number of species lumped together in each box. It should also be stated that the feeding habits in the field of most soil organisms are poorly understood, and any food web of this type relies heavily on guesswork. Omnivory, i.e., consumption at more than one trophic level, is more of a rule than an exception, and a more realistic food web would really resemble a well-cooked batch of spaghetti. However, with all its limitations, a food web may help to identify questions pertaining to the participation of different organism groups in decomposition processes and their possible influence as controllers of other organisms.

In this web the columns from left to right show the taxonomic groups classified into trophic categories. Plants are the only primary producers, and herbivores and saprovores are primary consumers. Microbivores and predators are consumers at higher trophic levels (third, fourth, and possibly higher) in the food web, their position depends on the trophic level of their prey. However, omnivory makes the concept of trophic levels somewhat misleading, since a predator may feed on herbivores, microbivores or other predators, and saprovores certainly can consume and assimilate microorganisms or any animal or egg that is small enough. The saprovores also create favourable conditions for microbial activity, e.g., by comminuting the litter, and then by consuming the microorganisms and their products. In soil animals this is often expressed as coprophagy, i.e., eating microorganisms grown in faeces. This strategy can be viewed as keeping an "external rumen" (Macfadyen 1963), which unfortunately makes neat classifications of trophic levels in the soil more or less impossible.

Strictly speaking, it is even doubtful if herbivores in general are primary consumers. More often than not, herbivorous animals rely on microorganisms to digest, e.g., cellulose (A. Pokarzhevski, pers. comm.), which should render them a position as secondary consumers. The cow's rumen is perhaps the most well-known application of this strategy, but it is also common among soil arthropods.

The microbivores in Fig. 5.36 should ideally be divided into bacterivores and fungivores. Nematodes can be divided into these groups, since the mouthparts differ considerably according to feeding habits (Nicholas 1984). For Acari and Collembola, this morphological division is not so clear. For example, the collembolan *Folsomia fimetaria* was concluded to consume fungi, bacteria and protozoa in a laboratory experiment (Andrén 1984, Andrén and Schnürer 1985). Thus the microarthropods were not divided into bacterivores and fungivores. Enchytraeidae and earthworms probably have a mixed diet (Persson et al. 1980), and the positioning of earthworms as saprovores and enchytraeids as 50/50 saprovores/microbivores is somewhat arbitrary.

Predators also show a low degree of specialization, and recent trials have shown that Acari predators consume prey as different as nematodes and Collembola. There is, however, some degree of specialization, larger Acari predators mainly attack arthropods, and smaller

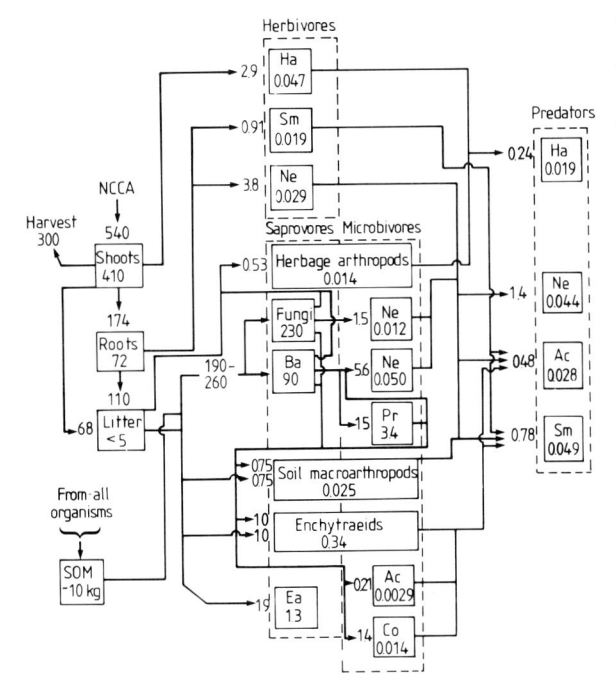

Fig. 5.36. Carbon food web of the Kjettslinge field, exemplified with data from fertilized barley (B120). Values in the boxes indicate biomass carbon (g C m^{-2}) at harvest for the plant; for other organisms mean values for September 1982–1983 are given. On the left side of each box, consumption (g C m^{-2} yr^{-1}) is indicated, calculated using energetic quotients as described in the text. All biomass values are given for the top soil, 0–27 cm depth. (NCCA = net carbon canopy assimilation, i.e., the net amount of carbon fixed by the plant, Bacteria = Ba, Protozoa = Pr, Nematodes = Ne, Herbage arthropods = Ha, Soil macroarthropods = Sm, Acari = Ac, Collembola = Co, Earthworms = Ea. Data partly from Paustian et al. 1990).

species attack nematodes (Karg 1971, 1983, Walter et al. 1988). Enchytraeids are also a potential food source, although investigations are scarce (Karg 1971). The soil macroarthropods are dominated by generalist predators such as beetles and spiders. They feed on any suitable prey of the right size – in the soil, on the soil surface and in the herbage. Microbivorous/detritivorous Diptera larva constitute another major component in the macroarthropod community.

In spite of all the reservations stated above and the restriction to one treatment, Fig. 5.36 can be used as a basis for discussing some of the results. Starting with herbivore consumption, it is clear that although there was a comparably high abundance of aphids in this crop (B120), the herbivore consumption was a negligible fraction of the net canopy carbon assimilation (NCCA), only around 0.5%. If the root consumers are included, the fraction consumed by herbivores rises to 1.5%. The root consumption expressed as a fraction of C allocated to roots was about 3%. These low percentages do of course exclude other damage to the plant by herbivory, since the indirect effects of herbivory such as transmission of pathogens and structural damage do not show up in a carbon flow analysis of this type.

A semi-parallel path in the food web leads from litter (and older soil organic matter) to saprovores, of which the herbage arthropods, soil macroarthropods and enchytraeids were doubly classified as saprovores/microbivores. A reasonable approximation is that these groups consume 50% from each food source. Using 200 g C m^{-2} yr^{-1} as an approximation for consumption by microorganisms, the saprovorous animals were responsible for almost 15% of the total consumption (by microorganisms + animals). The major contributors were earthworms, which consumed about 10% and enchytraeids, which consumed about 5%.

The microbivores in B120 consumed about 35 g C m^{-2} yr^{-1}, partitioned into 45% by protozoa, mainly naked amoebae, 30% by enchytraeids, 15% by bacterivorous nematodes, and 5% each by fungivorous nematodes and microarthropods. Assuming an effective growth yield of 20% for the entire microflora, the microbial production in barley was 40 g C m^{-2} yr^{-1}, and the entire gross microbial production during a year could have been consumed by the microbivores (cf. Clarholm 1985b, Paustian et al. 1990).

Predator consumption in the soil totalled 2.7 g C m^{-2} yr^{-1}, of which nematodes were responsible for about 50%, Acari around 20%, and soil macroarthropods for the remaining 30%. Soil animal production in B120, including predator production and excluding earthworms, was 4.1 g (Fig. 5.33). Using the production estimates for the potential prey groups shown in Fig. 5.33 it is possible to evaluate predation pressure. Assuming that Acari predators prefer soil Collembola as prey, they may consume the entire annual Collembola production (0.17 g C) and still be able to consume 25% of the nematode production. Nematode predators could consume the entire nematode production, and macroarthropod predators could consume almost half the enchytraeid production. Thus the predators would be able to consume most of the annual soil faunal production, especially if one considers that one additional trophic level, predators feeding on predators, is not included in the calculations. In addition predatory fungi trap and kill an unknown fraction of the nematode population (Mankau 1980).

There seems to be a widespread lack of specialization within the soil fauna, which contrasts with the specialization that can be found among, e.g., flying insects. This difference probably stems from the great differences in habitat size between the above- and belowground fauna. The strategy of a flying insect to search over great areas for the 'correct' plant or host would be impossible for a soil animal that may have difficulties in moving through the dark, narrow and erratic soil pore system, only guided by smell (Bengtsson et al. 1988) and tactile sensations from antennae, legs or bristlehairs. The compartmentalization of the soil has consequences for predator-prey relationships; if the prey can access small soil pores where the predator cannot go, the predator-prey balance is clearly altered in favour of the prey (Elliott et al. 1980). The pore size distribution at an even smaller scale can determine the size of microbial biomass protected from microbivores (van Veen et al. 1984).

Food webs can help in evaluating the contributions of various organisms to the nitrogen flows in an ecosystem. A nitrogen food web (Fig. 5.37) can be easily constructed from the carbon food web (Fig. 5.36) and from a few assumptions concerning C/N ratios of organism biomass, consumption and defecation (Paustian et al. 1990). The food web structure would be almost identical, with the exceptions that NCCA (net carbon canopy assimilation) would be excluded and root uptake added. Under each organism box, the amount of nitrogen mineralized by that group is indicated (Fig. 5.37).

The harvested nitrogen, 13 g N m^{-2} yr^{-1}, was fairly well balanced by the 12 g of added fertilizer. The non-harvested part of the plant nitrogen constituted 27% of the maximum amount of nitrogen present in the crop, so in spite of all efforts made to maximize harvests, a considerable part of the nitrogen accumulated in the plant over the growing season was recycled to the soil.

The N flows in the food web are to a great extent similar in proportions to the carbon flows, with the exception that herbivores and saprovores are relatively less important since their diet contains a lower nitrogen concentration than that of microbivores and predators. The fauna in general contributed about 25% to N mineralization, compared with only about 10% to carbon mineralization. Protozoa were responsible for over 50% of the N mineralization by fauna, and other groups with considerable contributions were enchytraeids and nematodes.

A detailed investigation of the detritus food web,

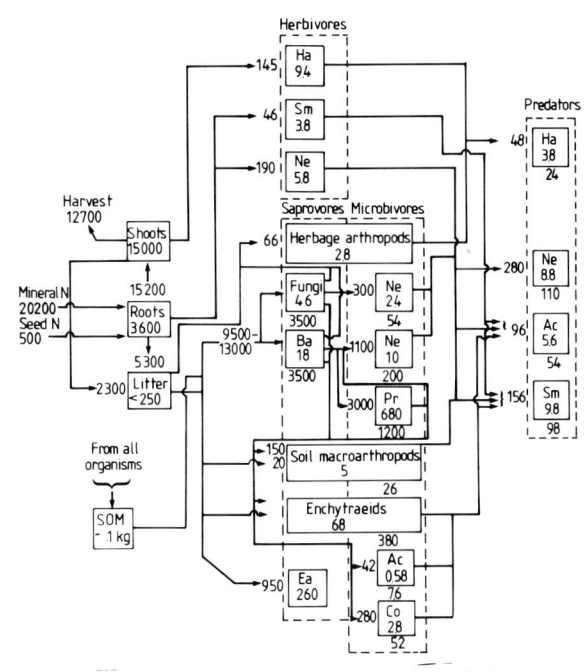

Fig. 5.37. Nitrogen food web of the Kjettslinge field, exemplified with data from fertilized barley (B120). Values in the boxes indicate biomass nitrogen (mg N m^{-2}) at harvest for the plant; for other organisms mean values for September 1982–1983 are given. On the left side of each box consumption (mg N m^{-2} yr^{-1}) is indicated, calculated using energetic quotients as described in the text. All biomass values are given for the top soil, 0–27 cm depth. See text for more information. Nitrogen mineralization is indicated by values under each box. (Bacteria = Ba, Protozoa = Pr, Nematodes = Ne, Herbage arthropods = Ha, Soil macroarthropods = Sm, Acari = Ac, Collembola = Co, Earthworms = Ea). Data partly from Paustian et al. 1990. Nitrogen mineralization was divided between bacteria and fungi for practical reasons; only the sum (7000) is relevant.)

under conventional tillage and no-till, was performed in Georgia, USA (Hendrix et al. 1986, 1987). The biotic components of the food web from Georgia are fairly similar to those in Fig. 5.36, e.g., bacteria, fungi, Protozoa, nematodes, enchytraeids, micro- and macroarthropods and earthworms are included.

Although both climate and treatments differ from our investigation, some comparisons can be made, assuming that our barley treatments were similar to the conventional treatment and our leys resembled the no-till situation. In the no-till treatment, earthworms and macro- and microarthropods showed higher biomass than in the conventional treatment, whereas enchytraeids and nematodes showed the opposite response. These results are consistent with the results from Kjettslinge, with the exception of nematodes. Hendrix et al. (1986, 1987) found the highest biomass of bacterial-feeding nematodes in the conventional tillage. In the Swedish investigation, this group's biomass was fairly similar between barley and leys (Fig. 5.26). Nematode fungivores and root feeders responded similarly in both studies, with lower biomass in the annually cultivated plots.

Another food web that may be comparable to that in our leys was compiled by Hunt et al. (1987) for a shortgrass prairie. The components in the prairie food web, e.g., bacteria, fungi, Protozoa, nematodes, Collembola and mites, are to a great extent similar to those at Kjettslinge. The main differences are that the prairie food web contains mycorrhizal fungi and lacks enchytraeids and earthworms. Amoebae and bacterial-feeding nematodes were calculated to account for over 83% of N mineralization by the prairie soil fauna. The corresponding figure for B120 was considerably lower, around 60% (Fig. 5.37).

The similarities between the food webs in different environments are apparent. Too literal interpretations of similarities or differences should, however, be avoided. Some of the similarities may be due to the fact that the same assumptions and techniques have been used, e.g., for calculating flows from biomass values. Differences between the sites could partly be due to differences in methodology or areas of competence in the research groups, i.e., if a nematologist is present nematodes are found – otherwise not. However, synthesis of results from different biomes indicate that even on a global scale the major organism groups in the terrestrial detritus food web are the same (Petersen and Luxton 1982). Naturally, the relative contributions from the groups, as well as the species representing the groups, differ between sites.

There are many proposed control mechanisms, by which soil fauna can affect microorganism activity to a greater extent than may be inferred from abundance and respiration data. In gnotobiotic (i.e., known organisms are introduced into a previously sterile environment) microcosms, Protozoa (Clarholm 1984), nematodes (Ingham et al. 1985) and microarthropods (Hanlon and Anderson 1979) have been shown to significantly influence decomposition and/or nitrogen mineralization rates.

However, these effects have not always been demonstrated in experiments with a more diverse organism community. *Folsomia fimetaria*, a common collembolan at Kjettslinge, was used as a model organism to test the influence of a microbial consumer on the decomposition rate of barley straw (Andrén and Schnürer 1985). The experiment was made using gas-tight containers (Fig. 5.38), and numbers of springtails corresponding to those found in litter-bags in the field. After 1000 h of incubation, the density of fungal hyphae on the straw surface was reduced by the presence of *Folsomia*, but no significant differences in CO$_2$ evolution, straw mass loss or total microbial biomass were found. This lack of significant effect can be attributed to the unspecified feeding habits of this specific collembolan, or, more generally, to the fact that we used unsterilized straw and soil solution, retaining the natural microflora including Protozoa. A hypothesis may be that the effects found in a simple experiment, e.g., without the natural microflora, are not applicable to field situations, where the

Fig. 5.38. Photograph showing the arrangement of gas-tight containers used for respiration studies with the collembolan *Folsomia fimetaria*.

complex food web compensates for the presence or absence of a certain organism group.

It is possible that most species of soil organisms are "unnecessary" from the system's viewpoint, i.e., decomposition and mineralization proceed at the same rate regardless of whether they are involved. This hypothesis is not very popular among soil zoologists, but it remains to be refuted. In more general terms, it may well be that most species in any food web are only weakly interacting with other species, and only a few species really matter as controllers (Paine 1988). However, food webs are not fixed structures, and it is possible that species that are weakly interacting under one set of conditions interact strongly under other conditions. For example, some nematode-consuming mites cannot eat anhydrobiotic nematodes, i.e., nematodes that have entered a resting stage due to dry conditions (Walter et al. 1988). A "dry" food web would thus show no interaction between the mites and nematodes, but a "wet" food web could show considerable interaction. Successional changes, e.g., in decomposing litter could also result in different food webs at different times (Lagerlöf and Andrén 1985).

The calculations from field data indicate that microbivores and predators, due to their relatively nitrogen-rich food, should have considerable influence on nitrogen mineralization. This was found to be true in a laboratory experiment, where plant uptake of nitrogen increased by 60–120% when Protozoa were present (Clarholm 1985a, see also Tab. 5.11).

The litter-bag experiment, in which barley straw in litter-bags was investigated during two years more closely approximated field conditions. One thousand litter-bags were buried 15 cm deep under B0. Straw mass loss and chemical composition (Wessén and Berg 1986, Andrén and Paustian 1987), nematode abundance and energetics (Sohlenius and Boström 1984), microarthropod and enchytraeid abundance and energetics (Lagerlöf and Andrén 1985) as well as microbial biomass and activity were monitored. The litter-bags rapidly built up a numerous and diversified decomposer community, and about 30 species of nematodes (Sohlenius and Boström 1984) and 30 of arthropods (Lagerlöf and Andrén 1985) were recorded. However, the total biomass of all organisms never exceeded 2% of the litter mass. When respiration was calculated from biomass and soil temperature for nematodes, microarthropods and enchytraeids, their contributions to the total carbon loss during the two-year period were 0.2, 0.2 and 0.8%, respectively (Sohlenius and Boström 1984, Lagerlöf and Andrén 1985).

Biomass dynamics of selected organisms are shown in Fig. 5.39. Total biomass increased gradually during the two years, when total fungal mass was included. When only the FDA-active fungal fraction was included, the increase was considerably smaller, indicating that inactive fungal hyphae were accumulating. The early peak in total biomass was due to a temporary increase in amoebal biomass in May the first spring, reaching around 5 mg per g initial straw. There was a continuous increase in bacterial biomass during most of the experimental period. This increase, as well as the increase in total organism biomass, indicated a negative correlation between microbial biomass and the absolute mass loss rate (i.e., g litter-bag^{-1} d^{-1}), which declined over time following a negative exponential function.

In other words, total organism biomass as well as bacterial and fungal biomass were poor indicators of biological activity. One reason for this may be the early high abundances of bacterivorous amoebae and nematodes (Fig. 5.39) and enchytraeids (not shown), which are at least party bacterivorous. During the initial period of rapid decomposition and presumably high bacterial production, the bacterial population may have been kept at a low level by bacterivore grazing. Predatory mites, including nematode-feeders then entered the litter-bags and may have reduced the grazing pressure on bacteria. Thus the bacterivores were better indicators of bacterial activity than bacterial biomass in itself.

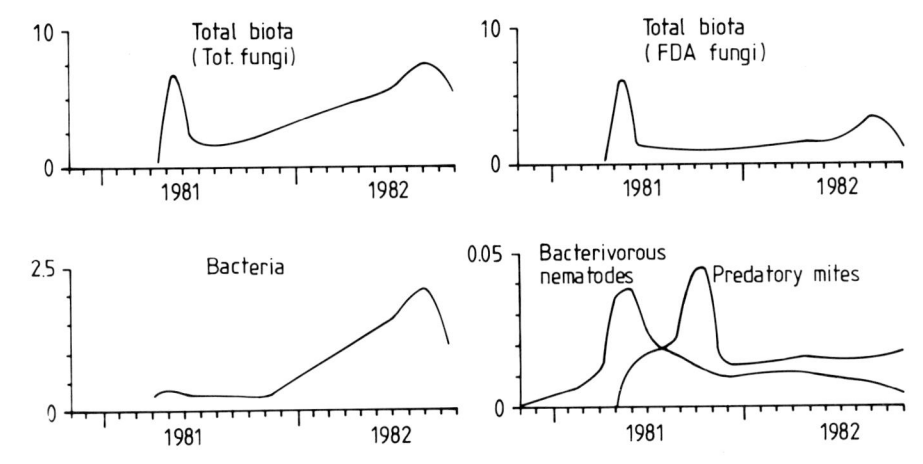

Fig. 5.39. Biomass dynamics (mg) in litter-bags with 1 g initial ash-free dry mass of barley straw: Total biota (bacteria, fungi – total mass, protozoa, nematodes, microarthropods and enchytraeidae), total biota with FDA-active fungal biomass only, bacteria, and microbivorous nematodes together with predaceous mites (Adapted from Andrén et al. 1988).

Several theoretical papers dealing with food webs have been published, attempting to formulate a general theory based on patterns found in published food webs (Pimm 1982). Much emphasis has been put on connectance, i.e., the fraction of possible trophic interconnections that are realized in a certain food web. From the preceding paragraphs it should be clear that the food web presented in Figs 5.36–37 is a highly subjective construction, especially considering connectance. We simply know too little about the feeding habits of organisms in the soil.

The difficulties in formalizing food webs are, however, not restricted to the soil. In a critical review of food web theory, Paine (1988) stated that connectance in itself is highly variable and subjectively defined, not at all fit for generalizations. He also suggested that "theoreticians should ask whether the web under consideration is a biologically realistic representation of that community". This has not always been the case, and neither has the field-oriented scientist always been aware of the subjectivity involved in food web construction. As in many other fields in ecology, the gap between theory and practice, or rather theoreticians and empiricists, is too wide and needs to be bridged.

Comparative and experimental studies in simple, replicated and clearly bounded food webs are one way of getting reliable data (Pimm and Kitching 1988), which can be used as fuel for theory-building. Concerning soil fauna, there is a great need for more information on "who eats what, where and when?", i.e., trophic interconnections from general, spatial and dynamic viewpoints. Only when this information is available, does the lumping of species into functional groups or guilds (Walter et al. 1988) become a truly meaningful exercise.

However, it is not self-evident that an increased understanding of ecosystems requires more knowledge about every possible organism interaction. For example, the decomposition models discussed earlier in this chapter would not gain much predictive power by explicitly including organism interactions. From a functional viewpoint, the ecosystem can often be viewed as a hierarchy, with different levels of organization (O'Neill et al. 1986). Lower organizational levels represent rapidly changing rates, and higher levels represent slower rates. The differences in rates isolate the levels from each other, and the result is that a complex ecosystem can be viewed as being composed of only a few hierarchical levels, segregated by differences in response times. Rapid fluctuations of lower levels are attenuated when being passed on to higher levels, which thus only receive integrated, or average, information. This theory can tentatively be applied to the results from the investigations of decomposition processes in this project. The observation that litter decomposition dynamics could be well described by temperature and moisture dynamics, in spite of a multi-species and rapidly fluctuating organism community, could then be interpreted as: "The fluctuations at the lower level, i.e., within the decomposer community, were attenuated when passed on to the higher level, which was represented by the litter mass loss observations". One may of course claim that this is only a rephrasing and not an explanation. This may be true for this example, but it is definitely not true for the hierarchical concept in general, which adresses a central problem of ecological theory – how a seemingly endless set of organism interactions can result in an ecosystem with quite simple and predictable properties.

Summary

– Photosynthesis measurements in the field showed that the low barley above-ground production in 1985 could partially be attributed to a reduced assimilation rate due to drought. Leaf area measurements indicated that the main cause for the low production was reduced leaf area.
– Primary production, calculated using repeated samplings of above- and below-ground parts and not including rhizodeposition, ranged from 500 g organic matter m^{-2} yr^{-1} in B0 to 1600 g in LL. Annual harvests

(grain + straw) in barley were 250 g (B0) and 650 g (B120). Both leys yielded similar harvests, around 800 g organic matter m^{-2} yr^{-1}.

- In unfertilized barley 20% of net primary production was allocated below ground and in fertilized barley 16% was allocated below ground. Thus input of organic matter to the soil increased less with nitrogen fertilization of barley than would be expected from differences in above-ground production or harvest. In both leys around 30% was allocated below ground.
- Laboratory investigations using ^{14}C-labelled CO_2 indicated that rhizodeposition and root respiration together were 19, 15 and 19% of net canopy carbon assimilation in B0, B120 and GL, respectively. Around 10% was lost as root respiration, and consequently rhizodeposition was in the range of 5–10% of net carbon canopy assimilation. Expressed as a proportion of the carbon allocated to root production, rhizodeposition constituted about 50%.
- Symbiotically fixed nitrogen found in above-ground lucerne biomass was 84, 240 and 320 kg ha^{-1} yr^{-1} for 1981, 1982 and 1983, respectively. Nitrogen in below-ground production was 140 kg ha^{-1} in 1983, of which about 80% can be assumed to come from nitrogen fixation. Therefore, more than 400 kg N ha^{-1} was fixed from the atmosphere in 1983.
- Nitrogen fertilization increased the fraction of plant nitrogen found in harvest from 55% (B0) to 70% (B120). Nitrogen in roots decreased from 26% in B0 to 18% in B120. In the leys, 50–60% of the N was harvested, and about 20% was used for root growth. The fraction of added fertilizer that was recovered in the harvest amounted to about 60% in all crops. Virtually all fertilizer added was taken up by the grass ley during the same growing season.
- Barley straw decomposition in litter-bags during two years could be satisfactorily modelled assuming first-order kinetics and cessation of activity during winter. To model a 5-yr decomposition experiment using ^{14}C-labelled plant material a double exponential model was necessary. Shoot and root decomposition rates during one year in litter-bags could be modelled with a single first-order function, provided that the initial water-soluble fraction was excluded. The first year's mass loss ranged from 28% (B120 roots) to 91% (LL roots). A four-component simulation model could reproduce the dynamics of total mass, nitrogen and water-solubles in barley straw decomposing during two years in the field.
- Litter with high N content mineralized N at higher rates than litter with lower N content, and thus the N concentrations of the various litters converged with time. During the first 8–10 months the change in N varied between an immobilization of 6 kg ha^{-1} and a mineralization of 13 kg ha^{-1}. Ploughing of the leys resulted in an input to the soil of 150 kg ha^{-1} in GL and 290 kg N ha^{-1} in LL, which gave a N mineralization of 24 and 185 kg ha^{-1} in GL and LL, respectively.
- Microbial biomass at Kjettslinge constituted about 1.5% of total soil C and 2.5% of total soil N. The turnover time for microbial biomass was around 1 yr as a mean for the top soil. Microbial and root biomass were in the same order of magnitude.
- Earthworms dominated the metazoan biomass in all crops, followed by enchytraeids, soil macroarthropods and nematodes. The general relationship of soil fauna biomass in the four cropping systems was B0 = B120 < GL < LL. The distribution of herbage fauna was B0 < B120 = GL < LL. Enchytraeids showed a deviating preference pattern, B0 > B120 > GL < LL. The differences in faunal biomass between cropping systems were attributed to differences in amount and quality of litter input and, to a lesser extent, moisture differences. Due to the high specific respiration of smaller species, respiration rates of enchytraeids and nematodes were more similar to that of earthworms, than corresponding comparisons of biomass.
- Herbivore consumption above and below ground was about 1.5% of net canopy carbon assimilation. Saprovorous animals accounted for 15%, and microorganisms, 85% of saprovore consumption in B120. Microbivores consumed about 35 g C m^{-2} yr^{-1}, partitioned into 45% by Protozoa, 30% by enchytraeids, 15% by bacterivorous nematodes, and 5% each by fungivorous nematodes and microarthropods. The consumption by microbivores was equal to the minimum estimate of total microbial production. Predator consumption was 2.7 g C m^{-2} yr^{-1}, 50% by nematodes, 20% by mites, and 30% by soil macroarthropods. The predators were able to consume most of the annual soil fauna production.
- The fauna, including Protozoa, contributed with 25% to N mineralization in B120, and 10% to C mineralization. Protozoa were responsible for more than 50% of the faunal N mineralization. In a two-year litter-bag experiment the contributions of nematodes, microarthropods and enchytraeids to litter carbon loss were 0.2, 0.2, and 0.8%, respectively.

6. Inorganic nitrogen cycling processes and flows

Thomas Rosswall, Per Berg, Lars Bergström, Martin Ferm, Christer Johansson, Leif Klemedtsson and Bo H. Svensson

6. **Inorganic nitrogen cycling processes and flows.** T. Rosswall, P. Berg, L. Bergström, M. Ferm, C. Johansson, L. Klemedtsson and B. H. Svensson

Introduction	129
The organisms	130
Introduction	130
Nitrifiers	130
Denitrifiers	134
Effects of roots on nitrification and denitrification	134
Nitrification	134
Denitrification	135
Water as a regulator of transformation rates	136
Introduction	136
Nitrogen mineralization	137
Nitrification	137
Nitrous oxide production	138
Nitric oxide production	139
Denitrification	140
Water as carrier of substances	143
Introduction	143
Drainage conditions	144
Nitrate in drainage water	144
Nitrogen losses from four cropping systems	145
Introduction	145
Leaching	145
Denitrification	146
Nitric oxide losses	148
Ammonia losses	150
Total losses	151
Management impacts on nitrogen losses	151

Introduction

In addition to its importance for plant nutrition, the production of inorganic nitrogen (Chapter 5) is also the initial step in inorganic-N transformations. That the inorganic part of the biogeochemical nitrogen cycle plays a crucial role in terrestrial ecosystems is attributable to the fact that the major losses of nitrogen, except at harvest, involve nitrate. Agroecosystems receive large additions in the form of fertilizers, but also sustain large losses, leading to pollution of water and air. It has been estimated that about 50% of the N added to Swedish agricultural land is eventually lost (Jansson and Simán 1978). This is a major societal concern, and efforts are underway to implement strategies aimed at decreasing fertilizer inputs and developing management techniques for minimizing losses.

In this chapter the transformations of inorganic-N in the four cropping systems at Kjettslinge will be dealt with. Leaching of nitrate and total denitrification losses are the most important processes from the point of view of the overall N balance of the systems. However, losses of nitrous oxide (N_2O), nitric oxide (NO) and ammonia (NH_3) are also discussed, even if they are not quantitatively important for the N-balance of the studied cropping systems. Nevertheless, their role in atmospheric chemistry and the lack of quantitative field data motivated that studies of their production rates be included in the project.

Bacteria and fungi are the major organisms involved in soil inorganic N-transformations, although plants are important in immobilizing both ammonium- and nitrate-N. The ammonium produced via nitrogen mineralization is rapidly transformed by biological immobilization or by nitrification to nitrate. It can also bind to soil minerals or organic matter. Most of the inorganic nitrogen transformations are of bacterial origin. Although fungi may take part in inorganic-N transformations, the rates of fungal nitrification and denitrification are at least 2–3 orders of magnitude lower than those of the bacterial processes (Focht and Verstraete 1977).

The nitrogen cycle is unusually complex in that nitrogen occurs in several valence states, and both oxidative and reductive processes are involved. The oxygen concentration is thus one of the major factors regulating nitrogen transformations in the soil environment (Fig. 6.1). In principle, nitrogen fixation and mineralization can occur in the presence of oxygen as well as in its absence, whereas nitrification only occurs in its presence and denitrification in its absence. The existence of aerobic denitrification has been reported, however (Robertson and Kuenen 1984). Although the oxidative and reductive pathways of inorganic nitrogen transformations operate simultaneously they are differentiated in space. Nitrification and denitrification can occur at the same time in different parts of soil aggregates, as shown by oxygen concentration measurements using microelectrodes (Tiedje et al. 1984). Soil water is a major constraint on oxygen diffusion, which is four orders of magnitude slower in water than in air. Water also regulates the rate of soil respiration and consequently, the oxygen concentration in soil air. Water is also important for root uptake and as the medium for transport of dissolved nitrogen out of the rooting zone and eventually to groundwater and surface waters.

It was hypothesized that the major differences affecting inorganic nitrogen cycling between the four studied cropping systems were related to the presence of a higher root biomass in the leys. Roots might (1) decrease soil oxygen concentrations through respiration, (2) increase oxygen diffusion into the soil through water uptake, (3) supply denitrifying bacteria with an extra supply of energy through exudation, (4) increase the total soil organic matter – and thus heterotrophic respiration per volume of soil – through biomass production, and (5) decrease nitrogen leaching through immobilization.

The studies of inorganic N-transformations within the project have concentrated on the quantification of nitrogen leaching through the use of different types of

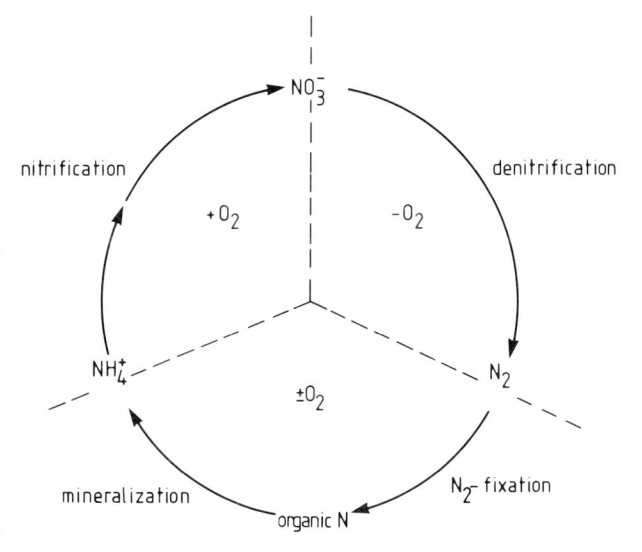

Fig. 6.1. A schematic diagram of the nitrogen cycle and its dependence on oxygen.

lysimeters and individually tile-drained plots (Bergström 1987a), and on quantifying losses of gaseous nitrogen through denitrification by using the acetylene blockage technique (Yoshinari and Knowles 1976, Klemedtsson et al. 1977). In addition, ammonia losses (Ferm 1983) and NO losses (Johansson and Granat 1984) were determined through measurements by cuvette techniques. Nitrification rates were estimated as both potential and actual rates using chlorate-inhibition techniques (Berg and Rosswall 1985). Studies were also made of the population dynamics of nitrifying bacteria (Berg and Rosswall 1985, 1987).

The organisms
Introduction

Apart from the differential influence of oxygen, there are many other differences between nitrification and denitrification in soil, owing to the fact that the former process is mainly maintained by chemolithotrophic autotrophs and the latter by chemoorganotrophic heterotrophs. Denitrifiers are thus dependent on available organic compounds both as an energy source and for biomass production. The respiratory oxidation of the reduced products from redox reactions during the primary conversions of the energy source is preferably via oxygen, but nitrate is a good substitute under anaerobic conditions. Nitrifiers in soil are probably never carbon (CO_2) limited, whereas they are dependent on either the N-mineralization of organic matter or the addition of ammonium-N through fertilization for their energy supply. Owing to the difference between autotrophic and heterotrophic growth as well as the aerobic vs anaerobic nature of nitrification and denitrification, different factors regulate the two processes. The choice of cropping systems and soil management practices will differentially affect the two processes. Even if, for example, nitrate does not directly stimulate nitrification, the application of nitrate fertilizer, which was used at Kjettslinge, will result in increased root growth and subsequent mineralization, which in turn stimulates nitrification. The relative importance of these factors will be discussed below.

In general, population studies of microorganisms provide little information on rates of energy flow or nutrient cycling, although there are important exceptions. Apart from the determinations of total numbers of bacteria and lengths of fungal hyphae (Chapter 4), functional groups of soil microorganisms were not studied in the project except for autotrophic nitrifiers.

In addition to denitrifiers, many other groups of soil microorganisms, including fungi, can produce N_2O (e.g., organisms reducing nitrate to ammonium via dissimilatory as well as assimilatory pathways; Bleakley and Tiedje 1982, Kuenen and Robertson 1988). Nitrous oxide formation through dissimilatory nitrate reduction by non-denitrifying bacteria was probably low at Kjettslinge, since the soil was low in organic matter and high in nitrate. Non-denitrifying bacteria only contribute substantially to N_2O formation when the C/NO_3^- ratio is high (Tiedje 1982). The possible role of fungi and heterotrophic nitrifiers for N_2O and NO production is not known (Rosswall et al. 1989).

Nitrifiers

Numbers of nitrifiers in the Kjettslinge soil were determined by the Most Probable Number (MPN) technique (Berg and Rosswall 1987). Nitrifier populations were quantified because they constitute a very specialized group of microorganisms as compared with denitrifiers, for example, and differences in population sizes were assumed to reflect differences in process rates. Population sizes and their seasonal fluctuations in the four cropping systems were determined, and attempts were made to relate population counts to process rates.

The nitrifier populations were largest in the lucerne ley, followed by the grass ley, B120 and B0 (Fig. 6.2), but differences between cropping systems were smaller than for total bacterial counts (Chapter 4). The higher counts in the leys, especially in the autumn samples, were probably due to higher ammonium concentrations in these cropping systems (Bergström 1986), resulting from higher amounts of organic matter, which can support higher rates of microbial N-mineralization. From the budget calculations presented in Chapter 7, it was estimated that relative nitrogen mineralization rates for the four cropping systems were 10:11:19:21 for B0:B120:LL:GL, while ratios of ammonium oxidizers for the same systems were 10:13:19:16 (based on median values). The agreement between mineralization rates and nitrifier population sizes is in line with the notion that sizes of these bacterial populations are determined by the production rate of NH_4^+ (Molina et al. 1979, Riha et al. 1986), although the reason for the comparatively low population numbers in the grass ley is not known.

The MPN counts of ammonium oxidizers in Kjettslinge were comparable to those made in other similar soils (Tab. 6.1), whereas counts of nitrite oxidizers were mostly higher in Kjettslinge. The nitrite-oxidizer counts were higher than expected based on the relative energy requirements of NH_4^+- and NO_2^--oxidizers. If microbial autotrophic ammonium-oxidation was the sole source of substrate production for the nitrite-oxidizing population, numbers of ammonium-oxidizing bacteria should have been three times higher than those of the nitrite-oxidizers, since three times more energy is available from ammonium oxidation per mole of oxidized substrate compared with nitrite oxidation. As the relevant genera (*Nitrosomonas, Nitrosolobus* and *Nitrobacter*) have comparable cell sizes (1.1–1.3 µm³; Watson et al. 1981), one would expect a 3 to 1 ratio between numbers of ammonium- and nitrate-oxidizers. If there had been a systematic underestimation of the numbers of ammonium-oxidizers relative to the numbers of nitrite-oxidizers, the two groups of organisms would have covaried, but this was not the case (p >0.1; Berg and Rosswall 1987). The reasons for the poor correlation between ammonium- and nitrite-oxidizer MPNs in our study and elsewhere (Steele et al. 1980) as well as the higher counts found for nitrite-oxidizers could be one or several of the following:

– Ammonium-oxidizers may locally inhibit their own growth in soil aggregates by acidifying their microenvironment (Molina et al. 1979)
– Heterotrophic growth of some strains of nitrite-oxidizers (Watson et al. 1981)
– Anaerobic growth of nitrite oxidizers on nitrate and organic substances (Bock et al. 1987)
– Nitrite-oxidizing bacteria obtaining substrate from nitrate-reducing bacteria (Belser 1979, Klemedtsson et al. 1987b).

Numbers of ammonium- and nitrite-oxidizers were 23 and 12 and potential oxidation rates 23 and 14 times higher in the plough layer as compared with the sand layers, respectively (Tab. 6.2). No significant differences were found between the two topsoil layers. This is in agreement with studies on the biomass of bacteria, fungi and protozoa (Schnürer et al. 1986a) and potential denitrification activities (Svensson et al. 1985).

Numbers of ammonium-oxidizers peaked in the spring and autumn, whereas the nitrite oxidizers only showed autumn peaks (Fig. 6.2). Higher soil moisture during spring and autumn stimulated mineralization, and higher ammonium-N contents were also observed during these periods (Bergström 1986). The seasonal variation was larger for nitrifiers than for total counts of bacteria (see Chapter 4). It is not very likely that slow-growing nitrifiers would fluctuate more than fast-growing heterotrophs. Berg (1986) has suggested that the higher variation in MPN counts could reflect the fact

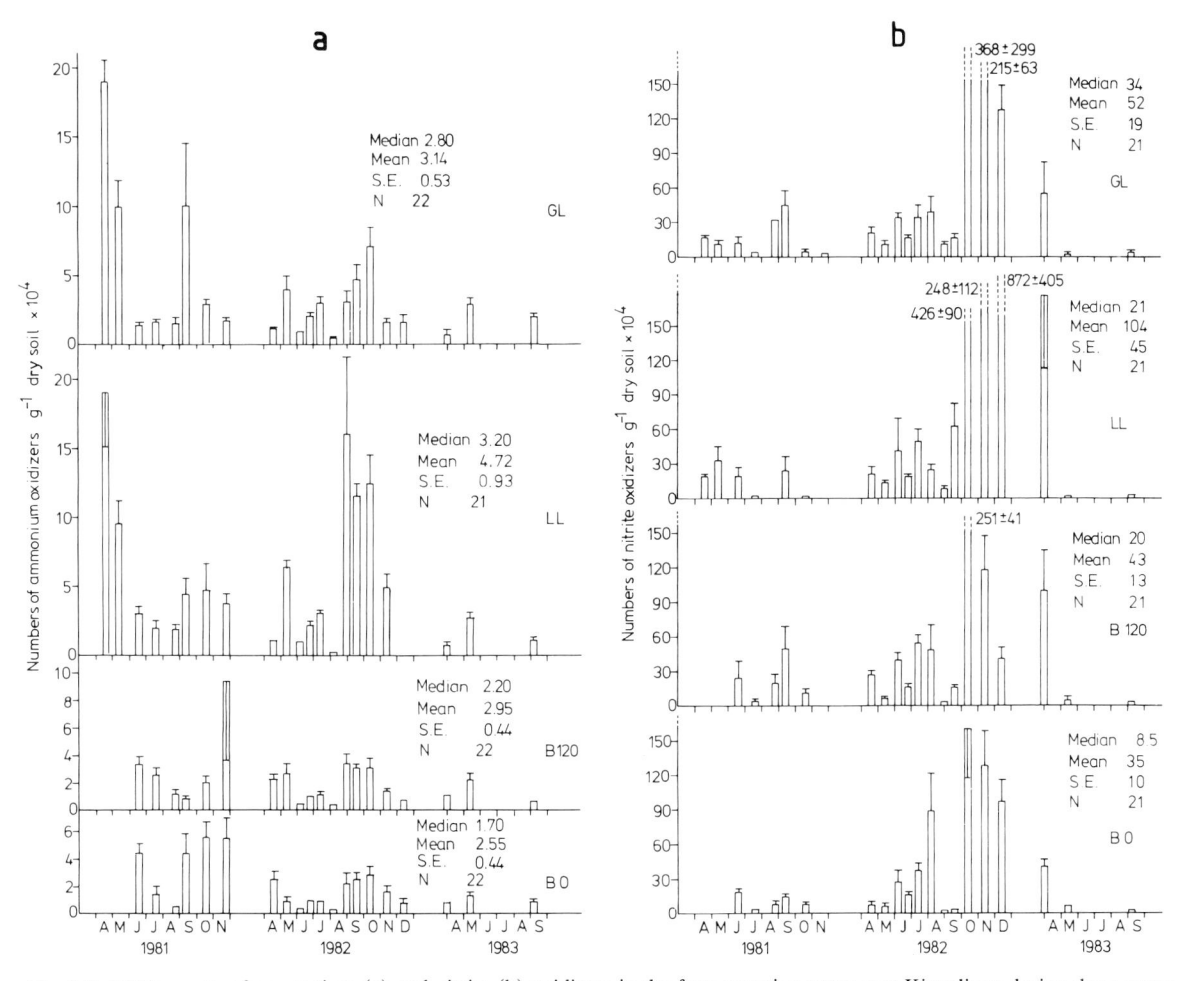

Fig. 6.2. MPN counts of ammonium (a) and nitrite (b) oxidizers in the four cropping systems at Kjettslinge during three years. Bars indicate standard error (SE). Median, mean, SE and N were based on all sampling occasions during 1981–1983 (Berg and Rosswall 1987).

Tab. 6.1. Most Probable Number (MPN) counts of ammonium and nitrite oxidizers in various topsoils. Bars means not reported.

Site	Soil type	pH	% org. C	% N	Plant cover	Abundance (No. 10^{-3} g^{-1} dw)	Reference
Ammonium oxidizers							
Kjettslinge, Sweden	loam	6.3	2.2	0.23	Barley or grass ley (NO$_3^-$-fertilized)	22–48	Berg and Rosswall (1987)
Ultuna, Sweden	sandy clay loam	6.6	1.13	0.15[a]	Bare fallow	1.4	Berg and Rosswall (1985)
Ultuna, Sweden	sandy clay loam	6.7	1.27	0.16[a]	Cereals (no additions)	4.1	Berg and Rosswall (1985)
Ultuna, Sweden	sandy clay loam	6.8	1.41	0.16[a]	Cereals (NO$_3^-$-fertilized)	2.1	Berg and Rosswall (1985)
Ultuna, Sweden	sandy clay loam	6.9	1.79	0.19[a]	Cereals (NO$_3^-$ + straw)	5.8	Berg and Rosswall (1985)
Ultuna, Sweden	sandy clay loam	6.8	1.86	0.22[a]	Cereals (Farmyard manure)	30	Berg and Rosswall (1985)
Rothamsted, England	silt loam	6–7	–	–	Spring barley (NH$_4^+$-fertilized)	17	McDonald (1979)
Lima, N.Y. U.S.A.	podzolic silt loam	7.0	6.0	–		3–4	Molina and Rovira (1964)
Wharekohe, New Zealand	silt loam	6.1	–	–	White clover grass	12	Steele et al. (1980)
Waimate, New Zealand	silt loam	6.1	–	–	White clover grass	55	Steele et al. (1980)
Kaitoke, New Zealand	silt loam	6.7	6.2	0.43	Clover ryegrass pasture	10–100	Stout et al. (1984)
Marua, New Zealand	clay loam	5.3	–	–	Permanent pasture	16	Sarathchandra (1979)
Wharekohe, New Zealand	silt loam	5.5	7.6	0.48	Permanent pasture	52	Sarathchandra (1979)
Horotiu, New Zealand	sandy loam	5.5	12.1	0.86	Permanent pasture	680	Sarathchandra (1979)

Site	Soil type	pH	% org. C	% N	Plant cover	Abundance (No. 10^{-3} g^{-1} dw)	Reference
Wakapuaka, New Zealand	sandy loam	–	–	–	Clover grass pasture	10–19	Belser and Mays (1982)
N Carolina USA (Coweeta watershed)		–	–	–	10-yr-old grass ley	3.3	Rowe et al. (1977)
Hutcheson Forest, New Jersey, USA	sandy silt loam	5.0	1.3	0.12	Annuals: 1-yr fallowed extensive agriculture	100	Robertson and Vitousek (1981)
		4.9	1.4	0.13	Perennials: 4-yr grass ley after extensive agriculture (no fertilizer)	107	Robertson and Vitousek (1981)
Nitrite oxidizers							
Kjettslinge, Sweden	loam	6.3	2.2	0.23	Barley (NO$_3^-$-fertilized)	200	Berg and Rosswall (1987)
Kjettslinge, Sweden	loam	6.3	2.2	0.23	Barley (Unfertilized)	85	Berg and Rosswall (1987)
Kjettslinge, Sweden	loam	6.3	2.2	0.23	Grass ley (NO$_3^-$-fertilized)	340	Berg and Rosswall (1987)
Kjettslinge, Sweden	loam	6.3	2.2	0.23	Lucerne ley	210	Berg and Rosswall (1987)
Kaitoke, New Zealand	silt loam silt loam	6.7	6.2	0.43	Clover and ryegrass pasture	3–30	Stout et al. (1984)
Wharekohe, New Zealand	silt loam	6.1	–	–	White clover grass	12	Steele et al. (1980)
Waimate, New Zealand	silt loam	6.1	–	–	White clover grass	22	Steele et al. (1980)
Marua, New Zealand	clay loam	5.3	–	–	Permanent pasture	2000	Sarathchandra (1979)
Wharekohe, New Zealand	silt loam	5.5	7.6	0.48	Permanent pasture	240	Sarathchandra (1979)
Horotiu, New Zealand	sandy loam	5.5	12.1	0.86	Permanent pasture	2500	Sarathchandra (1979)

[a] From Bonde et al. 1988.

Tab. 6.2. Depth distributions in fertilized barley (B120) of root biomass (Hansson 1987), soil organic carbon (Steen et al. 1984), hyphal lengths and bacterial biomass (Schnürer et al. 1986a), nitrifier numbers (Berg and Rosswall 1987), potential nitrification rates (Berg and Rosswall 1987) and potential denitrification rates (Svensson et al. 1985).

Soil	Root biomass (g dw m^{-2})	Hyphal lengths (m 10^{-3} g^{-1} dw)	Bacterial biomass (mg g^{-1} dw)	NH$_4^+$-oxidizers (no. 10^{-4} g^{-1} dW)	NO$_2^-$-oxidizers	Potential NH$_4^+$-oxidation (ng N g^{-1} h^{-1})	Potential NO$_2^-$-oxidizers	Potential denitrification (ng N$_2$O–N h^{-1})
Plough layer								
0–10 cm	95	1.0	0.55	7	11	230	550	1207
10–27 cm	60	0.7	0.45	5	11	190	490	1515
Loamy sand								
27–45 cm	7	0.4	0.17	0.3	0.8	10	40	16
Clay loam								
below 45 cm	20	0.3	0.07	0.2	0.5	10	90	5

that the cells are more efficiently extracted during the spring and autumn as a consequence of higher water contents of the soil.

As an alternative to the MPN method, Belser and Mays (1982) proposed that numbers can be calculated based on potential oxidation rates of the substrate, if the specific oxidation rates of the organisms in pure culture are known. Based on potential rates, estimated with the chlorate inhibition technique (Belser and Mays 1982, Berg and Rosswall 1985), Berg (1986) estimated the counting efficiency for ammonium-oxidizer MPN over one growing season to range between 0.4 and 3.7%. This discrepancy in determining nitrifier numbers could be the combined result of the inefficiency of the MPN-count procedure and the difficulties in determining numbers of bacteria based on specific growth rates obtained in pure culture studies.

No studies were made of heterotrophic nitrifiers, which are known to be important in some acid soils (Focht and Verstraete 1977, Killham 1986). Their oxidation rates are low, however, compared with those of autotrophic nitrifiers. It is unlikely that they were of any importance in the near neutral Kjettslinge soil (Kuenen and Robertson 1987).

Denitrifiers

Counts of denitrifiers by MPN in a large number of different soils have resulted in values ranging between 10^4 and 10^7 g^{-1} soil (dw) (Gamble at al. 1977). The MPN of denitrifying bacteria in a sandy loam (Denmark) ranged from 9 to 42×10^6 g^{-1} soil in the spring, but had decreased to 3×10^6 by August (Christensen and Bonde 1985). The denitrification activity was also higher in spring than in late summer, indicating that the population level reflected denitrification rates. Other observations also indicate that numbers of denitrifiers may vary over the season, and differences of one order of magnitude have been reported (Terry and Tate 1980, Valera and Alexander 1961). No similar studies were made on the Kjettslinge soil. Denitrifiers often account for a considerable proportion of the total number of soil bacteria that can be isolated (Rosswall and Clarholm 1974). Therefore, it is likely that the denitrifying bacteria varied over time, with depth and between the different cropping systems in a manner similar to that of total bacterial counts (Chapter 4), because the population sizes of denitrifying bacteria, like those of other heterotrophs, reflect their competitiveness for carbon (Tiedje 1988).

In the plough layer of B120 there were no significant differences in potential denitrification rates between the 0–10 and 10–27 cm depths, but potential denitrification rates in the sand and clay layers were at least two orders of magnitude lower than in the plough layer (Svensson et al. 1985; Tab. 6.2). The depth distributions of roots and soil organic matter are given in Tab. 6.2, illustrating the influence of roots on soil organic matter content and the degree to which they provide substrate for both nitrification and denitrification. It is probable that numbers of denitrifiers in the soil profile varied in a manner similar to that of the potential denitrification rates.

Effects of roots on nitrification and denitrification

Nitrification

Autotrophic nitrifiers are inactivated by many organic compounds, and they are poor competitors for ammonium and nitrite compared with heterotrophs (Schmidt 1982). These limitations render the environment close to roots rather unsuitable for autotrophic nitrifiers. Nevertheless, in B120 significantly higher numbers of nitrite oxidizers were found within rows compared with between rows (Tab. 4.11; Berg and Rosswall 1987), indicating that ammonium and nitrite concentrations were higher within rows. In a 5-wk growth chamber experiment with barley in Kjettslinge soil, Klemedtsson et al. (1987b) found a positive correlation between root biomass and number of ammonium oxidizers. This is consistent with the results of Molina and Rovira (1964), who showed that the growth of *Nitrosomonas* and *Nitrobacter* was stimulated by maize and lucerne roots within two weeks of planting. In their case, the growth promoting effect disappeared after five weeks.

Denitrification

Roots may selectively stimulate denitrifying bacteria, as has been shown by Jagnow (1983), who found that numbers of denitrifying bacteria in the rhizosphere of wheat (fertilized with 12 g N m^{-2} yr^{-1}) were more than two orders of magnitude higher than those in non-rhizosphere soil. This relationship, however, not only applies to denitrifiers but also to the total heterotrophic population of soil bacteria. Increased denitrification rates in the presence of plant roots were observed both in the field (Svensson et al. 1985, 1990) and in the laboratory (Klemedtsson et al. 1987a, b).

Plants can affect denitrification rates either positively or negatively. First, the presence of roots can increase denitrification as a result of an increase in anaerobicity, attributable to root respiration, and an increase in the utilization of root-derived carbon by heterotrophic organisms. Second, under anaerobic conditions, rhizodeposition can serve as a source of energy and carbon for denitrifying bacteria, thereby increasing the use of nitrate as an electron acceptor. Third, roots may decrease denitrification by taking up nitrate. Direct competition for nitrate between roots and denitrifiers need not occur, however, since denitrifiers utilize nitrate in anaerobic zones, while root-uptake will be low or non-existent under anaerobic conditions (Nye and Tinker 1977). Lastly, plant roots can facilitate oxygen diffusion into soil by taking up water, and root channels may contribute to increased aeration of the soil (Tiedje et al. 1984). The channels remaining after decomposition of dead roots also result in increased soil porosity (Leměe 1975).

The effects of roots on denitrification will operate on a micro-scale and extrapolating results obtained at the microsite level to the ecosystem level is difficult. It is thus easy to understand that the presence of plants has been reported to stimulate (Woldendorp 1963, Stefanson 1972 a-c, Bailey 1976, Klemedtsson et al. 1987a, b), have no effect (Haider et al. 1985, 1987) and even decrease denitrification (Smith and Tiedje 1979).

In a greenhouse pot experiment with barley (Klemedtsson et al. 1987a), denitrification was higher in planted pots than in unplanted ones. The amounts of denitrified N increased with increasing root biomass (Fig. 6.3). The N$_2$/N$_2$O ratio increased along with the root mass (Tab. 6.3), which also indicated a higher level of anaerobiosis at the higher root-mass contents. Although total amounts of denitrified nitrogen were higher in the presence of higher amounts of root biomass, only the N$_2$-production increased (Klemedtsson et al. 1987a), indicating that N$_2$O was produced in soil not influenced by plants.

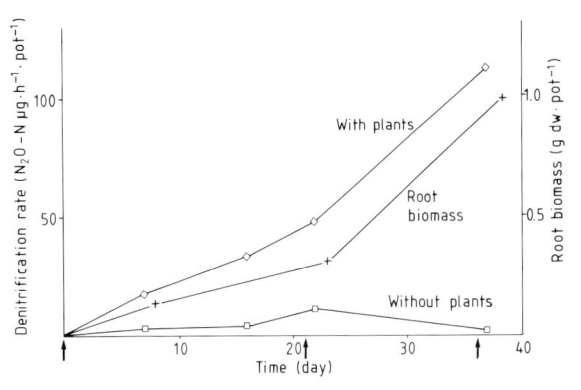

Fig. 6.3. Effects of roots on denitrifiction in a pot experiment (Klemedtsson et al. 1987a). Arrows indicate fertilization with 80 μg Ng^{-1} (dw) as NH$_4$NO$_3$.

The positive effects of roots on potential denitrification rates were smaller than their effects on actual denitrification rates, which also indicated that an increase in the anaerobic volume of the soil was the main cause of the increased denitrification (Klemedtsson et al. 1987a). The actual denitrification rates increased from 5 to 22% of the potential rates in the planted pots, whereas actual rates were only between 1 and 4% of potential in the unplanted pots (Tab. 6.4). If the increased denitrification observed with increasing root biomass had been caused by the addition of carbon and energy to the denitrifier population, population growth would have increased the denitrification potential.

In a cuvette experiment (Klemedtsson et al. 1987b), barley was grown with the roots confined to the middle (20 mm) layer of the cuvette. Three more layers of soil (4, 6 and 10 mm wide) were separated by the insertion of nets with different mesh sizes. Potential denitrification rates in the root layer co-varied with the amount of root biomass. The potential denitrification rates were generally largest in the root layer (Klemedtsson et al. 1987b).

Although there are many problems involved in determining field denitrification rates, the most serious one, if the aim is to quantify total yearly losses for a particular cropping system, is the high degree of spatial and temporal variability (Schimel et al. 1989). Folorunso and Rolston (1984) reported small-scale variability, with coefficients of variation for N$_2$O flux in 3 × 36 m plots ranging from 161 to 508%. For our plots (16 × 40 m) the C. V. varied between 10 and 469% (Svensson et al. 1990). The spatial variability may have been caused by several factors, but a large part was probably a reflection of the activity of plant roots. Field rates based on samples collected over plant rows were significantly different from those measured in samples collected between rows. Ratios for samples taken within and between plant rows were 1.7 for fertilized barley, 8.6 for grass ley and 15.5 for lucerne ley (Tab. 6.5).

Tab. 6.3. Ratios of N_2/N_2O production rates in pots planted with barley and ratios of $N_2 + N_2O$ production from denitrification in planted and unplanted pots (based on Klemedtsson et al. 1987a).

Days after planting	N_2/N_2O ratio from planted pots	$\dfrac{N_2 + N_2O \text{ planted pots}}{N_2 + N_2O \text{ unplanted pots}}$	Root biomass (g dw pot^{-1})
0	< 1	1	0
16	11	9	0.14
22	3	4	0.31
37	76	56	0.98

Water as a regulator of transformation rates

Introduction

Water is a major determinant of oxygen diffusion in soil. Concentrations of oxygen not only determine the overall process rates but also the relative proportions of the various end products (NO_3^-, NO and N_2O for nitrification; NO, N_2O and N_2 for denitrification). Useful determinations of soil oxygen concentrations are very difficult to make, mainly owing to the great spatial heterogeneity (Flühler et al. 1976). Because water affects both the overall respiration rate and the oxygen diffusion rate, soil water contents are often used as an indirect measure of oxygen availability. Published data on the effects of water on nitrification and denitrification are often given as percent of water holding capacity (WHC), which cannot easily be transformed to water potentials (Sommers et al. 1981). However, water potentials should be used for determining effects on microbial activities and make it possible to compare different soils.

The effect of water on nitrification and denitrification is related to oxygen diffusion, and the percentage of air-filled pores is the important variable to consider.

Tab. 6.4. Measured denitrification rates as a percentage of the corresponding denitrification potential rates in a pot experiment with barley in Kjettslinge soil (Klemedtsson et al. 1987a).

Treatment	Days after planting		
	7–8	22–23	37–38
Without plant	1.2	3.6	1.0
With plant	4.9	8.5	22

Tab. 6.5. Comparison of denitrification rates in soil cores taken within (W) and between (B) plant rows in Kjettslinge during 1983 (modified from Svensson et al. 1990; see also Tab. 6.9).

Cropping system	Ratio W/B	Number of sampling dates
B120	1.7*	26
GL	8.6***	56
LL	15.5***	32

Significance was determined using F-tests in which interaction variances were tested against residual MSS, and variance due to plant effects was tested against interaction variance.
* $p < 0.05$, *** $p < 0.001$.

The apparent diffusion coefficient of oxygen in soil is proportional to the gas-filled porosity, and a small change in water content near saturation will drastically reduce diffusion rates (Currie 1984).

Soil often contains both aerobic and anaerobic microsites (Fig. 6.4), and thus oxidative and reductive processes, both involving inorganic nitrogen compounds, may occur simultaneously. This is important to recognize in interpreting data on N_2O and NO losses, because these gases may be produced both by nitrification and denitrification. During nitrification, N_2O is produced when a decrease in oxygen concentration creates microaerophilic conditions (Goreau et al. 1980, Lipschultz et al. 1981). On the other hand, the relative proportion of N_2O in gaseous denitrification products increases as the oxygen concentration increases in an initially anaerobic environment (Firestone et al. 1979). Thus, both processes produce N_2O in the transition zone between aerobic and anaerobic areas.

Anaerobic zones in soil are formed when oxygen consumption rates (respiration) become higher than oxygen diffusion rates. The respiration rate is mainly due to microbial activity, which in turn is determined by the availability of energy sources, temperature and water. Diffusion rates are regulated by the soil structure (pore and aggregate sizes and their distribution) and the water content (fraction of water-filled pores in the soil; Smith 1977). Any management practice affecting the soil structure will also affect oxygen concentrations and subsequently the nitrogen flows. In Kjettslinge, major differences in texture only occurred between the soil layers, and because rates of inorganic nitrogen transformations are very low in the sand and clay layers (Tab. 6.2), these differences will not be further considered. The fact that the plots sown with barley are ploughed annually, whereas the ley plots are not, should also have led to differences in soil structure. Such differences may also have occurred as a result of differences in root growth in the two types of systems. Porosity was higher in the grass ley than in the fertilized barley immediately before the breaking of the leys (Chapter 2). In addition, soil moisture was generally lower in leys than in barley owing to higher evapotranspiration rates in the leys over most of the year (Ch. 3).

Anaerobic conditions at high moisture levels are generally regulated by heterotrophic respiration. The effect of a rainfall on the aerobic/anaerobic status of micro-

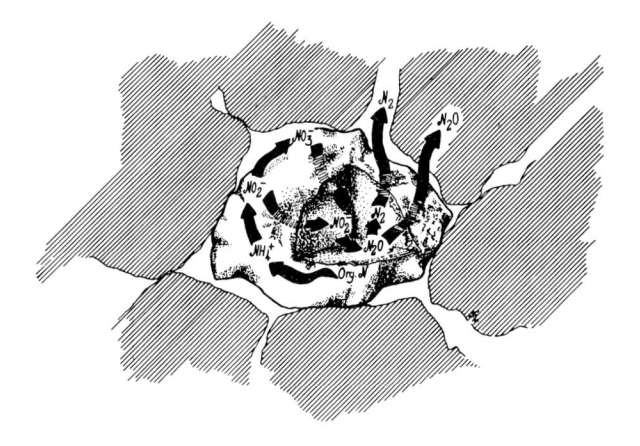

Fig. 6.4. Soil aggregate showing surface nitrification (aerobic) of ammonium to nitrite and nitrate. The ions diffuse into the anaerobic centers of the soil aggregates, where reduction occurs, producing N_2O and N_2 (Modified after Berg 1986).

sites depends on the availability of easily available energy sources, which can be liberated by sudden wetting (Birch 1960) or increased root growth. When Schnürer et al. (1986b) studied the effect of rain on microorganisms at Kjettslinge, they found that oxygen consumption rates were highest one day after a rainfall and that the bacterial population reached its maximum after three days. This should have resulted in lower oxygen concentrations in the topsoil, with conditions suitable for denitrification.

Thus rainfall directly affects rates of inorganic nitrogen transformations, and if process rates are to be determined in the field, a frequent sampling programme is needed based on rainfall events rather than on a regular daily, weekly or monthly measurement programme.

Nitrogen mineralization

Decomposition rates of organic matter are optimal at water potentials between –0.02 and –0.05 MPa (–0.2 to –0.5 bars; Sommers et al. 1981). Once the soil dries out, mineralization rates, which have been found to be negatively correlated with the logarithm of the water potential, will decrease (Stanford and Epstein 1974; Fig. 6.5). At water potentials above –0.02 MPa, mineralization may be inhibited by low oxygen concentrations. As nitrogen mineralization is often a rate-limiting step for nitrification, it will have a great influence on total inorganic nitrogen flows.

Nitrification

Belser (1979) considered soil moisture and temperature to be the main factors regulating nitrification rates. Nitrifiers are especially sensitive to low oxygen concentrations, and they also seem to be less effective in competing for available oxygen compared with heterotrophs (Focht and Verstraete 1977), since their K_m for oxygen is higher than that of heterotrophs. In the Kjettslinge soil, nitrification rates were optimal at –0.11 MPa (65%

of WHC; Berg and Rosswall 1987; Fig. 6.6). This agrees with the findings of Robertson (1982), who observed the highest nitrification rates for several soils to be around 70% of WHC (at 30°C). The optima may have been at higher water contents, since the highest rates were observed at the highest water contents studied. In laboratory studies, Linn and Doran (1984) found that microbial activity (respiration) was optimal at 60% of WFP (water filled pore space; approximately equal to WHC for the Kjettslinge soil). It has most likely been on only rare occasions that high water contents have limited nitrification rates by causing oxygen depletion at Kjettslinge. However, the often occurring dry conditions may have limited nitrification. Both short-term (days) and long-term (weeks) limitations of nitrification by low water potential (drought) were observed for the Kjettslinge soil (Berg 1986, Berg and Rosswall 1989).

Nitrification rates were generally determined with the chlorate inhibition method after addition of substrate, which gives potential rates (Belser and Mays 1982, Berg and Rosswall 1985). Potential ammonium-oxidation

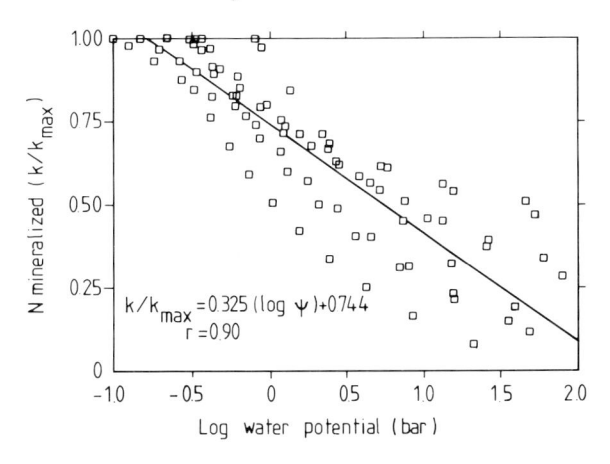

Fig. 6.5. Relationship between relative rates of N mineralization (k/k_{max}) and soil water potential. From Sommers et al. (1981), based on Stanford and Epstein (1974), with permission.

rates in Kjettslinge (Tab. 6.6) were similar to values presented by Belser and Mays (1982), who obtained between 300 and 1400 ng N g^{-1} (dry soil) h^{-1} in soils from two grazed Minnesota pastures. Groffman (1985) used potential oxidation rates to study tilled and untilled plots on a sandy loam in Georgia and obtained between 500 and 4000 ng N g^{-1} (dry soil) h^{-1} (0–5 cm), which was up to four times higher than the potential at Kjettslinge. For reviews on short-term nitrification measurements in general, see Belser (1979), Schmidt (1982) and Berg (1986). Actual rates were also determined without the addition of substrate (Berg and Rosswall 1985), although these determinations were sometimes difficult because of relatively high background concentrations of nitrite. The ratios of actual rates over potential rates were higher in July than in September (Tab. 6.6).

Annual nitrification rates calculated for the four cropping systems at Kjettslinge were based on short-term measurements of actual rates. The calculation was based on median values of actual ammonium oxidation rates (10–20 ng N g^{-1} (dry soil) h^{-1}) measured in B120 (Berg and Rosswall 1985 and unpubl.). These rates were assumed to occur during the 230 frost-free days at a mean soil temperature of 12°C (Alvenäs et al. 1986) over a top soil depth of 27 cm with a soil bulk density of 1.45 g cm^{-3} (Steen et al. 1984). Based on the above assumed values, actual nitrification in the B120 field over one year should range between 9.5 and 19.0 g N m^{-2} yr^{-1} (Tab. 6.6). Actual rates were then assumed to rank, on the average, in the same order as potential rates, i.e., GL>LL>B120>B0, with GL rates being 60% more than B0 rates. Values for B0, GL and LL plots were thus based on the average relative yearly differences in potential rates between the four cropping systems, with B120 as a reference.

The estimated nitrification rates can be compared with an estimated turnover of N via mineralization of 2% yr^{-1} of the organic material in the soil. With an organic-N pool of about 700 g m^{-2} in the Kjettslinge soil (Rosswall and Paustian 1984), about 14.0 g N m^{-2} should be mineralized annually. Estimated mineralization rates in nitrogen budget calculations ranked the

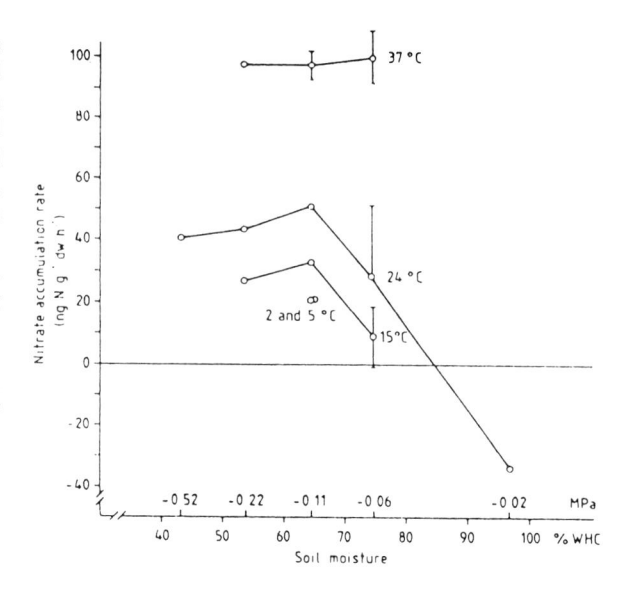

Fig. 6.6. Effect of soil moisture and temperature on nitrification rates in the Kjettslinge soil incubated in the laboratory (Berg and Rosswall 1989).

systems in the same order as the estimated nitrification rates and were always within the ranges given for nitrification rates (Tab. 6.6).

Nitrous oxide production

To be able to separately study N$_2$O produced by nitrifiers (N$_2$O^N) and that produced by denitrifiers (N$_2$O^D), it was necessary to inhibit one of the processes without affecting the other. Inhibition of autotrophic nitrification at low partial pressures of C$_2$H$_2$ has been reported by Hynes and Knowles (1978) and Berg et al. (1982). A method was developed taking advantage of the sensitivity of the nitrifiers to low acetylene concentrations. In the absence of acetylene N$_2$O will be produced by both nitrification (N$_2$O^N) and denitrification (N$_2$O^D). Thus,

Tab. 6.6. Potential and actual ammonium oxidation rates in the topsoil at Kjettslinge (0–10 cm; n = 4, mean ± SE) and estimated annual mineralization and nitrification rates.

Cropping system	Potential rates[a] (ng N g^{-1} h^{-1})		Actual rates[b]		Estimated annual nitrification[b] (g N m^{-2} yr^{-1})	Estimated annual net mineralization[c]
	July	Sept	July	Sept		
B0	150±4	206±6	34±6	0±0	7.6–15.2	8.0
B120	200±5	501±10	49±4	21±3	9.5–19.0	9.2
LL	131±13	261±24	42±3	11±2	11.1–22.2	14.7
GL	374±19	483±10	283±62	7.3±0.5	11.8–23.6	21.4

[a]Samples were taken in 1983 (Berg and Rosswall 1985).
[b]Berg (1986).
[c]Paustian et al. (1990).

by subtracting the N_2O^D obtained in the presence of 1–5 Pa of C_2H_2 from the total nitrous oxide production (N_2O^T), the N_2O^N can be estimated. Acetylene at low partial pressures completely inhibited N_2O^N production, while N_2O^D reduction to N_2 was almost unaffected (Klemedtsson et al. 1988a). The use of acetylene has distinct advantages over the conventionally used nitrapyrin in that the moisture content will not be changed by addition of the inhibitor. In view of the importance of oxygen concentrations for determining rates of nitrous oxide production from both nitrification and denitrification, it is essential that the method does not affect rates of oxygen diffusion, for example by water additions (Klemedtsson et al. 1988a).

In the field studies (see below), determinations of the relative rates of N_2O production from the two processes were generally not made. To evaluate the importance of soil moisture contents on nitrous oxide production rates, a laboratory experiment was set up in which the two sources of N_2O were determined (Klemedtsson et al. 1988b). During the initial period (0–2 d) considerable N_2O production from nitrification was only observed at 100% of WHC (Tab. 6.7), and there were no significant differences between the rates at the other three water contents (p <0.001). The N_2O produced from nitrification at the low water contents was assumed to have originated from microaerophilic aggregates, since nitrifiers produce N_2O when subjected to oxygen stress (Poth and Focht 1985). During the last period of measurement, the oxygen concentration at 100% of WHC was probably so low that nitrification ceased completely. At 90% WHC the increased respiration resulting from wetting of the soil led to a decrease in oxygen concentrations as nitrifier production of N_2O increased. The initial production in the two driest treatments was probably due to enhanced respiration caused by the water addition, an effect that disappeared during the last period of measurements. The highest total N_2O production was observed at 90% of WHC. Except for the last measurement period, N_2O^N was always greater than N_2O^D. The total losses via denitrification (N_2O^D + N_2) increased exponentially with increased soil water contents and time. Total nitrogen losses during the experiment were, however, low (ca 15 µg N as compared with the 80 µg N g^{-1} added). When considered from an agricultural point of view, losses associated with nitrification corresponded to 1% of the added NH_4^+, which is in agreement with earlier studies (Bremner and Blackmer 1978, Breitenbeck et al. 1980, Aulakh et al. 1983).

If these observations were to be interpreted in relation to the field conditions, both the intensity and frequency of rainfall events would be important in regulating N_2O production by nitrification and denitrification, but from an ecosystem point of view, the losses would be insignificant.

Nitric oxide production

Three different processes can lead to NO production in soils: nitrification, denitrification and chemodenitrification (Galbally 1985). Other suggested mechanisms for NO production include photolysis of nitrite in flooded rice fields (Galbally et al. 1987) and in ocean waters (Zafirou and McFarland 1981). Laboratory studies have shown the importance of anaerobic conditions for NO production. Johansson and Galbally (1984), using soil from the Kjettslinge site, observed that production rates in flow-through soil columns were 65 times higher under anaerobic conditions than under aerobic ones. Under aerobic conditions both biological production and uptake of NO were observed. It was tentatively suggested that the NO produced in anaerobic zones is consumed in aerobic zones near the surface. As yet, no aerobic NO consumption process has been identified (Rosswall et al. 1989). The emission of NO under field conditions may thus be regulated by microbial processes involving both production and consumption of NO and by physical factors regulating transport to the soil surface. No production of NO_2 was detected.

Galbally and Johansson (1989) presented a conceptual framework for relating laboratory measurements to diffusive flux from the soil to the atmosphere. Model calculations of the NO flux in the field based on the laboratory measurements of NO production and uptake under aerobic conditions agreed closely with those actually observed under the same conditions in the field. The nitrification rates were also similar under laboratory and field conditions (Galbally and Johansson 1989). Calculations based on oxygen consumption rates (measured on field samples by Schnürer et al. 1986b) and data on soil porosity showed that the top layers of the soil during the field measurements used in this com-

Tab. 6.7. Nitrous oxide production in the Kjettslinge topsoil, supplemented with 40 µg NO_3-N g^{-1} dw and 40 µg NH_4^+-N g^{-1} dw, incubated in the laboratory at different moisture contents. The rates are given for different elapsed times after the start of the incubations. The rates were determined with a selective inhibition technique using different concentrations of acetylene. Data from Klemedtsson et al. (1988b). N represents production from nitrification and D from denitrification. Nitrogen gas production during denitrification is also given as an indicator of total denitrification rates (N_2).

Water content (% of WHC)		0–2 d	3–6 d	9–15 d
		(ng N_2O-N g^{-1} dw d^{-1})		
60	N	3.3	3.3	0
	D	0.2	0.2	0.2
	N_2	0	0	0.7
80	N	3.0	3.0	0
	D	0.5	0.5	0.5
	N_2	2.0	5.5	8.8
90	N	2.3	33.4	158
	D	1.6	17.9	65.3
	N_2	0.8	66.7	1617
100	N	34.0	21.3	0
	D	7.3	15.4	0
	N_2	8.0	1100	1850

parison were fully aerobic. These results indicate that nitrification rather than denitrification was the main process regulating the NO production observed in the Kjettslinge soil.

Soil moisture has a substantial influence on NO emissions. Johansson and Granat (1984) observed increased NO emissions from wetted soil after a period of drought. As the soil moisture approached saturation, the NO emission decreased sharply. However, no simple relation between the flux of NO and soil moisture could be deduced. Similar relationships between NO emission and soil moisture have been reported by others (Slemr and Seiler 1984, Anderson and Levine 1987, Johansson et al. 1988). Anderson and Levine (1987), who made simultaneous measurements of NO and N_2O, observed an increase in the emission of N_2O and a decrease in the emission of NO as the soil water content approached or exceeded field capacity. In light of laboratory experiments that have shown NO production to be much higher under anaerobic conditions it appears as if NO is more readily consumed by the bacteria than is N_2O.

Denitrification

Denitrification measurements were concentrated on determinating loss rates in the field. Although it has been suggested that gaseous losses might be as important to the nitrogen economy of Swedish agroecosystems as leaching losses (Jansson and Simán 1978), no quantitative measurements had previously been carried out in the field.

The acetylene inhibition technique (Yoshinari and Knowles 1976, Balderston et al. 1976, Klemedtsson et al. 1977) was used for the denitrification studies. Efforts were also made to collect 15-N dinitrogen in temporary enclosures of the small lysimeters fertilized with highly enriched 15-N nitrate (cf. Bergström 1987b). Although collections were made for extended periods (24–48 h) after rainfall events, the excesses obtained were generally not above the detection level.

The limited areas available for the denitrification measurements in the field (16 m^2 for each of the four replicates in the four cropping systems, see Chapter 2) precluded the use of a field cuvette technique as used by Ryden et al. (1979), since repeated measurements cannot be made at a plot previously exposed to acetylene. Instead, a soil core method was developed for short-term incubations (Svensson et al. 1985). Soil cores were collected in the field with an auger (3.2 cm diameter) and placed in plastic containers having lids with rubber septa. After the cores had been brought back to the laboratory, acetylene was injected, the cores were incubated at 15°C and gas samples were withdrawn periodically for 15 h. For soil cores at 100% of WHC, it took less than 3 h for acetylene to completely inhibit the nitrous oxide reductase system (Fig. 6.7). Production was linear for 10–20 h (Svensson et al. 1985), indicating little or no increase in denitrification due to the presence of C_2H_2.

Because of the large spatial variability in denitrification rates there were rarely statistically significant differences between denitrification rates on consecutive sampling dates (Svensson et al. 1990), although mean values varied greatly. A log-normal curve of the measured denitrification rates seemed to best approximate their distribution (Fig. 6.8). Log-normally transformed data were used for all variance analyses performed on the results from core denitrification measurements. Peaks after fertilization were observed both in the grass ley and fertilized barley plots. Increased rates were also observed during rainy periods in 1983, i.e., late June – early July and mid-September (Fig. 6.9).

The relationship between denitrification rates and soil moisture was described by a biphasic curve with no effects of water on the rates occurring at low moisture levels, followed by a steeper increase at the higher water contents (Fig. 6.10). This was evident for between-row samples during the two seasons investigated. During 1983, there were two exceptions, however; samples from within rows of the grass and lucerne leys followed a monophasic curve over the entire moisture range. Within plant rows in the leys a general restriction in oxygen availability was obvious, giving rise to a continuous and more rapid response to increases in the soil

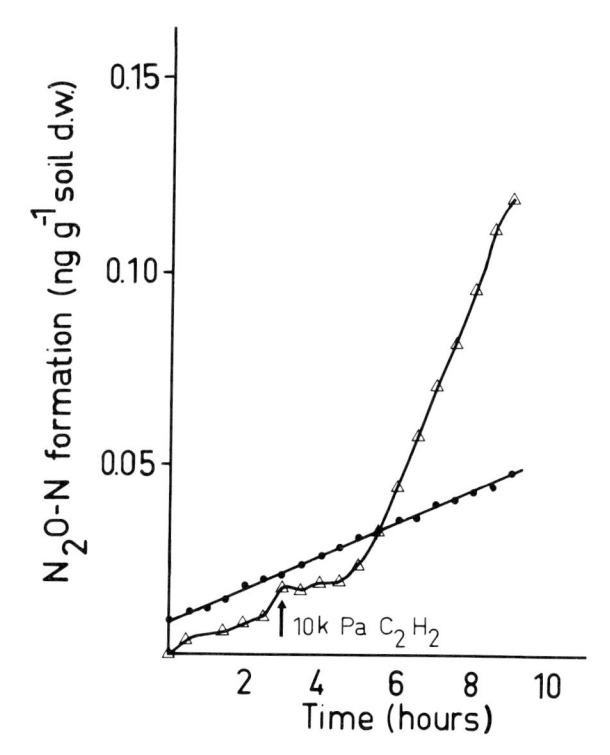

Fig. 6.7. Nitrous oxide production from typical soil cores without (●) and with (△) the addition of 10 kPa of acetylene addition, indicated by arrow). The soil cores were at 100% of WHC, incubated at 15°C and NO_3^- was added to a final concentration of 10 µg g^{-1} dry soil (Klemedtsson 1986).

water content compared with the between-row situation. The most probable reason for this response was the drought occurring that summer. The dry weather had especially severe effects on the meadow fescue grass, much of which wilted (Pettersson 1987). A source of easily metabolizable organic matter thus became available during the dry part of the season.

Moisture explained a significant fraction of the variance in denitrification rates, both within and between rows of all three crops, when calculations were based on an exponential increase at soil water contents above 0.2 g H$_2$O g^{-1} soil (Fig. 6.10). The effect of water explained 20% and 46% of the variation for within- and between-row samples, respectively, in barley and 42% of the variability between rows in the lucerne ley (Klemedtsson et al. 1990).

Nitrate content proved to be a useful variable in describing denitrification rates in the grass ley and in between-row samples of the lucerne ley (Klemedtsson et al. 1990). Higher NO$_3^-$-concentrations are often needed to half-saturate denitrification in field soils as compared with pure-culture studies, due to nitrate diffusion limitations (Myrold and Tiedje 1985). The nitrate effect on the log-normal denitrification rates was model-

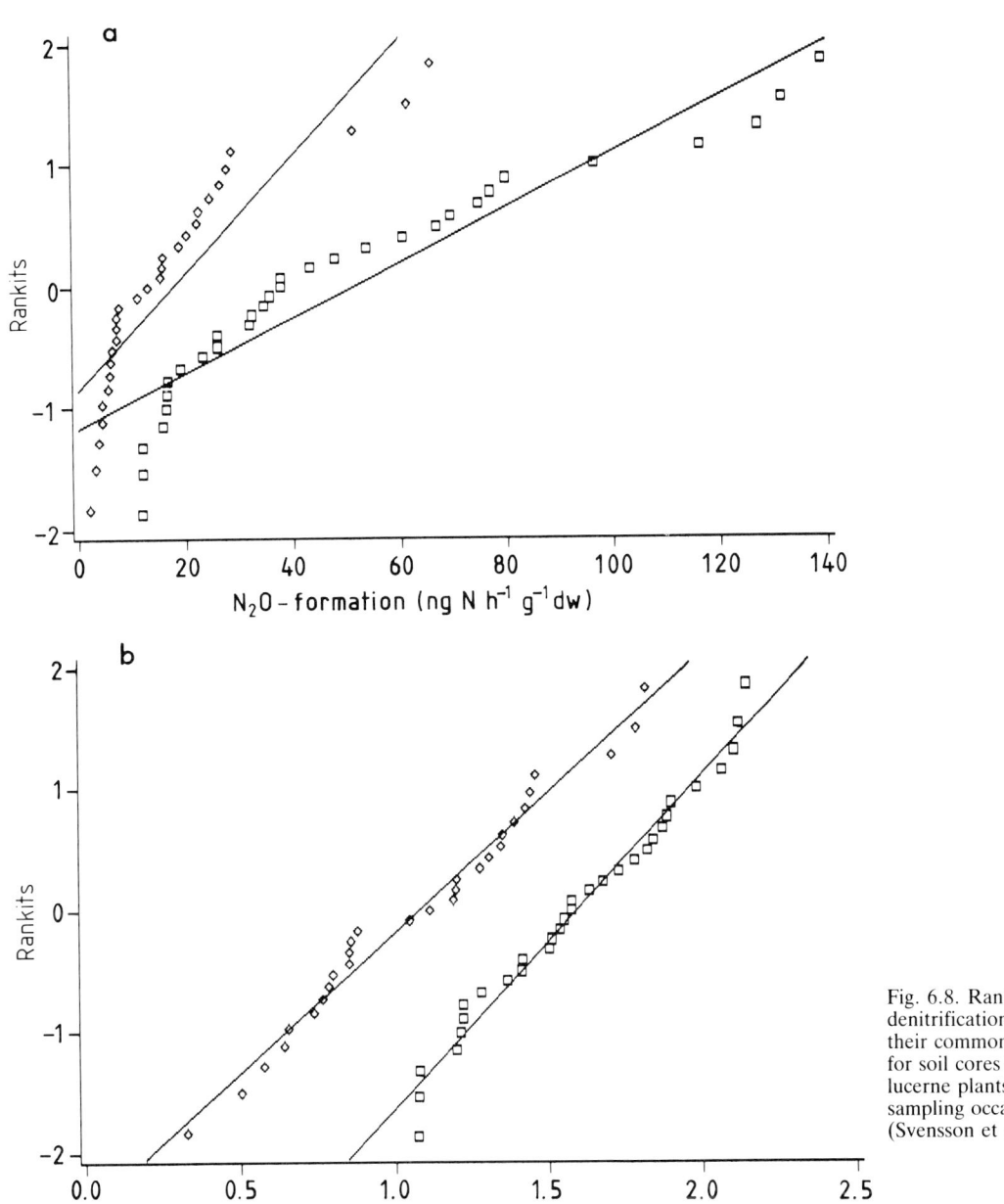

Fig. 6.8. Rankit plots of denitrification rates (a) and their common logarithms (b) for soil cores taken over lucerne plants on two sampling occasions (Svensson et al. 1990).

led using a Michaelis-Menten function. The K_m estimates for the field data are not true K_m values for the denitrifying population of the soil but a combined function of the K_m:s and soil chemical and physical reactions. Our studies indicated that K_m values were 0.12–1.40 mM (Klemedtsson et al. 1990) similar to those reported by Myrold and Tiedje (1985), i.e., 0.13–2.10 mM.

As an example of the relationship between denitrification rates, soil moisture and nitrate concentrations, a nonlinear regression analysis on samples collected between rows in the grass ley is shown in Fig. 6.11. The model construction did not allow separate determinations of the water and nitrate effects when both contributed significantly to the observed variability. When the nitrate effect was studied, the data were adjusted to take the influence of water into account. The combined effect of water and nitrate explained 37% and 24% of the variation in the within- and between-row data, respectively, in the grass ley and 65% of the variation for between-row samples in the lucerne ley.

With regard to the lack of a relationship between denitrification rates and soil moisture in the dry samples, it should be noted that the driest samples from the barley and lucerne ley plots were collected one or two days after a rain following an extended dry period (Kle-

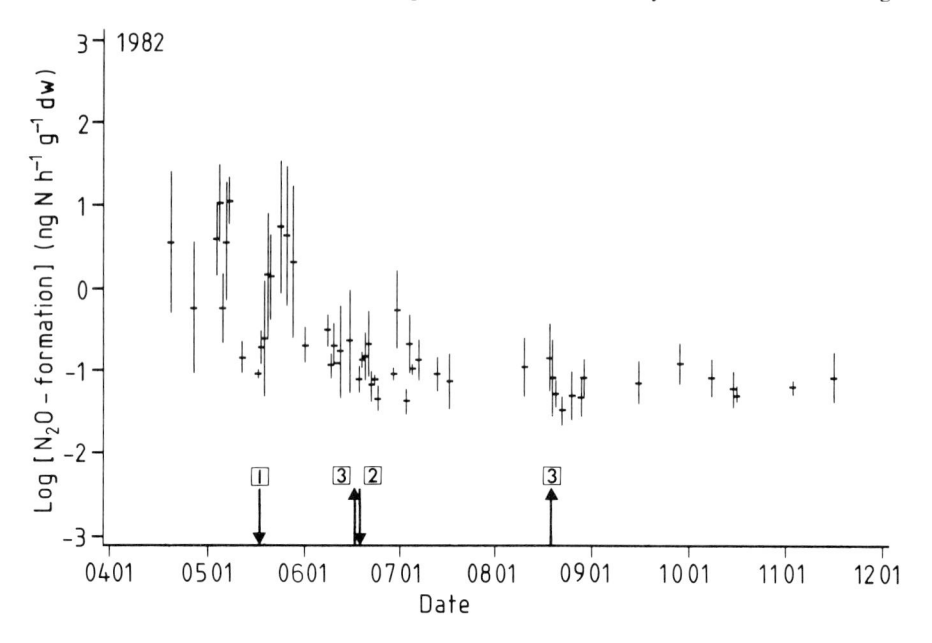

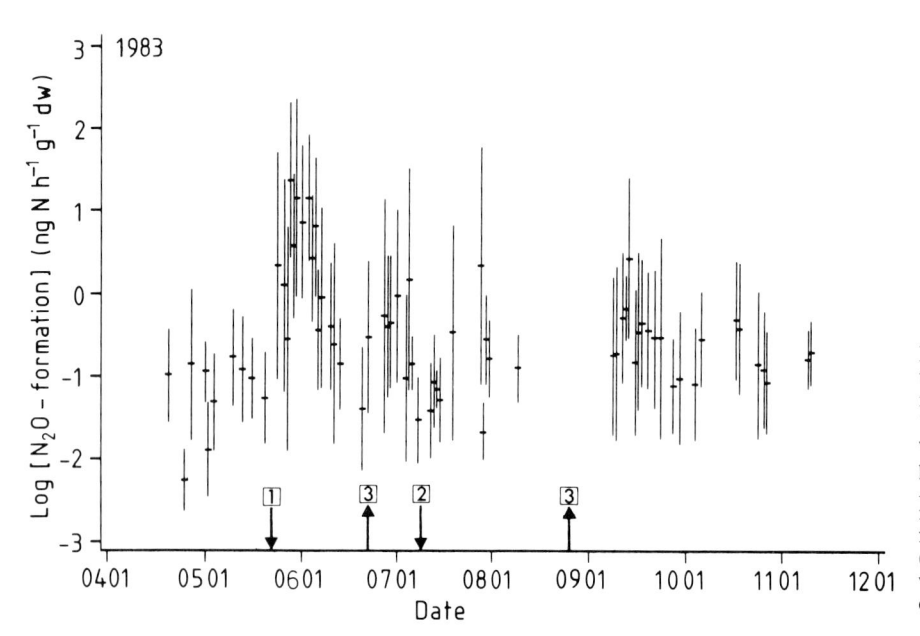

Fig. 6.9. Means of log-transformed denitrification rates of the grass ley during 1982 and 1983. The samples were collected between plant rows. Bars show S.D. Arrows indicate soil management treatments: fertilization with 12 g N m^{-2} (1), and with 8 g m^{-2} (2) as well as harvest (3) (Svensson et al. 1990).

medtsson et al. 1990). Although the soil was still quite dry after the rainfalls, it is possible that the partial remoistening of the dry soil resulted in a burst of respiration that could have consumed oxygen rapidly enough to give rise to anaerobic micro-niches. No such relationship was found between rainfall and denitrification rates in the dry samples of the larger data set collected from the grass leys.

The denitrification response after irrigation or rain events has been shown to be complicated. The rates may be dependent on irrigation frequencies (Rolston et al. 1982). Sextone et al. (1985) reported that denitrification rates are not always high following rainfall events, possibly as a result of nitrate or carbon depletion at the anaerobic sites after an earlier rain. Soil texture also influences denitrification response to rainfall. A non-aggregated sandy soil showed a more rapid increase as well as a more rapid decrease in denitrification rates than a clay soil (Sexstone et al. 1985). The sandy soil showed maximal rates already 3–5 h after rainfall, whereas maximum rates peaked in the clay soil after 8–12 h. The rates had returned to background levels after 12 and 48 h for the sand and clay soils, respectively.

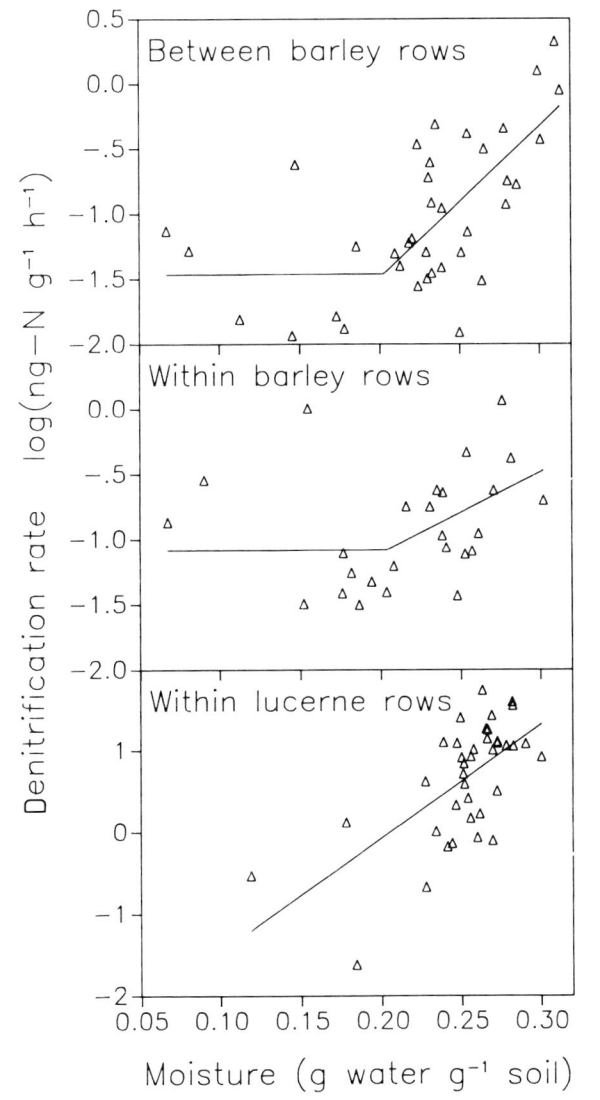

Fig. 6.10. Denitrification rates as a function of soil moisture between and within rows of barley and within lucerne rows. No relationship was found between denitrification rates and nitrate concentrations in these cropping systems (Klemedtsson et al. 1990).

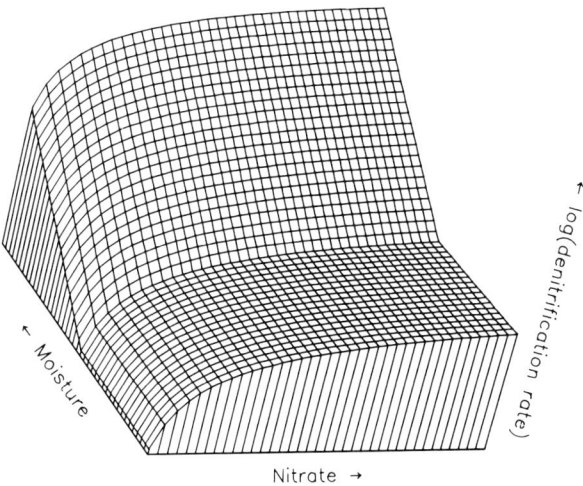

Fig. 6.11. The most general of the models used by Klemedtsson et al. (1990) to represent the combined effects of soil moisture and nitrate on denitrification rates.

Water as a carrier of substances

Introduction

In temperate soils, nitrate, unlike ammonium ions, is not bound to the clay minerals. This results in the nitrate-N being highly mobile, and it is the only nitrogen species transported by mass flow in the soil profile to any major extent. Typical drainage water from the Kjettslinge field contained about 10, <0.1, 1 and <0.01 mg N l^{-1} of nitrate, ammonium, organic-N, and nitrous oxide, respectively. When the soil is covered by vegetation, immobilization by root uptake is a major factor reducing nitrate transport from the topsoil, even when large amounts of nitrate are added as fertilizer. Furthermore, an established grass ley takes up nitrogen faster than a newly sown grain crop (Bergström 1986), thereby leading to a lower risk for leaching losses during early rains.

Although many of the causal connections between leaching of nitrate, cropping systems and environmental factors are known, it is still difficult to obtain proper quantitative estimates. Field measurements of leaching have been performed using a variety of methods. One important fact to keep in mind when comparing quantitative results is that important differences concerning surface area, water collection technique, etc., exist between methods. Lysimeters of 75 m^2 were used by Uhlen (1978) to determine nitrate leaching from a cultivated soil. Many experiments with small-scale lysimeters, both with disturbed and undisturbed soil profiles, have been reported (e.g., Cassell et al. 1974, Cannell et al. 1980).

The results from this project are based on measurements from both tile-drained plots (each 3600 m^2) as well as different types of lysimeters with both disturbed and undisturbed soil profiles. The lysimeter types are hereafter referred to as the large (27 m^2), medium (1.13 m^2) and small (0.07 m^2) lysimeters, of which the two former had disturbed soil profiles (see Chapter 2). Labeled fertilizers (^{15}N-NO$_3^-$) were also used to determine the amounts of fertilizer-derived nitrate in drainage water.

Drainage conditions

The distribution of rainfall over the year greatly influenced both the timing of the drainage discharges and their quantities. Heavy rains during autumn generally resulted in discharge peaks, while rainfall during late spring or summer, when a soil-water deficit normally existed, had far less direct influence on the amounts of discharge. Dry years, such as 1982 and 1983, resulted in low drainage flows during autumn as well.

The influence of the crop on drainage volumes was considerable and mainly depended on the degree to which the soil-water deficit developed during the growing season. In general, the barley cropping systems (B0 and B120) had considerably higher drainage volumes than the cropping systems with perennial leys (LL and GL). Much of this difference was due to the extended growing season and the consequent uptake of water early and late in the season, which is typical for leys. Thies et al. (1978) found that the drainage volume from non-cropped lysimeters was 26% higher than from lysimeters with a crop. Similarly, Hoyt et al. (1977) reported that drainage volumes from non-vegetated soils were slightly higher than those from cropped soils. The fertilization intensity of the barley and the type of ley also had a certain, although less pronounced, influence on the amounts of discharge. The lower growth rate of the unfertilized barley and the associated lower transpiration rates commonly led to soil water tensions being lower in this treatment than in the fertilized barley treatment (Jansson and Thoms-Hjärpe 1986). Consequently, drainage volumes from the B0 cropping system were often higher than those from B120.

In comparing the various methods used to measure drainage discharge, clear differences were found. Although the tile-drained plots and the different lysimeters showed similar temporal patterns of variation in drainage discharges, the lysimeters had considerably higher accumulated drainage volumes (326, 289 and 235 mm for the large, medium and small lysimeters, respectively) than the corresponding large-plots (141 mm), as exemplified by the B120 cropping system (Fig. 6.12). Much of this difference arose because all percolating water reaching the bottom of a lysimeter is measured, whereas only part of the inflow water is usually captured by a drainage system under natural conditions.

The amount of discharge from tile-drains depends on, for example, groundwater level and possible lateral water movements. The low hydraulic conductivity in the clay limited the percolation rate substantially. This indicates that during periods of high precipitation, lateral water movements commonly occurred in the topsoil and sand layers with high conductivity. The comparatively good crop growth on the B120 tile-drained plot during dry periods (R. Pettersson, pers. comm.) suggests that water contributions from the subsoil should also be considered. Measurements of soil water tension and calculations of water flows indicated that a capillary rise occurred during the growing season (Jansson and Thoms-Hjärpe 1986). Such conditions also require that groundwater storage in the subsoil be restored following dry periods, resulting in substantially reduced tile-drainage flows. Hood (1977) found that only 20% of the rainfall was recovered from tile-drains under field conditions compared with 38% for lysimeters. This suggests that measurements of tile-drainage volume in a field can give an unreliable picture of the total amount of water actually moving through the soil profile.

Drainage volumes measured from closed systems such as lysimeters may, in contrast, represent overestimations of actual drainage volumes under field conditions, since the water normally giving rise to surface runoff in a field situation is measured as drainage in lysimeters such as the medium-sized and small ones used in this study. An extrapolation of results from lysimeters to a field would therefore be inaccurate. Consequently, quantitative determinations of water transport in soils should be viewed with caution. For studies requiring thorough control of the total water balance of a soil profile and/or control of all solutes leaving the root zone, it is clear, however, that lysimeters are most suitable.

Nitrate in drainage water

Clear differences existed among the four cropping systems regarding the nitrate concentrations and the seasonal variations of nitrate in drainage water (Fig. 6.13). After an initial period of relatively unstable conditions, the nitrate concentrations were ordered hierarchically according to the cropping systems. The most obvious difference occurred between the fertilized barley and the perennial leys with ca. ten times lower concentrations in drainage water from the latter.

The stable conditions characterized by low nitrate levels in water draining from grass leys changed abruptly once the grass leys were ploughed in 1984.

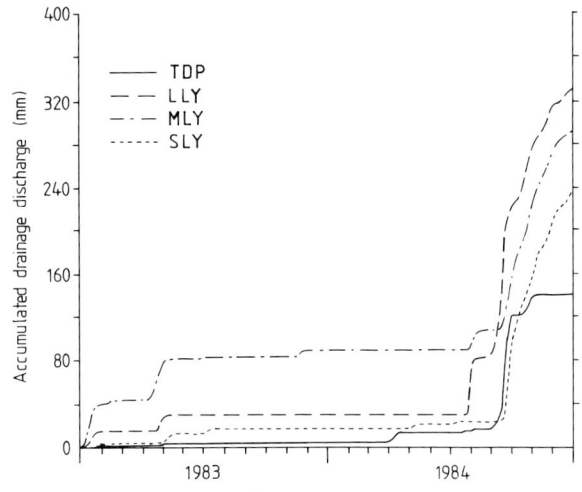

Fig. 6.12. Accumulated drainage discharges from the tile-drained plots (TDP) and the different lysimeters during 1983 and 1984. The large (LLY) and medium lysimeters (MLY) had disturbed soil; the small lysimeters (SLY) had an undisturbed soil profile (Bergström 1987c).

Nitrogen losses from the four cropping systems

Introduction

One of the main aims of the project was to quantify the losses of nitrogen from the four cropping systems and to compare their abilities to conserve nitrogen. In other studies on nitrogen cycling in agroecosystems, attempts to determine N budgets have generally been based on estimating losses as the difference between inputs and changes in the system storage over time (Frissel 1977). Although there are difficulties involved in making yearly estimates of all the loss processes, except harvest, the lack of quantitative data on leaching and denitrification is often a severe drawback in most other studies of nitrogen cycling in agroecosystems. We have, however, attempted to determine all major, and some minor, loss processes in the field.

Leaching

Since the nitrate concentrations in drainage water were fairly stable, with characteristic levels for each cropping system, the variations in yearly fluxes mostly reflected varying drainage volumes. Several other investigations showed a similarly strong interdependence between mass emission and drainage volume (e.g., Bolton et al. 1970, Letey et al. 1977).

Calculated nitrogen losses through nitrate leaching depended on which of the lysimeter systems were used for the determinations as discussed earlier. Two years after the large lysimeters had been installed, by which time the systems had stabilized, the B120 cropping system had the highest losses followed by B0, while GL and LL had the lowest losses (Tab. 6.8, Fig. 6.14). These differences arose mainly because in annual crops like barley there are usually considerable amounts of inorganic nitrogen in the soil at harvest in addition to that added by mineralization occurring during autumn.

From concentrations of commonly less than 1 mg N l^{-1}, nitrate increased to 10–20 mg N l^{-1} within a few months (Fig. 6.13). The nitrate concentrations in the drainage water even exceeded the concentrations of the B120 cropping system. Expressed as flux, the nitrate losses reached 4.2 g N m^{-2} during the 20 wk following ploughing of the grass leys.

A considerable increase in nitrate content of the soil owing to autumn mineralization of ley crop residues also occurred (Bergström 1986). Cameron and Wild (1984) found that about 10 g N m^{-2} of nitrate had leached down to levels below 0.9 m over two winters as a result of ploughing grassland, further supporting the belief that a large increase in mineralization is to be expected after ploughing a ley.

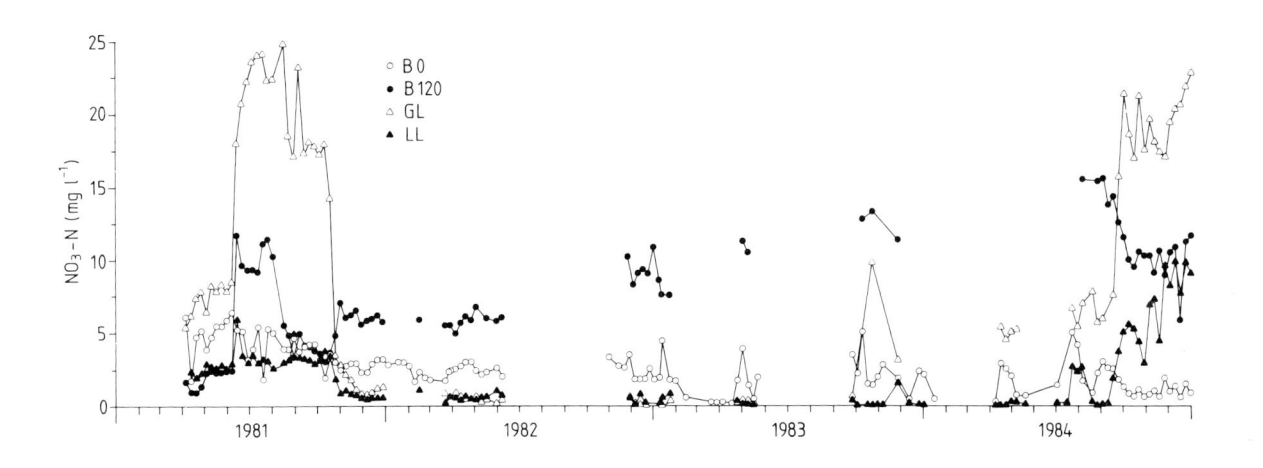

Fig. 6.13. Nitrate-N concentrations collected from the large lysimeters (Bergström 1987a).

These sources of nitrogen are usually very susceptible to leaching (Bergström and Brink 1986). Several leaching studies comparing different cropping systems have come to similar conclusions. Once a perennial ley is established, there is reason to expect a minimum of nitrate leaching (e.g., Bolton et al. 1970, Kolenbrander 1981, Gustafson 1983).

Although total nitrate leaching from the B120 cropping system could reach as high as 3.6 g m^{-2} during a single year, the part derived from fertilizer can still be low. The relative amount of the yearly fertilizer added that is leached out of the system is very variable and is mainly regulated by rainfall and percolation in the period following fertilization. A maximum of 1.2% of a single application of ^{15}N-labelled fertilizer to barley was recovered in drainage water during the three years after fertilization, which is equivalent to ca. 0.15 g m^{-2} (Bergström 1987b). However, extremely dry weather conditions following ^{15}N application limited leaching substantially, and a great deal of the labelled fertilizer was either taken up by the crop or immobilized in organic-N pools rather than being leached out (Lindberg et al. 1989).

Denitrification

Denitrification rates were measured in the field during 1982 (GL, B120) and 1983 (GL, LL, B120). To estimate annual losses, the rates measured at 15°C were recalculated based on the actual mean soil temperature for the days of sampling, using different Q_{10}-values for different temperature ranges (Svensson et al. 1990). The temperature-corrected rates were then subjected to a nonparametric resampling procedure ("bootstrapping") according to Efron (1979). Separate calculations were made for within and between row samples in each crop. Single rate measurements were randomly selected, for

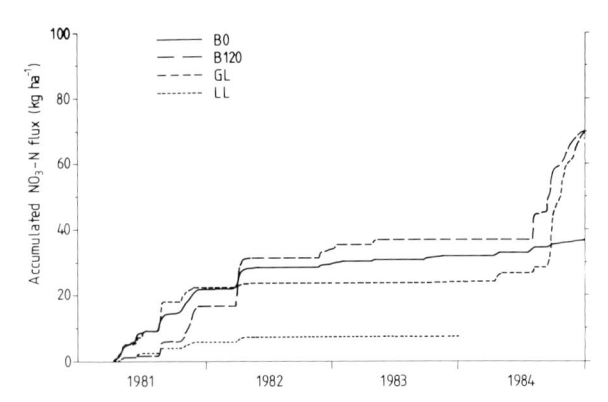

Fig. 6.14. Accumulated nitrate-N fluxes (kg ha^{-1}) from the large lysimeters during 1981–1984 (Bergström 1987a).

each sampling date, and then integrated over the season, assuming straight-line segments between the points. This procedure was iterated 10,000 times. From the frequency distribution of the integrated estimates, means and confidence intervals were calculated (c.f. Efron and Tibshirani 1986). To account for the denitrification occurring below the 10-cm depth (i.e., down to 27 cm, which is the mean border between the topsoil and the sand layer), the ratios between denitrification rates in the upper 10 cm and those below, obtained by ANOVA tests for differences in depth, were used (Svensson et al. 1990). The mean seasonal rates and the estimated total rates are given in Tab. 6.9 and 6.12, respectively.

Especially for the leys there was a pronounced difference between denitrification rates in samples taken within plant rows and those taken between rows, the former being higher. In laboratory studies (Klemedtsson et al. 1987a, b), it was shown that growing roots stimulated denitrification, whereas just the presence of

Tab. 6.8. Leaching losses from the different lysimeters in the four cropping systems during 1981–1984 according to Bergström (1987a): transport of NO$_3^-$ by drainage water measured in full-scale, tile-drained plots (TDP) and with three types of lysimeters (LLY, MLY, and SLY). The estimates for the medium and small lysimeters (MLY and SLY) are mean values ± SE for the replicates (Bergström 1987a).

Method	Treatment	1981	1982	1983	1984
			g N m^{-2} yr^{-1}		
Tile-drained plots (TDP)	B0	2.3	0.7	0.1	0.28
	B120	2.7	1.4	0.02	0.76
	GL	1.7	0.46	0.02	0.71
	LL	0.82	0.56	0.0	0.24[a]
Large lysimeters (LLY)	B0	2.3	0.81	0.22	0.48
	B120	1.5	1.6	0.28	3.6
	GL	2.3	0.14	0.04	4.5
	LL	0.61	0.16	0.01	1.4[a]
Medium lysimeters (MLY)	B120 (n=2)		0.18±0.04[b]	0.66±0.10	0.60±1.7
	GL (n=3)		0.01±0.01[b]	0.02±0.01	1.54±1.7
Small lysimeters (SLY)	B120 (n=3)		0.05±0.03[b]	0.44±0.14	1.3±1.0
	GL (n=6)		0[b]	<0.01<±0.01	0.50±0.13
	GLX[c] (n=3)		0[b]	<0.01<±0.01	<0.01<±0.01

[a]Changed to the B120 treatment during May 1984.
[b]Values refer to the period 1 Jul to 31 Dec 1982.
[c]GLX = grass ley that remained unploughed during the experiment.

a larger root biomass did not. The difference in denitrification rates between the annual barley crop and the perennial grass leys can largely be explained on the basis of this finding. When the N-fertilizer was applied to both crops at the end of May, the barley was just sown, while the grass was actively growing. A great increase in denitrification rates therefore occurred in the latter. The development of the root biomass in the barley coincided with dry periods in both 1982 and 1983 (June-July). This circumstance together with a smaller actively growing biomass in the initial phase of establishment of the barley may explain the less pronounced effect of fertilization in this crop as compared with the grass ley. The lower seasonal rate for barley during 1982 for within plant row samples as compared with 1983 (Tab. 6.9) was probably due to the drier conditions in 1982.

During 1982, total rates were about the same in the grass ley and the fertilized barley (0.5 and 0.3 g N m^{-2} yr^{-1}, respectively). The losses were about twice as high (1.2 g N m^{-2} yr^{-1}) for the grass ley in 1983 and the fertilized barley (0.6 g N m^{-2} yr^{-1}). The lucerne ley showed the highest denitrification losses of the three cropping systems (1.5 g N m^{-2} yr^{-1}).

In laboratory investigations (Klemedtsson et al. 1987a, b), an increase in biomass associated with active growth seemed to lead to an increase in denitrification rates. This was also observed to occur in connection with the fertilization of the perennial grass ley, especially in 1983 (Fig. 6.9). The lack of growing roots in the newly sown barley fields at the time of fertilization was probably the reason why the same effect was not seen there. Since the grass ley roots were concentrated in the surface layer (Hansson and Andrén 1986), differences in denitrification rates with depth in 1983, i.e., six times higher in the top 10 cm compared with the layer below (Svensson et al. 1990), could probably be attributed to the vertical root distribution. Thus, a relatively high level of heterotrophic activity should occur in the upper layer of this soil throughout the year. The growing period for roots is shorter and the rate of increase in root density lower in barley than in grass leys, and consequently, annual heterotrophic activity should be lower for the former crop. The upper layer will also be the one receiving precipitation, and thus anaerobic conditions should be more easily invoked within the perennial grass than in the annual barley. Although no direct correlation with rain events was observed during 1982/83, this effect may have caused the level of denitrification in the meadow fescue grass to be higher than that in the barley, which has a shorter growth period.

The difference in denitrification rates between the two successive years in the grass ley may partly have been due to the between-year difference in the soil water content. The moisture content 15 cm below the surface was about the same during spring and summer in 1982 and 1983, whereas moisture was higher in autumn 1983 (Alvenäs et al. 1986). The autumn values given by Svensson et al. (1990) for within-row samples seemed to be higher in 1983 than in 1982, although the variation was great and no statistical analysis was made. The same seemed to be valid for the period following the spring fertilization up until the drought during July and August, although no obvious differences were observed in soil moisture. The between-row rates tended to be lower in 1983 than in 1982, which is also reflected in the means (Tab. 6.9).

The gaseous losses corresponded to 3 and 5% of the nitrogen supplied as fertilizer or 14 and 43% of the total losses in the barley for 1982 and 1983. The grass ley showed a wider range of losses, i.e. 2 to 6% of the nitrogen supplied, while gaseous losses represented 50 and 71% of the total losses for the two years. The nitrogen fixed and built into the above-ground biomass of the lucerne ley was reported to be 31.9 g N m^{-2} yr^{-1} for 1983 (Wivstad et al. 1987). Another 13.0 g N m^{-2} yr^{-1} was calculated to have been incorporated in belowground biomass (Paustian et al. 1990). About 80% (or 36.7 g N m^{-2} yr^{-1}) was estimated to be biologically fixed nitrogen. The denitrification rate thus corresponded to less than 4% of the biologically supplied nitrogen. Denitrification accounted for about 75% of the total losses in the lucerne ley during 1983.

The denitrification rates in the different cropping systems and their relation to fertilizer supply for annual and perennial crops at Kjettslinge were about the same as those obtained for similar crops in other parts of the world (Tab. 6.10). The differences between the annual and perennial crops can be attributed to corresponding differences in several factors affecting the oxygen ten-

Tab. 6.9. Mean seasonal denitrification rates (g N m^{-2} yr^{-1}) and 95% confidence intervals (C. I.) during 1982–83 in three of the cropping systems in Kjettslinge, obtained by "bootstrapping" (Svensson et al. 1990). Note that within-plant row samples for barley refer to the time elapsed between appearance of the plant and ploughing of the fields while the between plant row samples and all samples of the grass ley and lucerne ley are based on the frost-free period of the year.

Cropping system	Year	Within-rows mean	C.I.	Between-rows mean	C.I.
B120	1982	0.109	0.038–0.252	0.189	0.076–0.371
	1983	0.286	0.094–0.578	0.202	0.065–0.692
GL	1982	0.325	0.084–1.049	0.165	0.066–0.362
	1983	1.718	0.842–3.093	0.061	0.030–0.114
LL	1983	2.279	1.185–4.923	0.276	0.135–0.543

sion in the soil, i.e., moisture, root density and distribution over time. These factors combined with the supply of nitrate will thus affect the rates of gaseous flow by denitrification differently.

The values above indicate that the losses of plant-available inorganic nitrogen (including organic-N mineralized) due to denitrification, are of minor importance from an ecosystem point of view. Denitrification may, however be of greater importance during years when moisture levels are higher throughout the unfrozen season. However, compared with the other types of gaseous losses, denitrification losses were the most important.

Nitric oxide losses

Nitric oxide production rates were studied at Kjettslinge using a cuvette technique (Johansson and Granat 1984). A cylindrical box equipped with a stirrer blade was placed over the soil. The emission rate was calculated as the linear increase in the concentration of NO within the first few minutes after installation over the soil. Several aspects concerning the validity of the method have been discussed earlier (Johansson and Granat 1984, Johansson et al. 1988, Johansson and Sanhueza 1989).

The spatial variability in the NO emission rate was around 25% (c.v.) for unfertilized areas (B0 and LL),

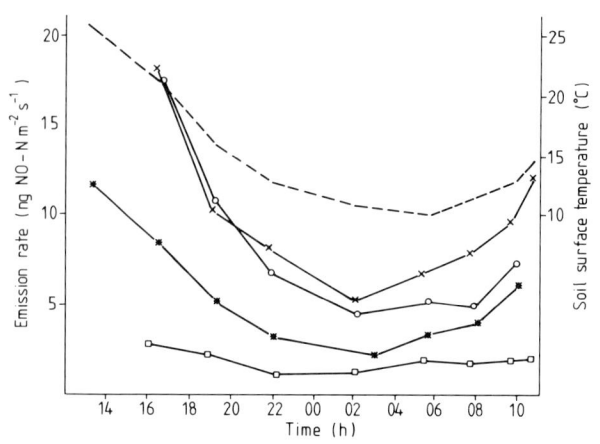

Fig. 6.15. Diel variations of NO emission rates in the four cropping systems. Dashed line shows the soil temperature (Johansson and Granat 1984).

whereas emissions from a recently fertilized grass ley varied more than a factor 10. The large spatial variability on recently fertilized areas was probably due to an uneven distribution of fertilizer pellets. However, much higher spatial variation was observed for denitrification in the Kjettslinge soil over the whole season (see above).

The NO fluxes were found to vary dielly (Fig. 6.15). The temperature dependence of emission rates from

Tab. 6.10. Total denitrification rates at Kjettslinge compared with annual estimates from other field studies and different N-fertilization practices. The time of integration is also included.

Cropping system	N-fertilizer used	N-application rate (g N m^{-2})	Time period (d)	Denitrification rate (g N m^{-2})	% of N added	References
Non-irrigated soils						
Corn	urea	0	85	0.16	–	Duxbury and McConnaughey (1986)[a]
		12	85	0.20	1.7	–
		12	85	0.24	2.0	–
Corn	(NH$_4$)$_2$SO$_4$	20	120	0.45	2.5	Mosier et al. (1986)[b]
Barley	(NH$_4$)$_2$SO$_4$	20	100	0.15	1.0	
Sugar beet	NH$_4$NO$_3$+urea	18	210	1.2	6.6	Benckiser et al. (1986)[a]
Ryegrass	NH$_4$NO$_3$	25–50	–	0.9–4.3	5–8	Ryden (1985)[a]
Grass sward	NH$_4$NO$_3$	25–50	365	1.1–2.9	4–6	Ryden (1983)[a]
Grassland	NH$_4$NO$_3$	21	21	0.21	1.0	Colburn and Dowdell (1984)[a]
Wheat+fallow[d]	urea	7.5	2 years	1.8	15	Aulakh et al. (1984)[a]
Fallow+wheat[d]	urea	7.5	2 years	1.9	16	–
Wheat+wheat[e]	urea	12	2 years	2.1	18	–
Wheat+fallow[e]	urea	7.5	2 years	4.9	60	–
Barley	Ca(NO$_3$)$_2$	12	200	0.3–0.6	3–5	Svensson et al. (1990)[a]
Grass ley	Ca(NO$_3$)$_2$	20	200	0.5–1.2	3–6	–
Lucerne	N$_2$-fix.	40	200	1.5	4	–
Irrigated soils						
Fallow	(NH$_4$)$_2$SO$_4$	20	60	2.0	10	Hallmark and Terry (1985)[a]
	Ca(NO$_3$)$_2$	20	60	1.9	10	–
Ryegrass	KNO$_3$	30	20	2.8	9	Colburn and Dowdell (1984)[a]
Vegetables	[c]	29–67	180	9.5–23	14–52	Ryden and Lund (1980)[a]

[a]Measured by means of C$_2$H$_2$-technique.
[b]Measured by means of N-15-technique.
[c]Various sources of nitrogen.
[d]Conventional till.
[e]Zero till.

fertilized soils, as described by an activation energy or Q_{10} value, was similar to that reported for N_2O production in the literature (see Johansson and Granat 1984). This strong temperature dependence suggests that production is a result of biological processes. It should be noted, however, that a clear covariation with temperature was only observed under favourable conditions such as those occurring in recently fertilized soil. The degree of temperature dependence was much less in the unfertilized barley (B0) and lucerne (LL). Similar diel patterns in NO and N_2O emissions have been reported by Slemr and Seiler (1984). The weak temperature dependence phenomenon has also been observed in natural, uncultivated soils (Johansson 1984, Johansson et al. 1988).

The relative importance of denitrification and nitrification for NO production was studied in a fertilization experiment at Kjettslinge (Fig. 6.16). It was shown that the application of nitrate (added together with glucose) and ammonium fertilizers produced an almost immediate increase in emission rates.

The highest fluxes were observed in the nitrate-fertilized areas and in an ammonium fertilized area where the plants had been cut. During the second day after fertilization the flux was larger from ammonium-treated soil, where the plants were cut, than the flux from the nitrate plus glucose treatment. The plant-covered NH_4^+ treatments resulted in NO losses half the size of those where plants were cut. Only about 0.3% of the applied nitrate-N was lost as NO and most of this loss occurred within the first few days after the fertilizer application. This is two orders of magnitude less than the fraction of nitrate lost as NO under anaerobic conditions in the laboratory study (Johansson and Galbally 1984), indicating that a substantial fraction of the NO produced within the soil is consumed and does not escape to the atmosphere. Similar losses have been reported for other nitrate fertilized soils, whereas larger losses, of up to 5.4%, have been reported for urea fertilization (Slemr and Seiler 1984). Because of the mineralization of urea, it should give rise to NO-fluxes similar to additions of NH_4^+.

It was also shown that emission rates of N_2O were about 100 times higher than those of NO (Johansson and Granat 1984). Slemr and Seiler (1984) found that. losses of NO plus NO_2 from an area fertilized with urea were about 30 times higher than the N_2O losses. The difference in the relative proportions of NO_x and N_2O lost between these two experiments is most likely due to differences in the type of additions: nitrate together with glucose, as used in our comparative study, will stimulate denitrification, whereas urea would stimulate nitrification. These results seems to favour nitrification as being responsible for NO production rather than denitrification in these soils.

Tab. 6.11 summarizes measurements of NO emission from cultivated land. Ranges are very large, especially on fertilized areas, with fluxes ranging from negative

Tab. 6.11. Measurements of NO emissions from cultivated soils as compiled by Johansson (1989).

Climate/System Location	Past use	Time period	Flux (ng N m^{-2} s^{-1}) range	mean	Reference
Temperate/Grassland Australia	grazed	Nov to May	1.5–7.3	3.5	Galbally and Roy (1978)
Temperate/Cropland Sweden (Kjettslinge)	unfertilized	Apr to Sep	0.3–17	0.6[a]	Johansson and Granat (1984)
Temperate/cropland Sweden (Kjettslinge)	fertilized	Apr to Sep	0.1–62	1.9[a]	Johansson and Granat (1984)
Subtropical/Cropland Spain		Sep Oct	–2–250		Slemr and Seiler (1984)
Temperate/Grassland Australia	grazed	Apr	0.6–124		Galbally et al. (1985)
Temperate/Cropland United States		June Jul	–9–28		Delany et al. (1986)
Temperate/Flooded rice Australia	fertilized	Dec	<0.2–0.95		Galbally et al. (1987)
Temperate/Cropland United States	unfertilized	May to Jun	0.003–67	1.7[a]	Anderson and Levine (1987)
Temperate/Croplands United States	fertilized	1 yr		6.7[a]	Anderson and Levine (1987)
Temperate/Pasture United Kingdom	fertilized	Jul to Aug	0–36	8	Colburn et al. (1987)
Temperate/Sward United Kingdom	unfertilized	Jul to Aug	–12–26	0.5[b]	Colburn et al. (1987)
Temperate/Cropland United States	unfertilized	Aug	0.2–3.8	1.2	Williams et al. (1988)
Temperate/Cropland United States	unfertilized	Aug	1.6–338	94	–

[a]Weighted yearly average.
[b]Calculated as NO_x–N (i.e. includes uptake of NO_2).

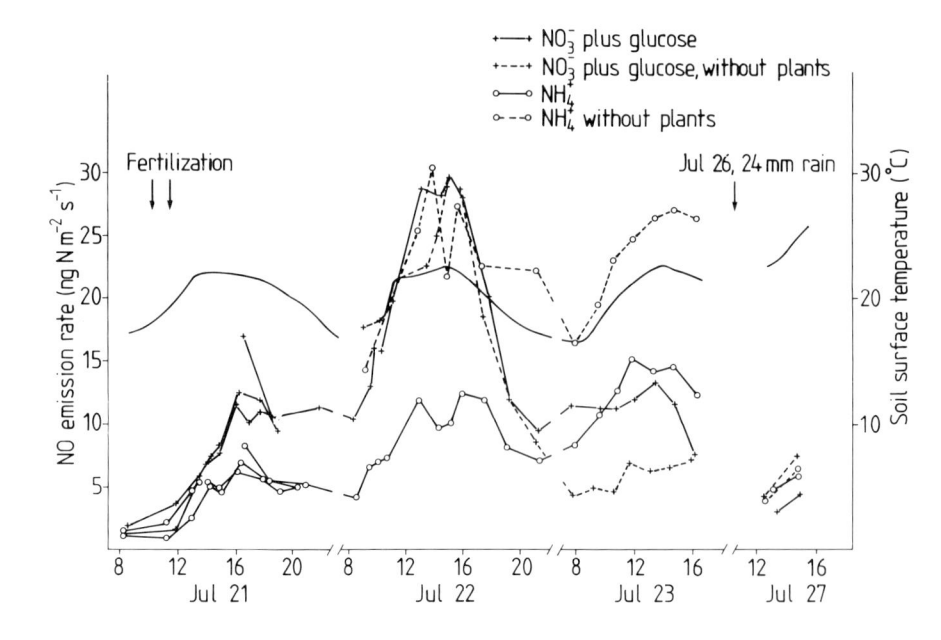

Fig. 6.16. Nitric oxide emission rates from field plots at Kjettslinge fertilized with NH_4^+ or NO_3^-. Glucose was added to the nitrate treated plots to stimulate denitrification. Soil temperatures are represented by the line without symbols (Johansson 1988).

vaues to 250 ng N m^{-2}s^{-1}. The highest fluxes occur within a day or two of the fertilizer application, whereas the negative fluxes of NO (i.e., uptake by soil or plants) are associated with relatively high NO concentrations in the atmosphere above the surface (Galbally et al. 1985).

It is difficult to extrapolate the data to obtain estimates of yearly losses for NO because of the relatively short measurement periods. Johansson and Granat (1984) obtained an approximate estimate of the yearly emission by using the weighted average based on day/night emission rates and different soil temperatures, assuming a zero emission during periods when the soil was frozen. It was thus estimated that the grass ley lost 0.06 g NO-N m^{-2} yr^{-1} and the unfertilized barley 0.02 g NO-N m^{-2} yr^{-1}. The fertilized barley was estimated to lose as much as the grass ley, while the lucerne ley lost only as much as the unfertilized barley.

Ammonia losses

Ammonia emissions were not considered to be important at the Kjettslinge field, because nitrate fertilizers were used and no manure was added. It was, however, felt that lucerne litter, with its high N content, might cause a local increase in pH during decomposition in the autumn, which could lead to gaseous ammonia losses.

It is difficult to determine the NH_3 emission from soil without disturbing the evaporation rate. A new technique, based on the measurement of the increase in the horizontal NH_3 flux downwind from the experimental plot, was therefore tried (Ferm 1983). The method does not affect the volatilization rate, but the flux increase must be high in comparison with the background flux (the flux of NH_3 entering the experimental plot from the surroundings). The last requirement was, however, not met at the Kjettslinge plots, since $Ca(NO_3)_2$ was used as fertilizer, and the NH_3 emission was small. The technique for measuring the horizontal NH_3 fluxes to estimate the NH_3 loss, which was initiated within the project, has later been improved (Ferm 1986). Two field measurements on areas to which cattle slurry had been applied have shown that the technique may be a good alternative to the "gradient" technique (Ferm and Christensen 1987).

The technique most commonly used to measure NH_3 losses to the atmosphere is the "chamber" technique. After initial attempts to use the horizontal flux technique, it was decided to determine NH_3 losses with a chamber method. The method was tested with respect to its influence on the volatilization rate from the surface with vegetation. It was found that the emission rate was proportional to the ventilation flow of air through the chamber. The NH_3 concentration inside the chamber was consequently constant and independent of the ventilation flow. This represents the equilibrium concentration of NH_3 in air above the soil, which is an important parameter necessary for the understanding of

Tab. 6.12. Removal of nitrogen (g N m^{-2} yr^{-1}) from the cropping systems at Kjettslinge. The denitrification rates are from 1983 (Svensson et al. 1990), NO emission rates from Johansson and Granat (1984), NH_3 emissions from Ferm (1983), leaching and harvest from Chapter 7.

Process	Cropping system			
	B0	B120	GL	LL
Harvest	3.4	12.4	24.1	24.6
Denitrification	0.5	0.5	1.2	1.5
NO-emission	0.02	0.06	0.06	0.02
NH_3-volatilization	0.06	0.12	0.26	0.45
Leaching	1.0	1.8	0.1	0.1
Total removal	5.0	14.9	25.7	26.7

the rate-determining process of the emission (Ferm 1983). Furthermore, the equilibrium concentration can be used together with wind speed measurements for making rough estimates of the NH_3 emission from an undisturbed soil surface.

Three transport mechanisms for NH_3 in soil have been described in the literature, viz., diffusion of NH_3 through the air-filled pores of the soil, diffusion of NH_3 and NH_4^+ in the aqueous phase and convection of the soil solution. Data from one measurement of the equilibrium concentration as a function of soil depth together with water content of the soil were used in equations describing transport rates by the different mechanisms. The NH_3 fluxes in the soil due to diffusion were directed downwards and set equal to 1.5 mg N m^{-2} h^{-1} (liquid phase) and 0.014 mg N m^{-2} h^{-1} (gas phase). The corresponding value for transport by convection was 4.8 (upwards) while the loss to the atmosphere was 0.5 mg N m^{-2} h^{-1}. Net transport in the soil to the surface was consequently higher than transport through the surface, indicating that NH_4^+ accumulated at the soil surface. The data clearly showed that NH_3 was mainly transported as NH_4^+ with the soil solution (convection) during the upward movement owing to evapotranspiration processes (Ferm 1983). The transport resistance in the soil is very low in comparison with the resistance at the air-soil interface, which determines the volatilization rate. The transport rate through the air-soil interface depends on wind speed (more specifically, the surface-transfer coefficient) and the NH_3 equilibrium concentration minus the NH_3 concentration in the air. Thus, the chamber technique does not directly measure emission rates, since the rates are also a function of meteorological variables.

The studies showed that NH_4^+ transport due to convection of soil water can be important and that there will be an accumulation of NH_4^+ towards the soil surface. The soil moisture will thus be important for regulating the NH_3 loss in two ways: the upward movement transports NH_4^+ to the soil surface, and the moisture content at the surface influences the NH_3 equilibrium concentration.

The neutral to slightly acid pH of the topsoil (pH 6.3) would not be expected to favour NH_3 losses, and only low emission rates were measured. Ferm (1983) assumed that convection of the soil solution was the main factor responsible for the ammonia transport to the boundary layer, and that this process is not rate determining. The rate-determining step is the diffusion of NH_3 through the laminar boundary layer. The emission of ammonia thus resembles the evaporation of water in this respect. Because water loss from the soil surface is much higher during daytime than at night, annual losses were calculated based on the period from one hour after sunrise to sunset (about 60% of the time between April and October). Ferm (1983) estimated losses of 0.06, 0.12, 0.26 and 0.45 g NH_3-N m^{-2} yr^{-1} for B0, B120, GL and LL, respectively. Ammonia volatilization thus accounted for 3–22% of the total gaseous losses. The relatively low contribution by ammonia can be attributed to the low NH_4^+ content in the soil compared with other forms of nitrogen. When manure, cattle slurry or urea is applied, NH_3 is rapidly lost to the atmosphere (1–3 d, e.g., Vallis et al. 1982) with losses on the order of 25–50% of the applied urea- or ammonium-N (Beauchamp et al. 1982, Catchpoole et al. 1983, Ferm and Christensen 1987).

Total losses

The total estimated non-harvest losses for the four cropping systems have been summarized in Tab. 6.12. The losses were small, with only 1.6, 2.5, 1.6 and 2.1 g N m^{-2} yr^{-1} for B0, B120, GL and LL, respectively. In the fertilized barley, losses amounted to 20% of the nitrogen added as fertilizer, while 8% was lost in the grass ley. Of the nitrogen transported from the B0 cropping system, 68% was in the form of harvest products, the remainder being lost from the system. It should be noted that in 1982 and 1983, when the most intensive field measurements were carried out, the weather was comparatively dry (Chapter 2), and because losses are regulated to a large extent by soil moisture, it can be assumed that considerably more nitrogen would be lost during wet years.

Management impacts on nitrogen losses

Management strategies in crop production aim at optimizing the plant uptake of mineral-N. However, the farmer also wants to minimize losses of nitrogen from the agroecosystem, which are regulated by inorganic nitrogen transformations. Nitrogen losses from areas under agricultural production are not only a waste of resources for the farmer, but are also a matter of serious environmental concern. Both losses by leaching and losses to the atmosphere must be minimized in present-day agriculture.

At the local and regional levels, contamination of surface water and groundwater is a common and increasingly serious problem. Fertilizer-N is often considered to be the most important factor contributing to the elevated nitrate levels in water, primarily due to the large surpluses of N applied (Baker and Johnson 1981, Bergström and Brink 1986). Under Swedish conditions, nitrate leaching from arable land occurs mainly during the autumn (September – November), a period characterized by high precipitation and low evapotranspiration (Bergström and Brink 1986). It has commonly been assumed that the nitrogen leached is mainly a consequence of commercial fertilizer applications, but in fact a considerable proportion of the losses may come from N mineralized during the autumn. In Kjettslinge, losses also occurred from the unfertilized barley (Bergström 1987a), and breaking of the unfertilized lucerne ley resulted in rapid mineralization and a subsequent increase in the soil mineral-N content (Bergström 1986).

Under the conditions occurring during 1983, which were dry, the total gaseous N-losses were estimated to be 28% for B120, 41% for B0 and 95% for the GL and LL of the total losses excluding removal by harvest. Denitrification, measured as N_2O- and N_2-release, contributed up to 20% in B120, 29% in B0, 68% in LL and 75% in GL. This process thus dominated the gaseous N-losses from the annual crops and the total N-losses from the leys. The gaseous N-losses for Swedish agriculture estimated to be 60% of the total losses by Jansson and Simán (1978) thus may be justified. From the variations shown between the crops of the Kjettslinge field, it is obvious that various proportions of crop-utilization and certainly soil conditions will influence such an estimate. It should also be noted that other soil systems may affect the estimate considerably.

In agriculture, any management practice that affects the soil oxygen status will have a fundamental impact on the fate of nitrogen in the ecosystem. Soil texture is also an important factor, since it influences both soil aeration and water flow rates. It is evident that sandy soils are more susceptible to nitrate leaching than heavier soils. Extensive studies in Denmark, reviewed by Hansen (1983) and Hansen and Aslyng (1984), showed that 6.5 g N m^{-2} yr^{-1} was lost from sandy soils compared with 4.0 g from clay soils (50% of the land was spring sown and the rest sown with autumn cereals). Denitrification losses, on the other hand, are generally smaller in coarsely textured soils (Nömmik 1956, Firestone 1982). Sandy soils may, however, denitrify heavily during simultaneous irrigation and fertilization (Ryden et al. 1979).

7. Ecosystem dynamics

Keith Paustian, Lars Bergström, Per-Erik Jansson and Holger Johnsson

7. **Ecosystem dynamics.** K. Paustian, L. Bergström, P.-E. Jansson and H. Johnsson

Introduction	155
Carbon and nitrogen budgets	155
Data base and calculations	155
Input/output balances	157
Internal fluxes	159
Carbon turnover	159
Nitrogen turnover	161
Comparisons with other agroecosystems	163
Simulation of nitrogen dynamics	165
Model description	166
Barley	168
Simulation results	168
Grass ley	171
Simulation results	172
Three crop rotations in southern Sweden	174
Simulation results	174
Long-term trends in nitrate leaching from an agricultural watershed	176
Simulation results	176
Summary	178

Introduction

The preceding chapters present an indepth view of the processes involved in carbon and nitrogen cycling in arable soil. In this chapter, the cropping systems are viewed at the ecosystem level of integration. Carbon and nitrogen flows through the plants, microflora, soil animals and non-living parts of the soil are analyzed using budgets and a simulation model.

One objective of the research project was to compile comprehensive C and N budgets for the four cropping systems. To accomplish this, data from the two most intensive years of field study, 1982 and 1983, were synthesized to derive annual budgets (Paustian et al. 1990). Besides quantifying fluxes into and out of the cropping systems, the budgets provided a concise format for making between-system comparisons.

While annual budgets provide an overview of mean flux rates, the influences of such factors as temperature, moisture and water flow on the dynamics of mineralization, immobilization, plant uptake, leaching and other processes are not addressed. To examine the dynamic behaviour of nitrogen in the cropping systems, an ecosystem-level simulation model was developed. Simulations were performed for unfertilized and fertilized barley and the grass ley (Johnsson et al. 1987, Bergström and Johnsson 1988).

The simulation model was also applied to other locations and cropping systems in Sweden. In one application, the model was used to analyze mineral N dynamics in different crop rotations at three research fields in southern Sweden (Jansson et al. 1987). In a second example, long-term trends in nitrate leaching from an agricultural watershed were evaluated in relation to changes in fertilization and cropping practices (Jansson and Andersson 1988).

Carbon and nitrogen budgets

A main objective in compiling the budgets was to quantify and compare total inputs and outputs of C and N from the four cropping systems. A unique feature of the budgets was that all major inputs and outputs of C and N from the cropping systems could be estimated from independent measurements and no assumptions of steady-state conditions for total C and N were invoked. In most other published agroecosystem budgets one or more input or output flows, for example gaseous N losses, have been estimated by mass balance assuming steady-state conditions (see reviews by Frissel 1977, Legg and Meisinger 1982).

Because of the large amount of C and N in most agricultural soils and its spatial heterogeneity, a reliable determination of changes in total soil C and N is generally feasible only in long-term field experiments. In the five year field study of the Kjettslinge cropping systems, trends in total C and N levels could be postulated from comparing inputs and outputs, but they could not be confirmed directly as significant differences between initial soil C and N levels and amounts present at the termination of the field experiment.

Another point to consider was the transitional nature of the leys. Ideally, budgets for the leys would have encompassed the entire four years from establishment through to ploughing and sowing of the new crop. In contrast, barley has a complete life-cycle within each year and is thus more suitable for representation on an annual basis. However, acquiring the full range of necessary data, for all the crops over a four-year time span was not feasible. Therefore, data from the most intensive period of field investigation, 1982 and 1983, were selected as the basis for annual budgets. By that time all treatments were well established and the leys had reached full maturity (Chapter 4).

Data base and calculations

The sources of data for the budgets are summarized in Fig. 7.1, which includes references to original papers for more detailed information. A complete description of the budget derivation is given by Paustian et al. (1990).

Organic C and N inputs and nitrogen uptake by the crop were estimated from field measurements of primary production, i.e., clipping and litter collection above-ground and root coring. An additional component of root production and turnover, i.e. rhizodeposition (root exudation, cell sloughing, etc.) could not be estimated from field studies. Therefore, the contribution of rhizodeposition to total root production and turnover was derived from results of a ^{14}C-labelling experiment performed in a growth chamber. The assumption used to extrapolate to field conditions was that the relative proportions of rhizodeposition to macro-root production measured in the growth chamber experiments were the same as in the field (where only macro-root production was measured). Nitrogen flows associated with primary production were derived from C fluxes and measured N concentrations in shoots, roots and litter.

Carbon output from soil as CO_2-C was measured several times per week in B120, GL and LL during

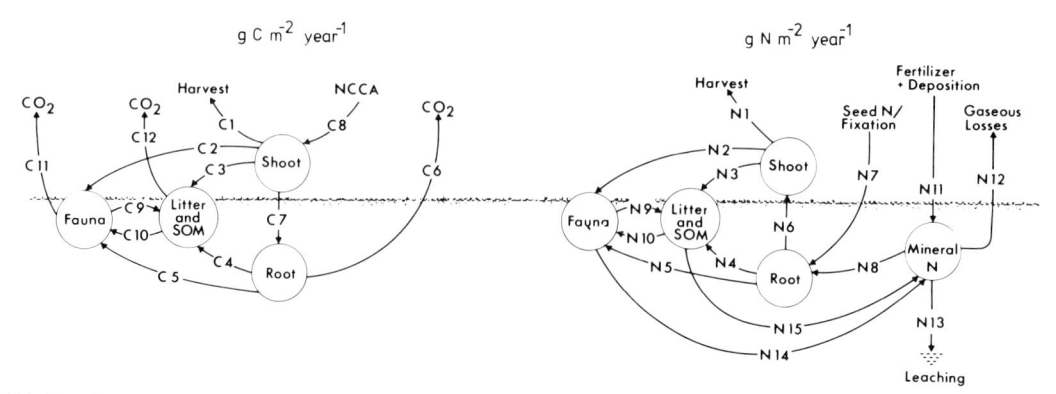

Fig. 7.1. Data base and methods used to calculate carbon and nitrogen budgets for the Kjettslinge cropping systems, shown in Figs 7.4–7.5. For clarity, more detailed calculations of fluxes (e.g., C and N flows through soil fauna and microorganisms, gaseous N losses) have been combined. Microorganisms have been included in the litter and soil organic matter (SOM) compartment.

Process	Flow no.	Determination method	Reference
Parallel carbon and nitrogen flows			
Harvest	(C1, N1)	Field measurement	Hansson and Andrén 1986
			Hansson et al. 1987
			Pettersson et al. 1986
Above-ground consumption	(C2, N2)	Respiratory metabolism and individual consumption rates	Curry 1986
Above-ground litterfall	(C3, N3)	Field measurement	Pettersson et al. 1986
			Pettersson 1987
Root litter input	(C4, N4)	Field measurement of macro-root input; ^{14}C labelling in growth-chamber	Hansson and Andrén 1986
			Hansson et al. 1987
			Pettersson et al. 1986
			G. Johansson unpubl.
Below-ground consumption	(C5, N5)	Respiratory metabolism	Sohlenius et al. 1988
			Paustian et al. 1990
Faunal production and defecation	(C9, N9)	Respiratory metabolism and earthworm life tables	Lagerlöf and Andrén 1988
			Lagerlöf et al. 1989
			Lagerlöf and Andrén in press
			Paustian et al. 1990
			Sohlenius et al. 1988
			Boström 1988a
Faunal consumption (excluding above-ground herbivory)	(C10, N10)	(Same as for C9, N9)	
Carbon flows			
Root respiration	(C6)	Extrapolated from root respiration with ^{14}C labelling in growth-chamber	G. Johansson unpubl.
C translocated below-ground	(C7)	C4 + C5 + C6	
Net canopy carbon assimilation	(C8)	C1 + C2 + C3 + C7	
Faunal respiration	(C11)	(Same as for C9, N9)	(Same as for C9, N9)
Microbial respiration	(C12)	Total soil respiration measurements minus root and fauna respiration (C6 and C11)	Paustian et al. 1990
Nitrogen flows			
N assimilation above-ground	(N6)	N1 + N2 + N3	
Seed N input/N$_2$-fixation	(N7)	Calculated from seeding rate (barley) fixation (lucerne) by ^{15}N dilution	Wivstad et al. 1987
Plant N uptake	(N8)	N4 + N5 + N6 − N7	
Fertilizer input plus deposition	(N11)	Field measurement (deposition)	Söderlund 1984
Gaseous losses	(N12)	Denitrification in soil cores NH$_3$ flux in field cuvettes NO$_x$ flux in field cuvettes	Svensson et al. 1990
			Ferm 1986
			Johansson and Granat 1984
Leaching	(N13)	Field measurement in lysimeters	Bergström 1987b
Faunal N mineralization	(N14)	From N consumption (N10) and group specific metabolic quotients	Paustian et al. 1990
Microbial net N mineralization	(N15)	Balance on mineral N pool (N8 + N12 + N13 − N11 − N14)	

April through November in 1982 and 1983, and the rates were integrated to derive annual estimates (K. Paustian, L. Klemedtsson and B. Svensson, unpubl.). Rates were measured in intact soil cores incubated (15°C) in the laboratory shortly after field sampling. The measured rates were adjusted for field temperatures, according to a series of Q_{10} functions, before integrating for annual estimates. Roots were present in the soil cores and thus the measurements represented total soil respiration, i.e., heterotrophic and root respiration. Root respiration and faunal respiration (Ch. 5) were independently estimated and thus microbial respiration was calculated as total soil respiration minus root and faunal respiration. Since soil respiration was not measured in B0, microbial respiration in B0 was assumed to be equal to that in B120.

Carbon and nitrogen flows through the fauna included consumption, respiration, production, defecation and N mineralization through excretion. Annual rate estimates for these processes were largely based on an energetic approach, whereby respiration was estimated from population variables and soil temperature data. The remaining flows were calculated on the basis of respiration rates, food selection and nitrogen content of the food consumed by the different soil fauna taxa. For earthworms it was possible to directly estimate production from population dynamics (including cocoon production) based on sequential samplings. Separate flow calculations were made for subgroups within the major categories of Protozoa, nematodes, micro- and macroarthropods, enchytraeids and earthworms (see Chapter 5).

The turnover of microbial biomass was estimated indirectly, by balancing net uptake from litter and soil organic matter with faunal consumption of microorganisms and carbon loss through respiration. Similarly, net N mineralization from microorganisms was determined as the difference between total net mineralization and N mineralization from the soil fauna.

Nitrogen losses via denitrification, NO_x emission and ammonia volatilization were estimated in the field studies of flux rates (see Ch. 6). Leaching losses of N, predominantly in nitrate form, were monitored continuously throughout the field experiment period, using a variety of methods (Chapter 2). Data from the large rubber lysimeters were used in the budgets.

Nitrogen inputs included atmospheric deposition (Söderlund 1984) and N added in seed and as fertilizer. Nitrogen fixation in lucerne was determined from ^{15}N-dilution studies where fixed-N was directly determined only in above-ground production. Budget values for total fixed N include N incorporated in below-ground production, where it was assumed that the measured fraction of N derived from N_2-fixation in above-ground plant parts was representative for the whole plant.

Input/output balances

There were clear differences between barley and ley crops in overall soil carbon balances (Fig. 7.2, Tab. 7.1). In barley, there was little net change in total C, indicating that C inputs from roots and crop residues roughly balanced soil respiration. In the leys, non-harvested C inputs by the plants exceeded respiration losses and there was a net accumulation of 140–150 g C m^{-2} yr^{-1}. This increase in total C mainly occurred as increased shoot, root, and surface litter standing crops, and there appeared to be little change in soil organic matter carbon (Tab. 7.1). Differences in soil carbon balances were mainly associated with crop type, i.e., annual vs perennial, and the amount and type of nitrogen supply, within crop types, had little effect on total soil C.

The major differences between the carbon balances of B0 and B120 were in net primary production allocated above-ground and in the subsequent harvest out-

Tab. 7.1. Carbon and nitrogen balances (g m^{-2} yr^{-1}) of the cropping systems at Kjettslinge for the top soil. Delta (Δ) values denote net annual changes in plant standing crop, soil organic matter and total C and N in the cropping systems. Seed N inputs in the leys were negligible. ND = not determined (From Paustian et al. 1990).

	Cropping system			
	B0	B120	GL	LL
Carbon balance				
Inputs				
Primary production	320	540	840	900
Outputs				
Harvest	110	300	360	350
Root respiration	60	60	100	110
Heterotroph respiration	170	170	250	290
Σ Outputs	340	530	700	750
Δ Total carbon	−20	+10	+140	+150
Δ Plant standing crop C[a]	0	0	+120	+180
Δ Soil organic matter C[b]	−20	+10	+20	−30
Nitrogen balance				
Inputs				
Atmospheric deposition	0.5	0.5	0.5	0.5
Seed N	0.5	0.5	negl.	negl.
Fertilizer	0	12	20	0
N_2-fixation	ND	ND	ND	38
Σ Inputs	1	13	21	39
Outputs				
Harvest	3.6	12.7	24.1	24.6
Gaseous losses	0.4	0.5	1.0	2.0
Leaching	0.5	1.0	0.1	0.1
Σ Outputs	5	14	25	27
Δ Total nitrogen	−4	−1	−4	+12
Δ Plant standing crop N[a]	0	0	+5	+9
Δ Soil organic matter N[b]	−4	−1	−9	+3

[a] Represents net annual accumulation of above-ground plant parts, surface litter and roots.
[b] Calculated as: Δ total C (or N) minus Δ plant standing crop.

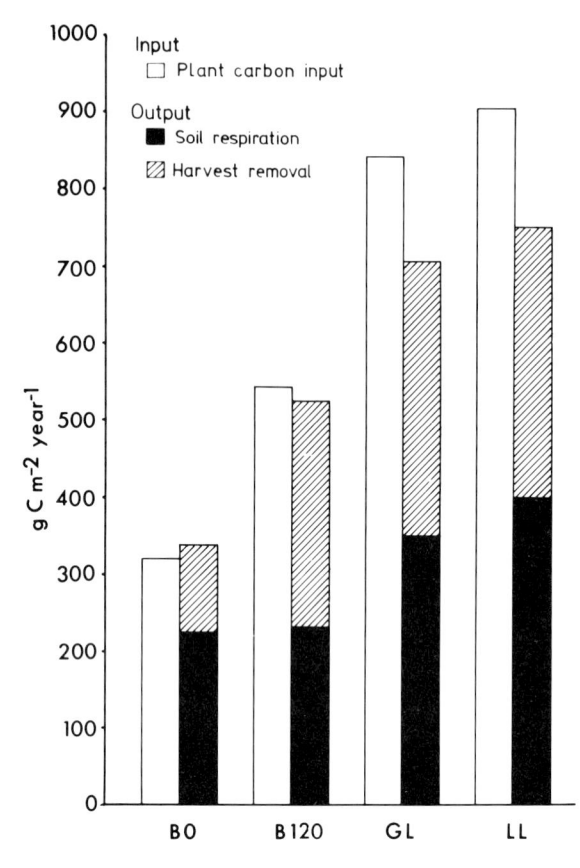

Fig. 7.2. Total carbon inputs and outputs in the annual carbon budgets for barley without N-fertilizer (B0), N-fertilized barley (B120), grass ley (GL) and lucerne ley (LL) at Kjettslinge. Values represent means for 1982–1983, for the top soil.

take (which included grain and straw). Otherwise, the organic matter inputs to the soil and the different components of soil respiration were similar (Tab. 7.1). The carbon balances of the leys showed a high degree of similarity, with nearly identical harvest out-takes and similar accumulations of total C (Fig. 7.2). Inputs of C to the soil were roughly equal, although the proportions of the input originating above-ground and below-ground were different in the two leys (see below). The main differences between the leys were that both annual net primary production and respiration were higher in LL than in GL (Tab. 7.1).

Overall nitrogen inputs and outputs and the resulting total N balances were less similar between cropping systems than was the case for carbon (Fig. 7.3, Tab. 7.1). Total annual nitrogen inputs, the main treatment variable along with crop type, ranged between 1 g N m^{-2} in B0 to almost 40 g N m^{-2} in LL. By far the greatest output of nitrogen in all the cropping systems was in the plant material removed at harvest. Total losses from gaseous efflux and leaching were modest, ranging from 1 to 2 N g m^{-2}. It is worth noting that in both of the fertilized systems, B120 and GL, nitrogen removed by crop harvest (including straw in B120) exceeded fertilizer input. Three of the crops, B0, B120 and GL, showed a net decline in total N, while nitrogen accrued in the lucerne ley (Tab. 7.1).

Not unexpectedly, total nitrogen declined more in the unfertilized barley than in the fertilized treatment (Tab. 7.1); numerous long-term field trials with cereals in Sweden (Jansson 1975, Persson 1980a) and elsewhere (Odell et al. 1984, Jenkinson et al. 1987) have shown similar results. Perhaps more significant was the high nitrogen recovery in the fertilized barley. According to ^{15}N studies, 59% of the applied fertilizer in B120 was recovered in the harvest alone (grain + straw). Using the difference method (i.e., comparing B120 and B0) fertilizer recovery was 59% in 1982 and 92% in 1983 (Tab. 5.5). Annual leaching losses during the same period (1982–83) averaged 1 g N m^{-2} and gaseous loss of N were less than 0.5 g N m^{-2}. Thus, the fertilized barley showed a high level of retention of the large fertilizer input during this period.

A comparison of nitrogen outputs reveals a number of similarities between the two leys. Nitrogen exported in harvest was nearly identical for GL and LL and total nitrogen losses were of similar magnitude (Fig. 7.3). However, other aspects of their nitrogen economies

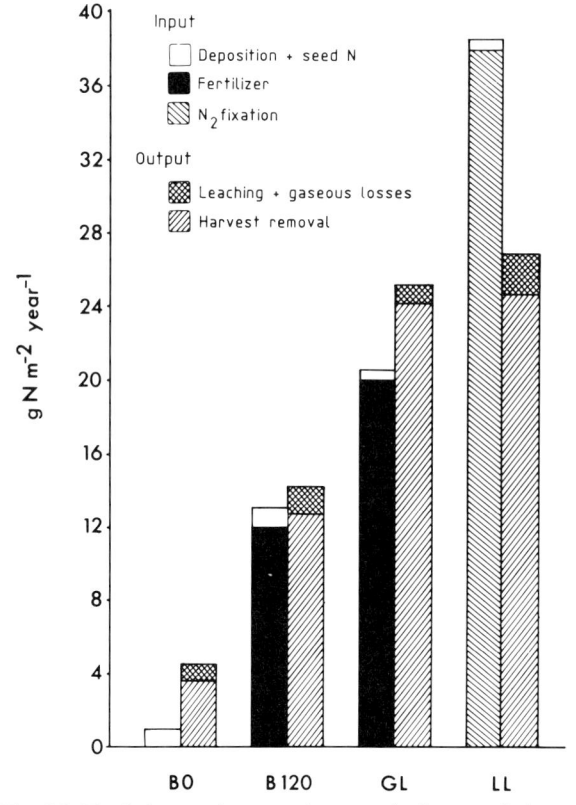

Fig. 7.3. Total nitrogen inputs and outputs in the annual nitrogen budgets for barley without N-fertilizer (B0), N-fertilized barley (B120), grass ley (GL) and lucerne ley (LL) at Kjettslinge. Values represent means for 1982–1983, for the top soil.

were very different. Input from N$_2$-fixation in lucerne substantially exceeded total outputs. However, most of the nitrogen accrual could be accounted for in the net increase in standing crops of roots, crowns, shoots and surface litter (Tab. 7.1), while the estimated increase in soil organic matter N was minor. In contrast, the grass ley showed a considerable deficit in the nitrogen balance between inputs and the amount exported plus that accumulated in plant biomass. Nitrogen in harvest exceeded fertilizer input by 4 g N m^{-2} and the total balance between inputs an outputs showed a deficit of about 4 g N m^{-2} (Tab. 7.1). As in the lucerne ley, there was a net accumulation in root, shoot and litter standing crops during 1982–1983. Considering the deficit in total nitrogen, this suggests that soil organic matter nitrogen in the grass ley declined by 9–10 g m^{-2} yr^{-1}, an amount equal to about 1% of the total soil nitrogen (Tab. 7.1).

Internal fluxes

Carbon turnover

When comparing crop species and levels of N input, there were important differences regarding the amounts, type and timing of organic matter input to the soil. These differences in inputs from the primary producers influence nitrogen mineralization and the supply of plant-available nitrogen, thereby acting as a feedback to primary production.

The total amounts of annual primary production made available to decomposers – in the form of dead roots, exudate/mucilage and sloughed cells from living roots, senescent leaves and stubble – were clearly divided into two groups: low C input (150–180 g m^{-2} yr^{-1}) in barley and high C input (260–270 g m^{-2} yr^{-1}) in the leys (Fig. 7.4). Carbon inputs originating from the roots made up 60% of the total inputs in barley and 75% in the grass ley. The relative contributions from above- and below-ground sources were approximately equal in lucerne.

There were also differences between the annual and perennial crops with respect to the timing of inputs. Organic matter input to the soil under barley occurred almost as a single pulse, at ploughing. With the exception of the small amount of senescent leaf fall and possible leaching of water-solubles from standing dead material, nearly all of the above-ground matter became accessible to the soil-borne decomposers only after harvest, and then mainly after ploughing, about one month after harvest. Inputs from barley roots, as sloughed cells, root exudate and dead fine roots, occur to some extent throughout the growing season (Martin 1987). However, the bulk of the barley root material became available after the plant matured and roots began to show senescence prior to harvest (Chapter 4).

The input of organic matter to the decomposers was somewhat continuous in the leys. Green leaves and stems were added to the soil surface as harvest spillage, which varied from negligible amounts to as much as 20 g C m^{-2} yr^{-1} (R. Pettersson, pers. comm.). Most above-ground material, however, became available to decom-

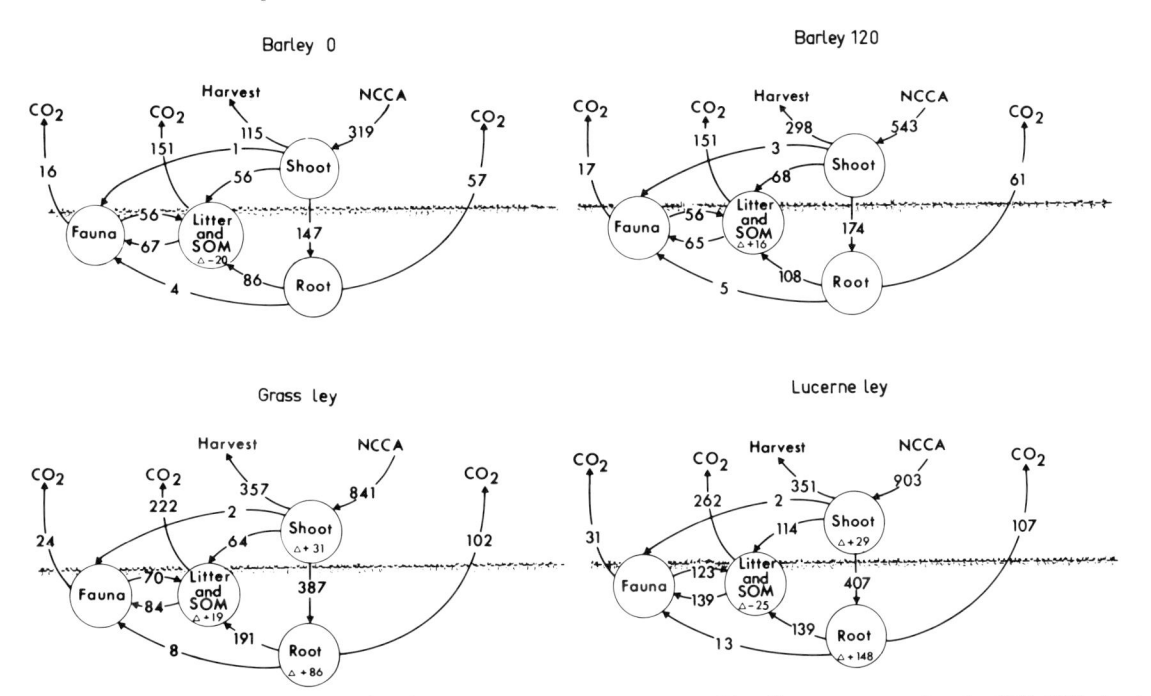

Fig. 7.4. Annual carbon budgets (g C m^{-2} yr^{-1}) for the four cropping systems at Kjettslinge, as mean values for 1982–1983, for the top soil. Net canopy carbon assimilation (NCCA) represents plant carbon uptake, excluding assimilated carbon which is respired above-ground. The delta symbol (Δ) denotes net changes over the year. See Fig. 7.1 for details on budget calculations (From Paustian et al. 1990).

posers only after plant senescence in autumn, and the major transfer of standing dead to the soil surface (followed by burial and comminution by soil fauna) occurred in late autumn and in the following spring (Chapter 4). The death and replacement of fine roots in the leys occurred at irregular intervals but increases in the dead root fraction were recorded primarily during the early part of growing season and to some extent from late autumn to the following spring (Chapter 4).

Decomposition rates determined from the litter-bag incubations (Chapter 5) and C mineralization rates estimated in ^{14}C growth chamber experiments (Chapter 5) can be used to estimate the relative contributions of the current year's input to total heterotrophic respiration (Tab. 7.2). The litter-bag incubations are most closely analogous to the pulsed input of organic matter in the barley, and are probably less representative for the leys. However, the litter-bag results incorporate differences in the decomposability of the litters and differences in abiotic factors between treatments and therefore provide a rough guide to first-year decomposition losses. The estimate for first-year mineralization of C in rhizodeposition was determined in conjunction with the labelling experiments designed to separate root respiration and microbial respiration components of the root-derived carbon measured in the growth chamber (See Chapter 5). The rapid turnover and low retention in soil for this part of the plant carbon input is supported by other studies of organic compounds introduced into arable soils (e.g. Kassim et al. 1981).

Despite the higher input of plant material in B120 compared with B0, similar amounts were respired, due to lower specific decomposition rates in B120 (Tab. 7.2). The drier conditions in B120 was the most likely reason for the difference in specific decomposition rates. In contrast, total inputs of plant material to the soil in the leys were similar, but measurements of soil respiration and budget calculations showed higher heterotrophic respiration in LL than in GL (Tab. 7.2). The higher respiration rates in LL were probably due to the higher moisture levels in the top soil, the greater input of surface litter and high decomposability of the lucerne root litter (Tab. 7.2). Very high first-year mass loss (91%) were found in the litter bag incubations of lucerne roots, which was attributed to their high content of water-solubles. Considerable amounts of these solubles may have been leached from the litter bags to the surrounding soil and thus the mass loss values may over-estimate carbon mineralization rates if some of the leached solubles were stabilized and not lost as CO_2. In addition, such high concentrations of water-solubles are probably not typical for lucerne roots during most of the growing season (Andrén 1987). Therefore, the more conservative estimate for first-year C mineralization of root material, derived from the incubation studies, was used in the calculation (see Tab. 7.2).

The partitioning of heterotrophic respiration and organic matter decomposition suggests that the CO_2-C flux from older organic matter, i.e., litter remaining from preceding years and soil organic matter, was nearly the same in all treatments (Tab. 7.2). These carbon mineralization rates corresponded to 0.7–1.1% of the mean total soil C contents (Tab. 2.5).

In relative terms, CO_2-C derived from current year inputs were very similar between the barley crops and between the leys, respectively (Tab. 7.2). Overall, it appears that differences in the amount and decomposition rate of organic matter additions and the prevailing moisture conditions can explain most of the differences in heterotrophic respiration in the different treatments.

The measurements of soil respiration and the estimates of heterotrophic respiration provide evidence for higher microbial activity in the leys compared with barley. However, there were no clear trends in direct-count measurements of biomass that could be used to estimate microbial production and turnover (Chapter 4). The difficulty in estimating production from microbial biomass measurements is attributable to the dormant state of much of the biomass (Clark and Paul 1970, Jenkinson et al. 1981) and when growth does occur, production and destruction of biomass (through consumption or lysis) can occur simultaneously. However, indirect estimates of microbial production can be made on the basis of the respiration carbon fluxes calculated in the bud-

Tab. 7.2. Partitioning of heterotrophic respiration (Q) into that derived from current year organic matter input and that derived from older soil organic matter, expressed as g C m^{-2} yr^{-1}. Values in parentheses denote fractional mass loss after one year.

Cropping system	Organic matter input (and fraction 1st year loss)			Total Q (Q_T)	Q from current year's input (Q_{CY})	Q from older organic matter (Q_T–Q_{CY})	% of Q from current year's input (Q_{CY}/Q_T) · 100
	Above-ground litter	Root litter	Rhizodeposition				
B0	57 (0.63)[a]	52 (0.40)[a]	38 (0.80)[b]	167	87	80	52%
B120	71 (0.55)	72 (0.28)	41 (0.80)	168	92	76	55%
GL	66 (0.73)	130 (0.54)	69 (0.80)	246	174	72	71%
LL	116 (0.67)	81 (0.80)[c]	71 (0.80)	293	200	93	68%

[a] Fraction 1st year mass losses estimated from in situ litter-bag incubations (Andrén 1987).
[b] Fraction 1st year C mineralization estimated from ^{14}C-labelled materials in laboratory incubations (G. Johansson unpubl., Chapter 5).
[c] Value for C mineralization from rhizodeposition used (see text).

gets. A reasonable estimate of effective growth yield for the microflora as a whole, including maintenance respiration and cryptic growth, might be between 20–40% (McGill et al. 1981, Kjöller and Struwe 1982). Using this range of growth yields, annual microbial production can be roughly estimated (Tab. 7.3). These estimates can be compared with microbivore consumption which was derived on the basis of animal abundance, trophic group classification and temperature influences on animal activity (Chapter 5). In all cropping systems a substantial portion of the microbial production appeared to have been consumed by soil fauna (Tab. 7.3). In addition, the close proportionality between the faunal consumption and the estimates of microbial production, across all treatments, suggest that the population data on soil fauna provided a good indicator of microorganism activity and production.

Nitrogen turnover
With respect to the different treatments, nitrogen turnover and net mineralization showed an overall pattern similar to carbon; that is, higher rates in the leys than in barley (Fig. 7.5). However, there were some significant differences in the behaviour of nitrogen, particularly in the grass ley, which merit closer examination.

Annual net mineralization in barley was 8.0 g m^{-2} in B0 and 9.2 g m^{-2} in B120, which represented about 1% of total soil N (Tab. 2.5). The ratio between C mineralized from the soil organic matter (Tab. 7.2) and net N mineralization was about 10 to 1, which is roughly equal to the C:N ratio of the soil organic matter. Since the crop residues in barley have relatively low N contents, most of the N mineralized would be expected to originate from organic matter already present in the soil. This was confirmed by the measurements of nitrogen dynamics in the various litter-bag incubations in the field (Chapter 5).

Net annual mineralization in the leys was estimated to be 21.4 g m^{-2} in grass and 14.7 g m^{-2} in lucerne, or about 1.6–2.3% of total soil N (Tab. 2.5). In some respects

Tab. 7.3. Microbial production (P) in the cropping systems at Kjettslinge, derived from estimates of microbial respiration (R), assuming an effective growth yield (Y) of 20–40% for soil microorganisms, i.e., P = Y/(1 − Y)R. Consumption of microorganisms by soil fauna was based on soil fauna population data and energetic assumptions. Values for respiration and microbivore consumption from Fig. 7.4 (From Paustian et al. 1990).

	Cropping system			
	B0	B120	GL	LL
Microbial respiration (g C m^{-2} yr^{-1})	150	150	220	260
Microbial production (g C m^{-2} yr^{-1})	40–100	40–100	55–150	65–175
Microbivore consumption (g C m^{-2} yr^{-1})	36	35	44	58
% of production consumed	40–90%	40–90%	30–80%	30–90%

Tab. 7.4. Carbon to nitrogen ratios pertaining to organic matter turnover (g m^{-2} yr^{-1}) in the cropping systems at Kjettslinge. Organic matter input is that which is available to the decomposer organisms and carbon mineralization rates are synonomous with heterotrophic respiration. Values from Figs 7.4 and 7.5.

	Cropping system			
	B0	B120	GL	LL
Organic C input	147	184	265	268
Organic N input	4.5	8.0	11.6	17.5
Ratio C$_{input}$/N$_{input}$	32	23	23	15
C mineralization	167	168	246	293
Net N mineralization	8.0	9.2	21.4	14.7
Ratio C$_{min}$/N$_{min}$	21	18	11	20

these higher mineralization rates parallel the higher carbon turnover and microbial and faunal activity in the leys. However, there are disparities in the relationship between C and N turnover, particularly in the grass ley, that suggest a basic difference in the pattern of nitrogen mineralization as compared with the barley.

Carbon to nitrogen ratios for litter input and for mineralized organic matter are summarized in Tab. 7.4. The overall C:N ratios of the litter inputs were highest in B0 and lowest in LL, reflecting the differences in nitrogen contents in the litter produced in the different cropping systems. More interesting is the comparison between C and N mineralization. Three of the cropping systems. B0, B120 and LL, have C:N ratios around 20. If C mineralization is used as a measure of general decomposer activity, then the results suggest a relatively constant proportionality between decomposer activity and the net release of nitrogen. However, for the grass ley the ratio between C and N mineralization is about one-half that for the other treatments, implying twice as high net nitrogen release in relation to decomposer activity.

There are two sources of mineralized nitrogen, fresh plant residue (litter) and the soil organic matter. The above-ground litter produced in both leys had higher nitrogen contents (2–3%) than in barley (0.5–1.0%). Net mineralization of the N in above-ground litter in the leys undoubtedly occurred and a rough estimate of 2–3 g m^{-2} yr^{-1} in lucerne and 1–2 g m^{-2} yr^{-1} in grass can be calculated from the decomposition rate and N contents (cf. Tab. 5.8). Although they had a lower N content, some N release from decomposing roots was also likely, particularly from lucerne roots where a large part of the total N in the roots was found in soluble compounds (Andrén 1987). The greater N mineralization from fresh litter can partly explain the higher throughput of mineral N in the two leys compared with barley. However, it does not explain the discrepancy between N removed and N inputs in the grass ley, which resulted in an overall deficit in total soil N. Possible explanations for an enhanced net mineralization from soil organic matter

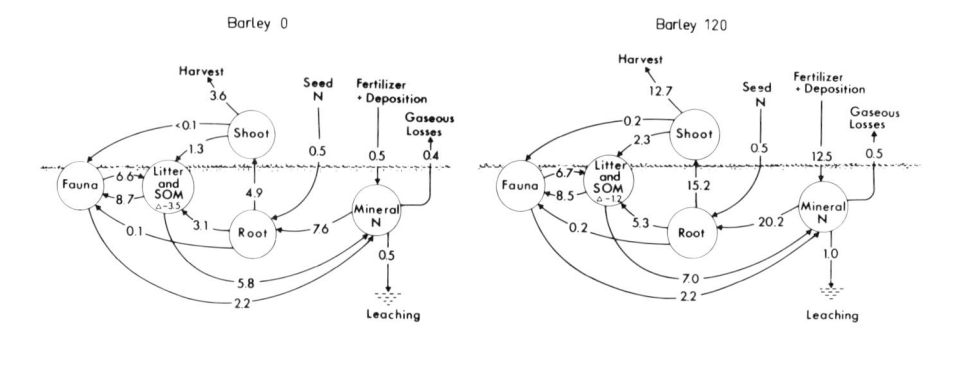

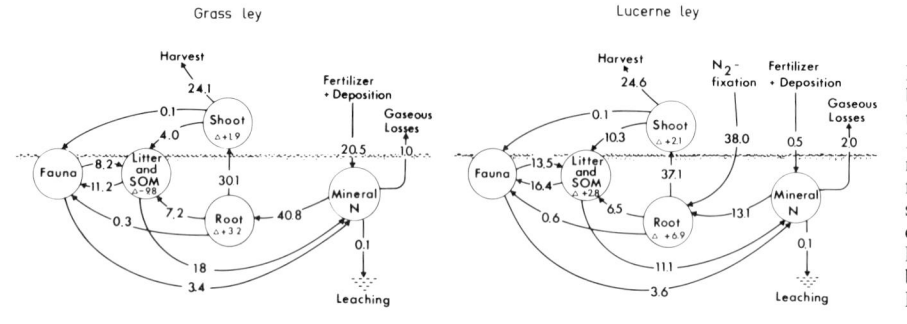

Fig. 7.5. Annual nitrogen budgets (g N m^{-2} yr^{-1}) for the four cropping systems at Kjettslinge, representing mean values for 1982–1983, for the top soil. The delta symbol (Δ) denotes net changes over the year. See Fig. 7.1 for details on budget calculations (From Paustian et al. 1990).

nitrogen in the leys, especially the grass ley, are examined below.

Since net N mineralization was not directly estimated and was largely dependent on measurements of nitrogen incorporated in primary production, one might speculate that the higher mineralization values merely reflect the longer growing period of the leys. For example, N uptake in GL before and after the growing season for barley (June-August) amounted to 8–10 g m^{-2} yr^{-1} (Hansson and Pettersson 1989), which is about equal to the difference in net N mineralization between barley and GL shown above. However, equivalent amounts of N did not appear to have been mineralized during the fallow period in barley. If this had been the case, greater accumulations and/or losses of mineral N, or substantial immobilization of N in crop residues, would have been expected. There were no year-to-year accumulations of mineral nitrogen (measured to 1 m depth) in the barley treatments and total nitrogen losses from leaching and gaseous losses in barley were not higher than in the leys. Furthermore, the results from litter-bag incubations showed a small initial release of nitrogen from soluble fractions of the harvest residues and a subsequent net immobilization of only 0.7 g N m^{-2} during the first 8–10 months after ploughing (Chapter 5).

Net mineralization and net plant uptake in the leys would be overestimated if there was a substantial internal recovery of nitrogen from senescing roots. However, even a 50% recovery of the nitrogen in dying roots by reabsorbtion into live parts would decrease the estimate of N mineralization (and plant uptake) in GL by only 3 g m^{-2}, and the overall input/output balance for N would be unaffected. Furthermore, results from the ^{15}N-labelling in GL and the N content of dead roots suggested that any internal recycling of root N was of minor importance (Hansson and Pettersson 1989).

Finally, an additional (unmeasured) source of N input which might account for the nitrogen deficit calculated in the grass ley is non-symbiotic N$_2$-fixation. However, results from other systems in temperate regions suggest that these inputs would be small, i.e., probably less than 1 g m^{-2} yr^{-1} (Lethbridge et al. 1982, Lethbridge and Davidson 1983), and much greater inputs would be necessary to alter the N balance of GL.

Considering the arguments stated above and the size of the deficit (9–10 g N m^{-2} yr^{-1}) in soil N required to match harvest removal, N losses and net accumulation of plant N standing crop in the grass ley, the data strongly suggest that there was a true enhancement of gross N mineralization, or alternatively a lower gross N immobilization, in the grass ley. Several workers have reported a stimulatory effect of plant roots on mineralization of soil or organic N (e.g., Hart et al. 1979, Clarholm 1985b, Billes et al. 1986, Mosier et al. 1987), which has been mainly attributed to a more active microbial flora in the vicinity of the roots and perhaps physical disturbance of soil aggregates by growing roots (Helal and Sauerbeck 1987). In contrast there are several reports of negative effects of roots on C mineralization from soil organic matter and litter, attributed to decreased soil moisture in the presence of plants (e.g., Shields and Paul 1973, Jenkinson 1977) or plant uptake of organic compounds (Sparling et al. 1982).

One explanation for contrasting effects of roots on C and N mineralization would be if a greater mineralization of N-rich organic compounds occurred in the rhizosphere. In contrast to the usually C-limited conditions in bulk soil, carbon supplied from the roots and the demands for N by roots and microflora may create a more N-limited environment and favour microorganisms utilizing the more N-rich organic compounds. Protein-rich constituents of soil organic matter have C:N ratios lower than the overall soil organic matter C:N ratio, and thus a greater utilization of these compounds under N-limiting conditions, induced by roots, could shift the ratio between N and C mineralization. However, evidence for such selectivity is presently lacking and McGill and Cole (1981) argue against the existence of a control mechanism linking organic matter mineralization to N-limitation of microorganisms. They cite pure culture studies where bacterial deaminase enzymes are instead repressed in the presence of labile carbon sources (Epps and Gale 1942).

An alternative explanation for contrasting responses of C and N mineralization in the presence of roots was suggested by Mosier et al. (1987). Their experiments included ^{14}C- and ^{15}N-labelled plant residue and soil organic matter and ^{15}N-labelled fertilizer in planted and unplanted soil. They found that C mineralization was unaffected by the presence of plants but that N mineralization was almost three times higher in the planted soils. They postulated that the decoupling of N and C mineralization was associated with the competition between roots and microorganisms for available N, and that N mineralized in the vicinity of the root was removed from the mineralization-immobilization turnover cycle by active plant uptake. Consequently re-immobilization by microorganisms of the liberated N was reduced. This implies that decomposition products and microbial metabolites would have a higher N content in the absence of plant roots. In the presence of plants more of this N would be removed by uptake and subsequently microbial products stabilized in the soil would have a lower N content. This hypothesis is consistent with (but not confirmed by) results showing that ^{15}N immobilized in microbial biomass had higher turnover rates in GL than in B120 (Lindberg et al. 1989).

Either of these two possibilities, a more selective mineralization of N-rich organic matter or a relatively lower re-immobilization and stabilization of N, could account for the altered C:N stochiometry in net mineralization from soil organic matter in the grass ley (Tab. 7.4). Although the mechanism is uncertain, we believe that the major factor contributing to the higher N mineralization rate in the grass ley, compared with barley, was the presence of a substantially larger root mass throughout the growing period, and consequently a more significant rhizosphere influence on N mineralization. In other words, with the high N demand of the grass, N-rich organic matter was effectively "mined" from the soil with a replacement of organic matter (of a lower N content) from the plant. It should be emphasized that the situation in the grass ley probably represented a transitional state, where in effect the ley was under-fertilized, in that harvested N and losses exceeded external inputs. Over a longer period of time under continuous ley cropping, the overall N and C balances would be expected to come into phase (cf. Clement and Williams 1967).

Comparisons with other agroecosystems

A variety of methods have been used to derive nutrient budgets for agricultural systems. The majority of studies employing budgeting methods are those where the main interest has been to quantify the fate of nitrogen applied as fertilizers. Earlier studies using exclusively difference methods and non-labelled fertilizer (cf. Allison 1955) have for the most part been supplanted by the use of ^{15}N-labelled fertilizers (Allison 1973, Legg and Meisinger 1982). Such budgets are, however, limited to quantifying the fluxes of the applied nitrogen and unless used in conjunction with other measurements, they do not quantify the total nitrogen fluxes in the system (Legg and Meisinger 1982). A fertilizer balance study of this type was conducted in the project (Lindberg et al. 1989) and results from this and other ^{15}N labelling experiments were presented in Chapter 5.

Analogously, carbon budgets can be determined from experiments using ^{14}C-labelling to follow the distribution of assimilated carbon through the plant and soil. This has been done most frequently in growth chamber experiments (e.g., Martin 1977, Sauerbeck and Johnen 1977, Merckx et al. 1985), as was done in our project (see Chapter 5). Fewer carbon budgets using ^{14}C have been derived from field experiments (e.g., Warembourg and Paul 1977, Martin and Kemp 1986).

Another methodology for estimating C and N balances is the use of long-term field experiments where at least some inputs (e.g., fertilizers, organic manures) and outputs (e.g., harvest) are routinely measured. Changes in soil organic matter are monitored over many years, from which the net result of unknown or poorly quantified inputs and outputs (e.g., N_2-fixation, gaseous losses, leaching) can be calculated by difference.

Still another methodology involves synthesizing results from a number of process studies conducted at a single site, as was done in deriving the C and N budgets for the Kjettslinge cropping systems. Budgets of this kind have been made for several native ecosystems, such as forests (e.g., Bormann et al. 1977, Sollins 1980), grassland (e.g., Woodmansee et al. 1978, Jones and Woodmansee 1979) and tundra (e.g., Rosswall and Granhall 1980, Svensson and Rosswall 1980). However, at the present time there are few budgets for agroecosystems which are based on integrated studies of this kind.

From investigations including measurements of primary production, litter decomposition, and soil respiration, Buyanovsky et al. (1987) calculated carbon budgets for winter wheat (fertilized with 50 kg N ha^{-1} yr^{-1}) and native tallgrass prairie in Missouri, USA. In constructing their budgets, they assumed that soil carbon was at equilibrium and therefore heterotrophic respiration should equal the input of plant biomass to the soil (on a carbon equivalent basis) over the year. Root respiration was then calculated as the difference between measured total soil respiration and the estimate of heterotrophic respiration. The main parts of their carbon balances together with analogous information on the fertilized barley and grass ley at Kjettslinge are shown in Tab. 7.5.

Inputs of plant residues were much lower at Kjettslinge than in the Missouri soil (Tab. 7.5), mainly due to greater harvest removal in the Swedish systems. In the winter wheat crop, all biomass produced, except for grain, was returned to the soil, whereas most of the straw in B120 was removed. The native prairie system was not harvested or grazed by livestock, whereas in GL much of the net primary production was removed in the harvests twice per year.

Total respiration was 2–3 times higher in the Missouri soil than in Kjettslinge (Tab. 7.5). In absolute amounts, root respiration from the perennial systems exceeded that of the annual crops, in both studies. However, there was a lower relative contribution from roots in the winter wheat compared with the other systems. Results from other field studies estimating the composition of soil respiration (i.e., % root respiration vs % heterotrophic respiration) include: 19% vs 81% in Canadian mixed grass prairie (Warembourg and Paul 1977), 15% vs 85% in old fields in South Carolina (Coleman 1973) and 27% vs 73% for wheat and 21% vs 79% for maize in India (Singh and Shekhar 1986).

In Tab. 7.6, the nitrogen budget for B120 is compared with three other ecosystem-level budgets: grain sorghum (plus a winter rye cover crop), with no-till or moldboard plough tillage, in Georgia, USA (Stinner et al. 1984) and winter wheat in the Broadbalk plots at Rothamsted, England (Powlson et al. 1986). Thus, in this comparison, all the budgets are for cereals receiving similar amounts of nitrogen fertilizer.

In the winter wheat N budget (Rothamsted), losses and non-fertilizer inputs were not directly measured but were estimated on the basis of ^{15}N recoveries in microplot experiments. The soil nitrogen content was assumed to be at equilibrium, considering that the treatments have been in place for 140 yr (Powlson et al. 1986). In the cropping systems in Georgia annual rates of gaseous losses of N were not estimated but occasional measurements of denitrification showed negligible activity (Stinner et al. 1984).

Despite the differences in crop species and climate regimes, there are a number of close similarities between the budgets (Tab. 7.6). Amounts of nitrogen removed in the harvest were similar and approximately equalled the amounts added in fertilizer. All budgets included estimates of total plant uptake (i.e. including N allocated to below-ground production). Sorghum showed lower amounts of total uptake recovered in harvest (ca. 50%) than the wheat and barley crops (ca. 75%). Net nitrogen mineralization varied between 90–150 kg ha^{-1} yr^{-1} and was equivalent to about one-half of total plant uptake in all systems.

An example of budgets derived from long-term field experiments is shown in Tab. 7.7. The results are from two localities in southern Sweden with two different crop rotations, with and without nitrogen fertilizer (Jansson 1986b). Crop rotation I consisted of barley, grass-clover ley, winter wheat and sugar beets, where cereal straw and beet tops were removed with the harvest and farmyard manure (FYM) was applied to the sugar beets. In rotation II, oil seed crops were substituted for the ley, straw and beet tops were returned to the soil and no manure was applied.

Soil nitrogen content decreased in all but one of the treatments and the rate of decline was significantly greater in rotation II than in rotation I (Bjarnason 1988). Organic matter N inputs from above-ground sources (i.e., stubble plus FYM in rotation I and stubble plus straw plus beet tops in rotation II) were similar in all treatments (50–70 ha^{-1} yr^{-1}), with the exception of the unfertilized treatments in rotation II (20–30 kg N ha^{-1} yr^{-1}) (Jansson 1986b). The slower decline in soil N for rotation I may reflect the influence of the ley, with reduced tillage and perhaps higher below-ground organic matter inputs. Below-ground production was, however, not estimated. Net N losses, determined by balance, were generally higher in the fertilized treatments, and when comparing rotations, losses appeared to be higher in rotation II. Part of the apparent lower losses in rotation I may have been due to an underestimation of N$_2$-fixation, which was based on measurements of the above-ground crop only. The unaccounted for N input (i.e., negative N losses; Tab. 7.7) in the non-fertilized treatment in rotation I at Fjärdingslöv supports this. For comparison, approximately 25% of the total N fixed went to below-ground production in the lucerne ley at Kjettslinge.

Tab. 7.5. Comparison of carbon budgets (g C m^{-2} yr^{-1}) determined from primary production and respiration measurements under field conditions, in winter wheat and native tallgrass prairie in Missouri, USA (Buyanovsky et al. 1987) and carbon budgets for B120 and GL at Kjettslinge (Paustian et al. 1990).

	Tallgrass prairie	Winter wheat	B120	GL
Litter inputs				
Above-ground	250	220	70	70
Below-ground	290	240	110	200
Soil respiration	610–630	600–640	230	350
% from heterotrophs	85–88%	70–75%	70%	70%
% from roots	12–15%	25–30%	30%	30%

Tab. 7.6. Comparison of ecosystem-level nitrogen budgets (kg N ha^{-1} yr^{-1}). Data for cropping systems in Georgia from Stinner et al. (1984), in England from Powlson et al. (1986) and in Sweden from Paustian et al. (1990). ND = not determined.

	Sorghum/winter rye (Georgia)		Winter wheat (England)	Barley (Sweden)
	conventional	no-till		
Inputs				
Fertilizer	131	131	146	120
Other (deposition, seed, N$_2$-fixation)	10	10	48	10
Outputs				
Harvest	131	127	140	127
Leaching	13	7	54[a]	10
Gaseous losses	ND	ND		5
Δ Total N	−3	+7	0	−12
Net N mineralization	150	128	105	92
Plant uptake	284	257	197	202

[a] Total non-harvest N losses (leaching + gaseous losses) were estimated indirectly from ^{15}N experiments.

Tab. 7.7. Nitrogen budgets (kg N ha^{-1} yr^{-1}) estimated from long-term field plots for the period 1957 to 1981, at Fjärdingslöv (Fj.) and Orup (Or.) in southern Sweden, by Jansson (1986b). Top soils were loam at Fjärdingslöv and silt loam at Orup, with carbon contents of 1.6% and 2.6–2.9%, respectively, at the beginning of the experiment (Jansson 1975). Nitrogen fixation was estimated as the difference in total N above-ground at harvest between the ley in rotation I and the unfertilized crop in rotation II. Δ Soil N is the mean annual change in total soil N from 1957 to 1981. See text for a description of the crop rotations.

	Rotation I				Rotation II			
	No N		Plus N		No N		Plus N	
	Fj.	Or.	Fj.	Or.	Fj.	Or.	Fj.	Or.
Fertilizer	0	0	100	100	0	0	100	100
Deposition	15	13	15	13	15	13	15	13
N$_2$-fixation	36	36	30	18	0	0	0	0
Σ Inputs	51	49	145	131	15	13	115	113
Net crop removal[a]	85	64	137	114	40	29	70	65
Δ Soil N	−5	−28	+6	−17	−33	−43	−27	−41
N loss (by balance)[b]	−29	+13	+2	+34	+8	+27	+72	+89

[a] Represents the difference between the amount of above-ground plant biomass at harvest and the amount of crop residues (and manure applied to sugar beets) returned to the soil.
[b] Nitrogen losses are calculated as Σ Inputs − net crop removal − Δ soil N, therefore a positive value represents a net loss of N and a negative value represents an unaccounted for addition of N.

Strictly quantitative comparisons between these long-term budgets and other budgets, including those for Kjettslinge, should be viewed with caution since much of the data on amount and N content of harvest residues were extrapolated from other field experiments (Jansson 1986b). However, the relative influences of N fertilized vs non-fertilized crops on nitrogen balances and the influence of nitrogen-fixing leys on soil organic matter N levels support the budgets calculated for the cropping systems at Kjettslinge.

Simulation of nitrogen dynamics

The C and N budgets presented earlier give an overall picture of fluxes on an annual basis, where the values have been derived for the most part directly from field measurements. In contrast, a simulation model provides a dynamic description of process rates in the form of time-dependent differential equations. That is, process rates vary continuously in response to changes in environmental conditions (e.g., soil temperature and moisture) and changes in the state variables (e.g., amounts of nitrogen in different forms). Thus the simulation approach focuses on how different abiotic factors control process rates and how multiple processes interact.

A number of simulation models of nitrogen in agricultural soils have been developed elsewhere, having different objectives and levels of resolution (reviewed by Frissel and van Veen 1981, Tanji 1982, Hadas et al. 1987, Jenkinson et al. 1987, Parton et al. 1987). Initial modelling work within the project utilized a model of heat and water fluxes developed for forest soils (Jansson and Halldin 1979) and a detailed model of nitrogen transformations developed for grassland soils (McGill et al. 1981). Based on our initial experiences with these models and from reviewing other models in the literature, we generalized the heat and water model to make it applicable to agricultural soils (Chapter 3, Jansson and Thoms-Hjärpe 1986) and developed a simplified nitrogen model (Johnsson et al. 1987).

The emphasis of the model was on nitrogen transformations in soil, particularly N mineralization and immobilization, and nitrate leaching. Primary production and plant uptake of N were thus treated as model drivers and empirical formulations, based on field measurements, were used for these processes. The ability to easily apply the model to different sites was another criterium used in model development. Therefore, the data input requirements and the model structure were simplified to a level compatible with information generally available in agricultural field research.

Model description

The model (Fig. 7.6) includes the major processes determining inputs, transformations and outputs of nitrogen in agricultural soils. Model inputs of nitrogen are fertilizer and/or manure added to the top soil and atmospheric deposition to the soil surface; outputs of nitrogen are harvest, leaching and denitrification. The model has a one-dimensional vertical structure with the profile divided into layers, which may vary depending on soil physical and biological characteristics and on the desired accuracy in the water flow calculations. The driving variables used in the nitrogen model, such as soil temperature, soil water content, vertical water flow and horizontal water flow to drainage tiles or directly to surface water, are generated by the heat and water model.

Organic N fractions include litter, faeces and humus. The litter fraction represents undecomposed material (e.g., crop residues, dead roots, bedding material incorporated in manure) and microbial biomass and metabolites. The faeces component represents the digested fraction in manure, i.e., excluding bedding material. The humus component represents stabilized organic material derived from litter decomposition. Organic carbon pools are included for both litter and faeces and are part of the formulation for nitrogen mineralization and immobilization.

The equations in the nitrogen model are fully described by Johnsson et al. (1987) and only a brief description is given below. Flow rates (shown in italics) are subscripted using an arrow symbol to indicate the origin and direction of flows between state variables, using the following abreviations: litter (l), humus (h),

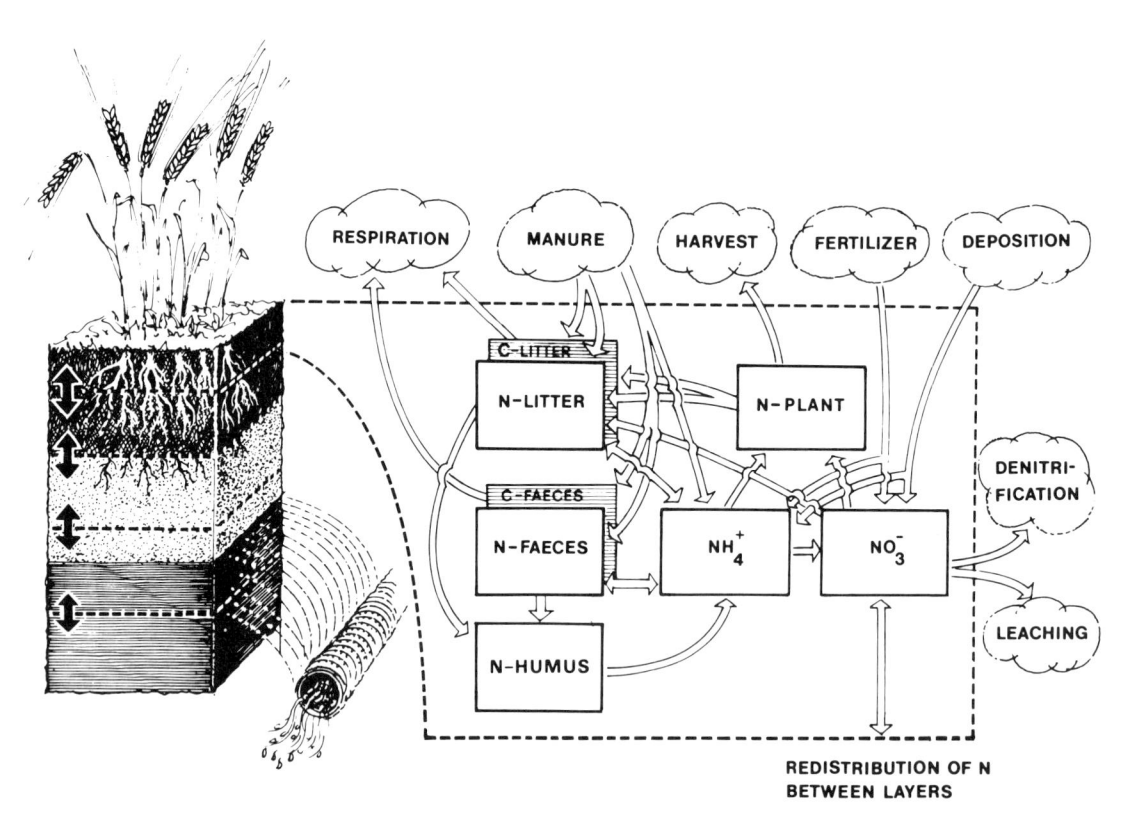

Fig. 7.6. Structure of the soil nitrogen model, showing state variables (boxes) and flows (arrows) included in the model. The model structure is replicated for each soil layer (From Johnsson et al. 1987).

ammonium (am) and nitrate (ni). Flows to a sink (e.g., denitrification, respiration) are subscripted by the source of the flow followed by an arrow.

Mineralization of humus nitrogen (N_h) is calculated as a first-order process.

$$N_{h \to am} = k_h \cdot e_t \cdot e_m \cdot N_h \quad (1)$$

where k_h is the specific mineralization constant and e_t and e_m are response functions for temperature and moisture (Eqs 4 and 5). Similarly, decomposition of the two organic carbon pools (litter and faeces) are calculated as first-order processes with specific mineralization constants k_l and k_f, respectively, and the same abiotic response functions as above. Decomposition products are partitioned into three fractions according to a microbial synthesis efficiency (f_e) and a humification fraction (f_h). One fraction is lost as CO_2–C, a second fraction is assimilated and recycled within the pool, and the remainder is stabilized as humus (Fig. 7.7). Corresponding nitrogen flows are calculated assuming a constant C:N ratio (r_o) of decomposer biomass and humification products.

The net mineralization of litter nitrogen, i.e.,

$$N_{l \to am} = (N_l/C_l - f_e/r_o) \cdot C_{l \to} \quad (2)$$

is determined by the balance between the release of nitrogen during decomposition and the nitrogen immobilized during microbial synthesis and humification, where N_l and C_l are N and C litter mass and $C_{l \to}$ is the litter carbon decomposition rate (see Fig. 7.7). Mineralization from faeces is handled in the same fashion. Negative values for Eq. 2 indicate net immobilization (i.e., $N_l/C_l < f_e/r_o$) and mineral nitrogen is transferred to litter. Both ammonium and nitrate can be immobilized but nitrate is taken up only if the ammonium supply is exhausted. If there is insufficient mineral N, the immobilization rate is reduced by assuming a maximum fraction (f_{ma}) of the mineral N is available.

Nitrification is modelled as a first-order process, modified by the inclusion of a threshold level defined by an equilibrium nitrate: ammonium ratio (n_q), which is assumed to be characteristic for a particular soil. The transfer rate of ammonium to nitrate,

$$N_{am \to ni} = k_n \cdot e_t \cdot e_m \left[N_{am} - \frac{N_{ni}}{n_q} \right] \quad (3)$$

further depends on a potential rate coefficient (k_n) and the abiotic response functions (Eqs 4 and 5).

As shown above, decomposition, N mineralization and nitrification are assumed to have similar soil temperature and moisture responses. A Q_{10} relationship is used to express the effect of temperature,

$$e_t = Q_{10}^{(T-T_b)/10} \quad (4)$$

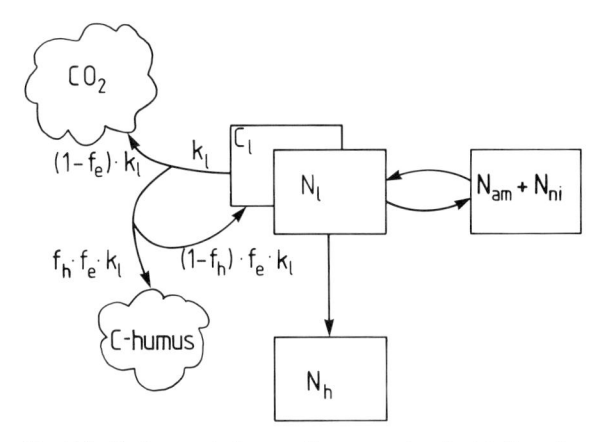

Fig. 7.7. Carbon and nitrogen flows associated with litter (or faeces). Decomposition is modelled as a first-order process, where decomposition products include respired CO_2-C, humified materials. See text and Tab. 7.8 for definitions of parameters (Modified from Johnsson et al. 1987).

where T is the daily mean soil temperature (°C) for the layer, T_b is the base temperature at which e_t equals 1 and Q_{10} is the factor change in rate with a 10° change in temperature. The soil moisture function for a given soil layer decreases on either side of an optimal region of soil moisture,

$$e_m = e_s + (1 - e_s) \left[\frac{\Theta_s - \Theta}{\Theta_s - \Theta_h} \right] \quad \Theta_s \geq \Theta \geq \Theta_h \quad (5a)$$

$$e_m = 1 \quad \Theta_h \geq \Theta \geq \Theta_l \quad (5b)$$

$$e_m = \left[\frac{\Theta - \Theta_w}{\Theta_l - \Theta_w} \right] \quad \Theta_l \geq \Theta \geq \Theta_w \quad (5c)$$

where Θ is the volumetric water content, Θ_s is the saturated water content, Θ_l and Θ_h are the low and high water contents, respectively, for which the soil moisture factor is optimal, and Θ_w is the minimum water content for process activity. A coefficient (e_s) defines the relative effect of moisture when the soil is completely saturated. The threshold water contents Θ_l and Θ_h are defined as,

$$\Theta_l = \Theta_w + \Delta\Theta_1 \quad (6a)$$

$$\Theta_h = \Theta_s - \Delta\Theta_2 \quad (6b)$$

where $\Delta\Theta_1$ is the range in water contents over which the response increases and $\Delta\Theta_2$ is the range over which the response decreases, with increasing soil moisture. Denitrification rate for each layer,

$$N_{ni \to} = k_d \cdot e_{md} \cdot e_t \left[\frac{N_{ni}}{N_{ni} + c_s} \right] \quad (7)$$

is a function of potential denitrification rate (k_d) and nitrate concentration, according to a hyperbolic function where c_s is the half-saturation constant (i.e. the concentration at which the rate is 50% of maximum). The temperature response function is the same as shown above. The moisture response function,

$$e_{md} = \left[\frac{\Theta - \Theta_d}{\Theta_s - \Theta_d} \right]^d \quad (8)$$

increases from a threshold point Θ_d and is equal to 1 at saturation (Θ_s), and the slope of the curve is determined by d.

Nitrate transport is calculated as the product of water flow and the nitrate concentration in the soil layer from which it originates. Nitrate is assumed to be totally in solution while ammonium is assumed to be immobile with respect to water flow.

Since the main focus of the model is on soil nitrogen transformations and mineral N transport, plant nitrogen demand is treated as a driving variable. A logistic function is used to define a cumulative curve for potential N demand during the growing season. Root distribution, which is allowed to change during the growing season, is an input variable that controls relative uptake from each layer. Actual N uptake of nitrate (or ammonium) is calculated from the relative proportion of roots in each layer, the proportion of nitrate (or ammonium) of total mineral N and the potential uptake demand. If N is insufficient to meet potential demand, then a fraction of the total mineral N, f_{ma} (see immobilization description above), can be taken up from the layer. Compensatory uptake from other layers may also occur and is implemented by adding the difference between potential and actual uptake rates to the potential demand from the soil layer below.

Barley

The first application of the model was to the barley treatments at Kjettslinge. The simulations included the period from 1981 to 1983 for which complete sets of abiotic driving variables, estimates of plant uptake and evaluation data sets were available. The data sets used in the model evaluation were mineral N profiles plus drainage flow and nitrate concentrations measured in the tile-drained plots (see Chapter 6).

Parameterization of the model (Tab. 7.8) was based on information from a variety of sources, including literature data and field measurements. A few parameter values which strongly influenced model behaviour were adjusted by trial-and-error to improve the model fit to measured soil mineral N values. The adjusted parameters were the nitrogen availability fraction, litter and humus mineralization rates, Q_{10}-value and the microbial activity coefficient in saturated soil (see Tab. 7.8). This model-tuning process was restricted to the B120 treatment and no adjustments were made in subsequent simulations of B0.

Simulation results

The two organic matter pools in the model, litter and humus, contributed approximately equal amounts to the annual net N mineralization in barley (Fig. 7.8). The "litter" pool, which includes both fresh plant debris, microorganisms and unstabilized decomposition products, showed two brief periods of net immobilization during the year. These occurred around harvest time, when the model simulated an influx of newly dead roots to the litter pool, and again when above-ground residues were incorporated after ploughing. The main contrast between the two barley treatments was the decline over the three-year period in mineralization from the litter pool in B0 (Fig. 7.8). This indicates that organic matter input was not sufficient to maintain the pool at its initial level, which is in accordance with the overall decline in soil C calculated in the budget for B0.

Simulated mineralization rates roughly paralleled the annual temperature cycle, with highest rates during summer and low activity during winter (Fig. 7.8). Soil moisture was seldom an important rate-limiting factor during the summer months and most of the fluctuations in simulated mineralization rates were associated with temperature fluctuations. Mineralization during winter periods was probably overestimated by the model due to the use of a single Q_{10} value over the entire soil temperature range, including sub-zero temperatures. According to this response function, specific mineralization rates could still be significant at 0°C. However, measurements of litter decomposition and N mineralization showed little or no activity in frozen soil (Andrén and Paustian 1987) and other studies indicate that Q_{10} values increase at low temperatures (cf. Svensson et al. 1990). Therefore, in later simulations of the grass ley (see below) the temperature response function was modified to incorporate a more rapid decrease in activity below 5°C.

The largest fluctuations in the mineral N levels in the soil were associated with the input of fertilizer (in B120) and the uptake of N by the plant (Fig. 7.9). Since fertilizer addition and plant N demand were specified input variables, the degree of model agreement with the mineral N measurements was not necessarily a good indication of model performance. However, the overall comparison between mineral N profiles for B0 and B120, gave some confidence to the model's treatment of the effect of nitrogen availability on actual plant uptake, since the parameters determining potential plant uptake were the same in both simulations. The areas which did allow for a more critical evaluation of the model's performance were the dynamics of mineral N outside the growing season and in the sequential changes in nitrogen amounts with depth, as mineral N from the top soil is moved down through the profile.

Tab. 7.8. Parameter values used for simulations of cropping systems at Kjettslinge (Johnsson et al. 1987, Bergström and Johnsson 1988) and additional sites in southern Sweden, including the long-term watershed study at Råån (Jansson et al. 1987, Jansson and Andersson 1988). Parameters used in the simulation of the grass ley (GL) were the same as for barley (B0/B120) unless otherwise indicated, and similarly the parameter values for the last four sites in southern Sweden were the same as listed for Lönnstorp, unless otherwise indicated. Parameter values which differed from those in the simulations of B120, when the model was first parameterized, are shown in bold type.

Parameter definition	Symbol	Unit	Kjettslinge B0/B120	Kjettslinge GL	Lönnstorp	Tönnersa	Ugerup	Råån
External inputs								
N-fertilization		g m^{-2} yr^{-1}	0/12	20	(see Tab. 7.9)			9/12
Dry deposition		g m^{-2} yr^{-1}	0.36		0.36			
Wet deposition in precipitation		mg l^{-1}	0.8		0.8			
Fertilizer dissolution rate day		d^{-1}	0.15		0.15			
Crop and management								
Harvested fraction	f_{hp}		0.6		*			**0.4**
Above-ground residue fraction	f_{ar}		0.1		*			0.1
Live root fraction	f_{lr}		0		*			0
C-N ratio of above-ground residues			50	**25**	*			50
C-N ratio of roots			25		*			25
Ploughing depth		m	0.27		**0.3**			0.3
Mineralization and immobilization								
Humus mineralization rate	k_h	d^{-1}	7.0·10^{-5}	20.5·10^{-5}	**1.0·10^{-5}**			**1.0·10^{-5}/2.1·10^{-5}**
Litter decomposition rate	k_l	d^{-1}	0.035		0.035			
Efficiency constant	f_e		0.5		0.5			
Carbon humification fraction	f_h		0.2		**0.15**			**0.2/0.63**
C-N ratio of humified products	r_o		10		10			
Fraction of available mineral N	f_{ma}	d^{-1}	0.08		0.1			
Specific nitrification rate	k_n	d^{-1}	0.2		0.2			
Ammonium-nitrate ratio	n_q		8	**1.0**	**6.0**	**6.0**	**5.0**	**15**
Moisture and temperature response								
Porosity	Θ_s	vol %						
layer 1			45		54	46	47	45
layer 2			45		46	50	41	44
layer 3			40		40	40	47	43
layer 4			53		40	35	53	50
layer 5			53		40	35	53	53
Minimum H$_2$O for decomposer activity	Θ_w	vol %						
layer 1			15		4	7	8	5
layer 2			15		4	6	5	6
layer 3			4		3	3	2	2
layer 4			27		3	3	2	10
layer 5			27		3	3	2	20
Interval for increasing activity	$\Delta\Theta_1$	vol %	10		10			
Interval for decreasing activity	$\Delta\Theta_2$	vol %	8		8			
Relative activity at saturation	e_s		0.6		0.0			
Response to a 10°C change	Q_{10}		3		2			
Base temperature for Q_{10}	T_b	°C	20		20			
Denitrification								
Potential rate	k_d	g m^{-2} d^{-1}	0.1		**0.01**			0.2
Half saturation constant	c_s	mg l^{-1}	10		**6**			
Soil moisture effect – activity range	Θ_d	vol %	10		10			
Soil moisture effect – coefficient	d		2		1			2

* Crop related parameters varied for different crops in the rotation. See Jansson et al. (1987) for a complete list of values.

In general, comparisons with the ammonium and nitrate contents measured in the top soil showed that soil N dynamics outside of the growing season were accurately reproduced by the model (Fig. 7.9). Ammonium levels were uniformly low (< 3 g m^{-2}) throughout the simulation period with small increases occurring primarily in early summer and occasionally in autumn. Mineralized nitrogen was rapidly nitrified and consequently the observable changes in mineral nitrogen content of the soil were almost totally limited to nitrate. Fluctuations in nitrate in the spring prior to the growing season and in the autumn following harvest were associated with net N mineralization and the absence of plant uptake. The model predicted N mineralization in the

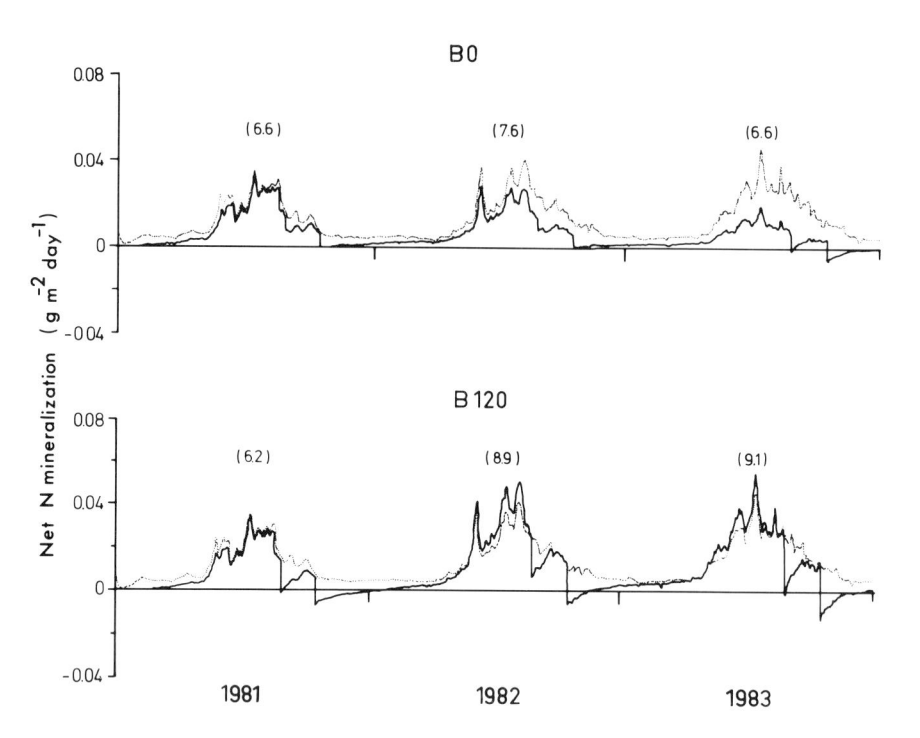

Fig. 7.8. Simulated net N mineralization rates in the barley cropping systems at Kjettslinge for the top soil. Mineralization from litter (i.e., crop residues, microbial biomass and unstabilized decomposition products) (—) and from humus (···). Total annual values are given in parentheses.

autumn to be higher in 1982 and 1983 than in 1981 and this agreed with the relative changes in nitrate-N found in the mineral N profile measurements (Fig. 7.9). The larger increases in nitrate in autumn 1982 and 1983 coincided with higher soil temperatures, during those two years, which stimulated mineralization (Fig. 7.8). However, there were also higher leaching losses in the autumn of 1981.

The most obvious discrepancies between simulated and measured nitrate contents were in the deeper clay layers, where measurements showed greater fluctuations than simulated values. These results suggested that water flow rates within the clay layer and transport of nitrate into and out of the clay were underestimated. Later simulations of the water flow paths, assuming a higher hydraulic conductivity in the clay (see Chapter 3), confirmed this.

Nitrate leaching was dependent on two model outputs: soil nitrate concentration in the leachate and drainage water dynamics. To separate the influence of these two variables on predicted nitrate leaching losses, measured and simulated values were combined into three different time series. One time series ("partly-simulated") was calculated from measured nitrate concentration in drainage water and simulated drainage flow; the second ("measured") represented observed leaching, calculated from measured drainage flow and nitrate concentrations; the third ("simulated") was the model output (Fig. 7.10). In this way, predictions of drainage water flow were evaluated by comparing the "partly-simulated" and "measured" time series and predicted nitrate concentration in the leachate were evaluated by comparing the "simulated" and "partly-simulated" series.

Nitrate losses occurred primarily over short time intervals during spring and autumn (Fig. 7.10). Peak water flow events and accompanying nitrate losses were generally well portrayed by the model, with the exception of the first half of 1981 when water flow was underestimated by the model. However, in 1982 and 1983, simulated nitrate concentrations in the leachate were higher than measured values, which can be most clearly seen in B0 (Fig. 7.10). This discrepancy was due to either an overestimate of bulk soil nitrate concentrations (see Fig. 7.9) or the existence of lower concentrations of nitrate in drainage effluent than in bulk soil. The latter situation can occur if the proportion of water flow in macropores is high (i.e. infiltrating precipitation does not have time to equilibrate with the soil solution having a higher nitrogen concentration). On the basis of our measurements the primary causal factors for the discrepancy in drainage water nitrate concentrations could not be determined.

Annual means of nitrogen fluxes were calculated for the simulation period (Tab. 7.9) and can be compared with the values determined in the nitrogen budgets (Fig. 7.5). The relative differences in flux rates for B0 and B120 were the same in the simulations and the budgets. Simulated mineralization and plant uptake were about 1 g N m^{-2} yr^{-1} lower than corresponding values in the budgets. While plant uptake in B120 cannot be used for comparison, since model parameters were specifically adjusted to conform with field measurements, the comparisons between budget and simulation values for plant

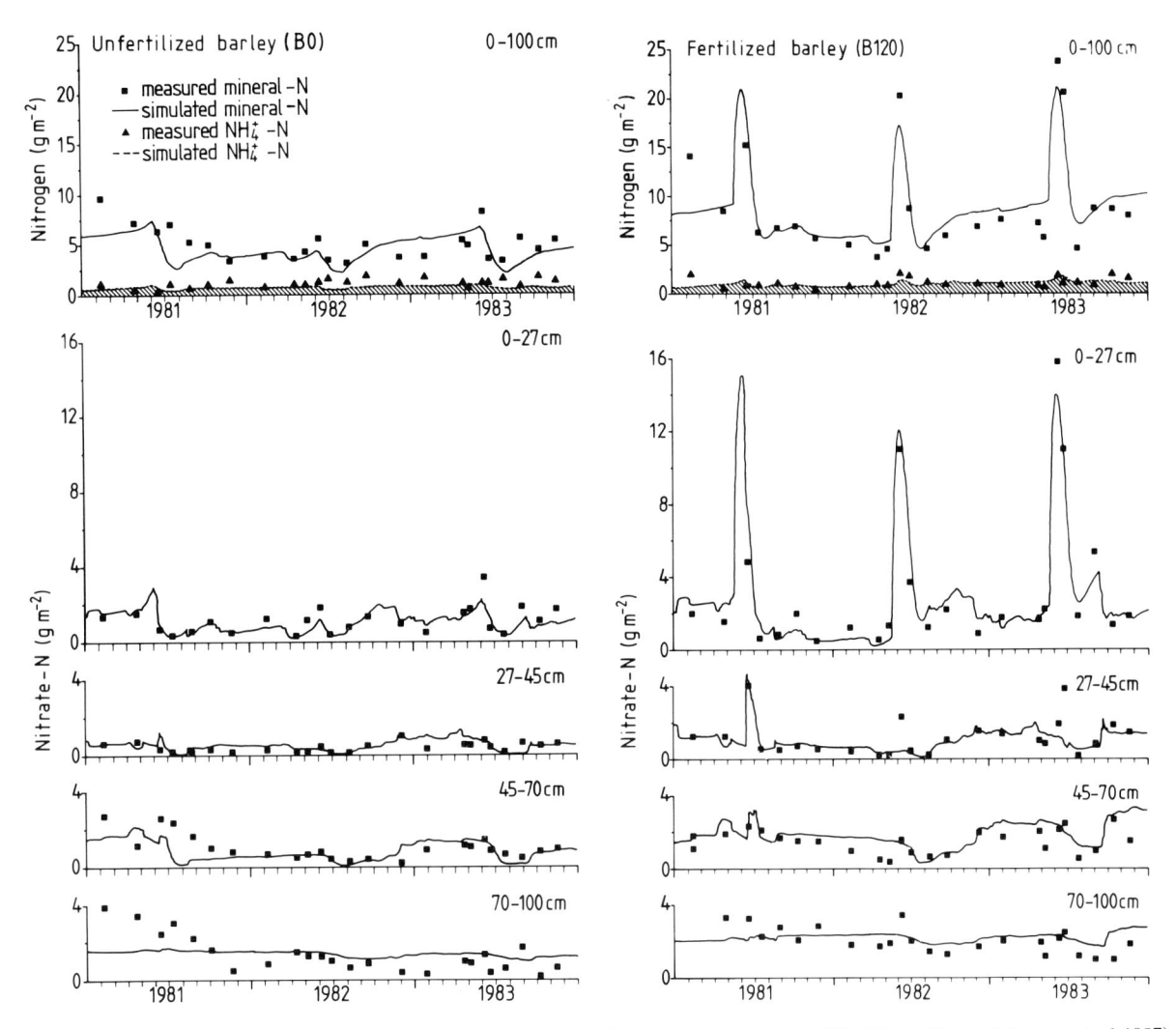

Fig. 7.9. Simulated and measured mineral N profiles in the barley cropping systems at Kjettslinge (From Johnsson et al. 1987).

uptake in B0 and net mineralization rates in both barley crops are valid criteria for assessing the performance of the model. Mean values for leaching were higher in the simulation than in the budget, but this was mainly due to the inclusion of 1981 (which had the highest leaching during the three years). Denitrification losses were approximately equal to the gaseous losses in the budgets (which were primarily as denitrification; Chapter 6).

Grass ley

The model analyses of the grass ley (Bergström and Johnsson 1988) encompassed the period from the first full year of ley production, 1981, through to the ploughing of the ley in 1984. A major objective was to examine the interactions between mineralization of added plant material, soil mineral N dynamics and nitrogen leaching in the period immediately following the ley ploughing.

The most important changes in model parameters from those used in the barley simulations were those involving plant N demand. Since primary production and N uptake of the grass ley increased as the ley matured, the parameters determining potential plant uptake were set separately for each year. It also proved necessary to adjust the humus mineralization rate constant to obtain a reasonable agreement with primary production data and measurements of soil inorganic N. This resulted in a specific rate constant for humus N mineralization about 3 times higher than that used in the barley simulations (Tab. 7.8). Another change was in the parameter determining the equilibrium ratio of nitrate and ammonium (n_q). The proportion of total mineral N in ammonium form was consistently higher in the grass ley than in barley, so the value of n_q was reduced (Tab. 7.8).

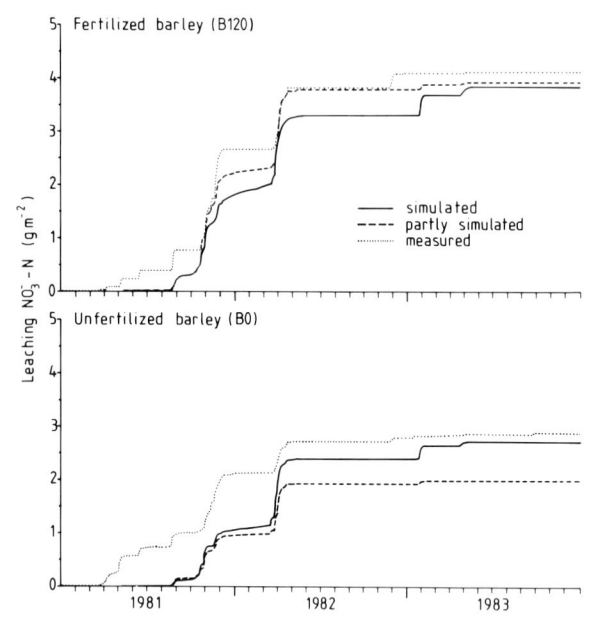

Fig. 7.10. Simulated, partly simulated (i.e. measured nitrate concentrations multiplied by simulated drainage flow) and measured leaching in the barley cropping systems at Kjettslinge. Values represent cumulative leaching over the time period (From Johnsson et al. 1987).

Tab. 7.9. Annual means of simulated nitrogen flows (g N m^{-2} yr^{-1}) for the period 1981–1983, for model simulations reported by Johnsson et al. (1987) and Bergström and Johnsson (1988). Partial flows are given in parentheses.

Flow	Cropping system		
	B0	B120	GL
Fertilization and deposition	0.8	12.8	20.8
Mineralization	6.9	8.0	16.7
Net litter mineralization	(2.3)	(3.4)	(6.2)
Humus mineralization	(4.6)	(4.6)	(10.5)
Humification	1.7	3.3	5.3
Plant uptake	6.8	18.3	39.4
Crop residues to litter	(2.7)	(7.3)	(13.0)
Harvest	(4.1)	(11.0)	(21.9)
Leaching	0.9	1.3	0.4
Denitrification	0.4	0.4	0.2

Simulation results

Net N mineralization was predicted to reach nearly 20 g m^{-2} yr^{-1} after the ley became established (Fig. 7.11). The initial period of net immobilization in the litter fraction was due to the large amount of plant residue originating from the winter die-back in 1980–81 (see Chapter 2). The only other net immobilization occurred for a brief period following the ploughing of the ley in 1984 (Fig. 7.11). Similar results, i.e., a short period of immobilization after ley ploughing, were obtained in incubation experiments using intact soil cores taken from the field (T. Lindberg, unpubl.). The net results of N mineralization and immobilization for the entire period (1981–1984) were that humus N declined by 19 g m^{-2} and the litter N increased by 7 g m^{-2} (Fig 7.12).

Simulated mineral N dynamics during 1981–1983 were generally consistent with field observations (Fig. 7.13). For example, the decline in inorganic N as the ley became established, the fluctuations during the growing season and the partial recharge of the mineral N pool in the autumn were all reproduced in the simulation.

After the ley was ploughed, mineral N content of the soil increased substantially due to the mineralization of the crop residues ploughed into the soil and the lack of plant uptake (Fig. 7.13). To separate the contributions of residue inputs vs cessation of crop N uptake to increased mineral N levels, two additional simulations were run (not shown). In these simulations, either residue addition or removal of the crop (i.e. ploughing) were omitted. By comparing these simulations, it was found that about 4 g m^{-2} of the increase in mineral N could be attributed to the lack of plant uptake, whereas mineralization of crop residues only increased the soil mineral N storage by 2 g m^{-2} by the end of the year (Bergström and Johnsson 1988).

Both measurements and simulations indicated that there was little nitrate leaching in the intact, mature leys (Fig. 7.14). Significant leaching losses were restricted to the establishment year of the ley (1981) and to the period after ploughing. The model underestimated leaching losses in 1981, probably due to errors in water flow calculations, as was the case in the barley simulations (see above). However, the lack of correspondence between predicted and observed values of nitrate leaching, after the ley was ploughed in 1984, is more difficult to evaluate.

After ploughing the ley and with the onset of autumn rains, a rapid drainage discharge was observed in September (Fig. 7.15). After this period, drainage flow was low during the remainder of the autumn, despite continued high precipitation. Simulated outflow from the sand layer to the upper clay layer showed a similar pattern as measured drainage discharge (Fig. 7.15). However simulated drainage discharge began first in October and continued at irregular intervals through December (Fig. 7.15). Examination of the hydrological properties of the different soil layers (Chapter 3) offers a possible explanation for the discrepancy between measured and simulated values.

Before the September rains, the soil was dry and piezometer measurements indicated that the groundwater table was below the depth of the drainage tiles. With the onset of rains there was a rapid percolation through the top soil and sand layer. Due to the low conductivity for matrix flow in the clay layer, there was potential for lateral flow in the sand layer with subsequent macropore flow through the clay horizon to the drainage tiles (Bergström and Johnsson 1988). Because the hydrological model accounted for only vertical matrix flow, simulated drainage would have been expected to show a more delayed response, as was the case. A

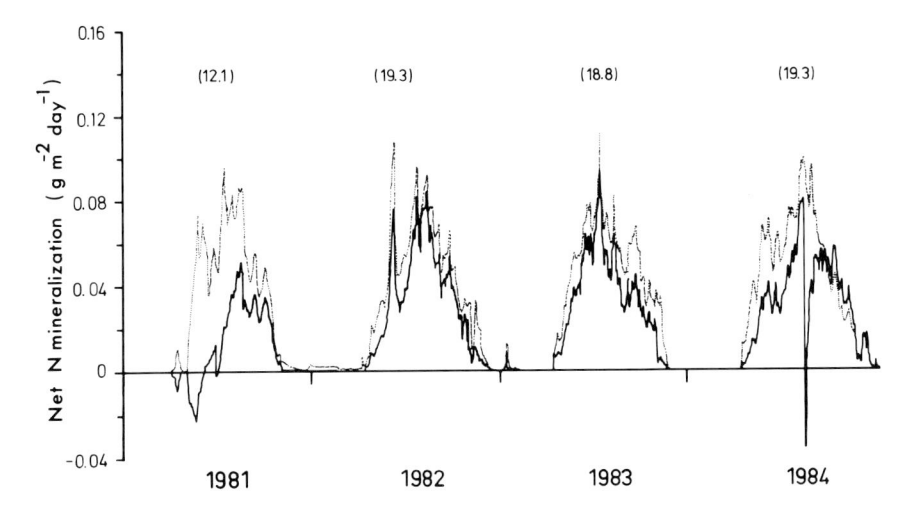

Fig. 7.11. Simulated net N mineralization rates for the top soil in GL. Mineralization from litter (i.e., crop residues, microbial biomass and unstabilized decomposition products) (—) and from humus (···). Total annual values are given in parentheses.

subsequent swelling of the clay as moisture content increased would have closed macropores and restored vertical matrix flow. With vertical matrix flow predominating, groundwater storage would have been recharged before further tile-drainage could occur. Cumulative water flow and nitrate leaching measured in lysimeters were also higher than in the tile-drained plots and corresponded well to simulated values (Bergström 1987c), suggesting that drainage and nitrogen leaching were underestimated in the tile-drained plots due to water flow bypassing the tiles.

Model calculations of mean annual N fluxes in the grass ley are summarized in Tab. 7.9. Values for 1984 were excluded to facilitate comparison with the barley simulations and the nitrogen budget for GL (Fig. 7.5) Simulated outputs exceeded inputs, resulting in a net decline in total nitrogen of about 2 g m^{-2} yr^{-1} for the period. The deficit was less than that determined for the 1982–83 budget, largely due to the lower harvest in 1981 (17.6 g m^{-2}). Net mineralization was about twice that simulated in the barley treatments. The higher N release from stabilized organic matter (humus) was a result of the higher value used for the specific rate constant for humus mineralization. The parameter change led to a close agreement with measured mineral N contents of the soil and most of the measured N fluxes, lending support to the hypothesis that the nitrogen mineralization potential was greater in the grass ley than in barley. The model predicted that denitrification would be lower in GL than in B0 and B120, due primarily to lower soil water contents in GL. While the soil was generally drier in the ley (Chapter 3), measured denitri-

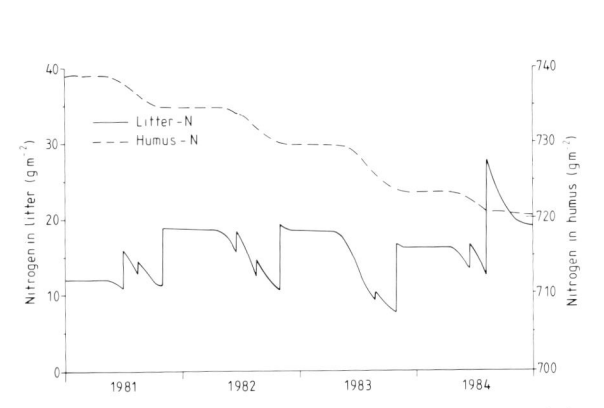

Fig. 7.12. Simulated changes in the litter and humus N pools in GL (From Bergström and Johnsson 1988).

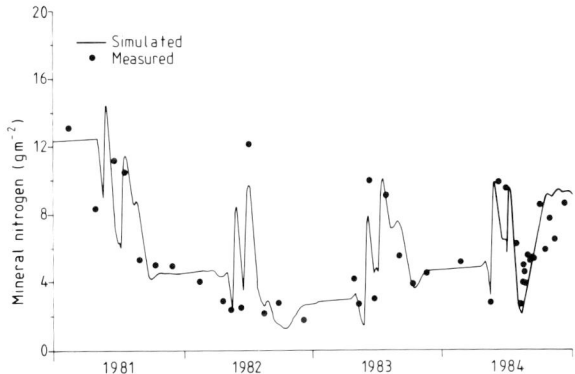

Fig. 7.13. Simulated and measured mineral N contents to a depth of 1 m in GL. Bergström and Johnsson (1988) included two simulated time series for 1984 (when the ley was broken), one based on preliminary data on N uptake and one assuming higher plant uptake. Final estimates of plant uptake were closer to the values for the high plant uptake simulation and these results are shown here.

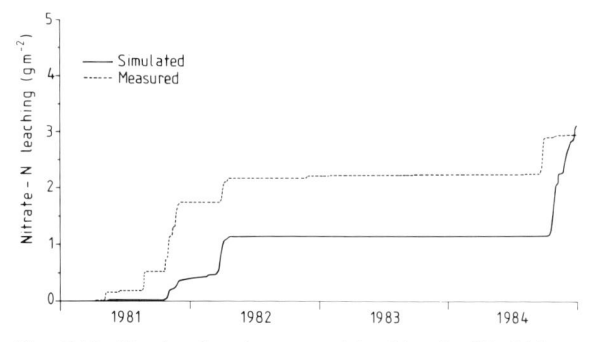

Fig. 7.14. Simulated and measured leaching in GL. Values represent cumulative leaching over the time period (From Bergström and Johnsson 1988).

fication rates were higher (Chapter 6). The influence of roots on denitrification was found to be important (see Chapter 6) and this factor was lacking in the rudimentary formulation of denitrification controls in the model.

Three crop rotations in southern Sweden

Although the model was originally developed for and first applied to the field experiment at Kjettslinge, our ambition was to apply the model to data from other experimental sites having different soil types, climatic regimes, and crops. While no other agricultural research site in Sweden has as detailed an information base as Kjettslinge, a number of other agricultural field stations have basic information such as crop yields, soil characteristics and meteorological records.

The model was applied to three experimental fields, Lönnstorp, Tönnersa and Ugerup in southern Sweden (Tab. 7.10, Jansson et al. 1987). The fields are located in a region characterized by high-intensity agriculture. The soils are often light-textured and generally remain unfrozen for much of the winter. Thus, there is considerable risk for nitrate leaching, which is why analyses of mineral N dynamics and nitrate leaching were of particular interest.

A number of parameters used in the calculation of soil heat and water flows and crop N uptake were site-specific (Tab. 7.8). Climatic driving variables were obtained from nearby meteorological stations or from on-site measurements. Soil physical parameters were derived from on-site measurements and from comparisons with similar soil types for which more detailed data were available (see Jansson et al. 1987).

Crop species varied from year to year (Tab. 7.11) and therefore parameters describing crop development, i.e., onset of crop growth, harvest time, plant N demand, etc., were assigned values for each specific year, based on field observations. The timing of root development was based on observations of crop phenological stages. Values for canopy surface resistance, a parameter used in calculating evapotranspiration were selected by fitting the model to measured soil water content.

Parameters determining soil N transformations were largely the same as in the simulations for Kjettslinge, with some exceptions. Two important parameter changes were the rate constant for humus mineralization and the equilibrium ratio between nitrate and ammonium (n_q), which was based on measured averages of nitrate and ammonium at each site.

Simulation results
Simulated mineralization and immobilization patterns reflected variations in temperature and moisture and the influence of crop residue inputs (Fig. 7.16). A common feature of the three southern sites, compared with Kjettslinge, was the relatively high activity during winter months, since the soils were unfrozen for much or all of the winter. The amount and quality of crop residues added in the autumn affected mineral N dynamics the following year. For example, after incorporation of sugar beet tops, which had an N content of 2%, only small amounts of mineral N were immobilized and there was a rapid and large mineralization of N the following year, as seen at Lönnstorp in 1981 and Ugerup in 1982. In contrast, the high input of turnip-rape straw gave a strong immobilization response after harvest and ploughing at Tönnersa in 1981.

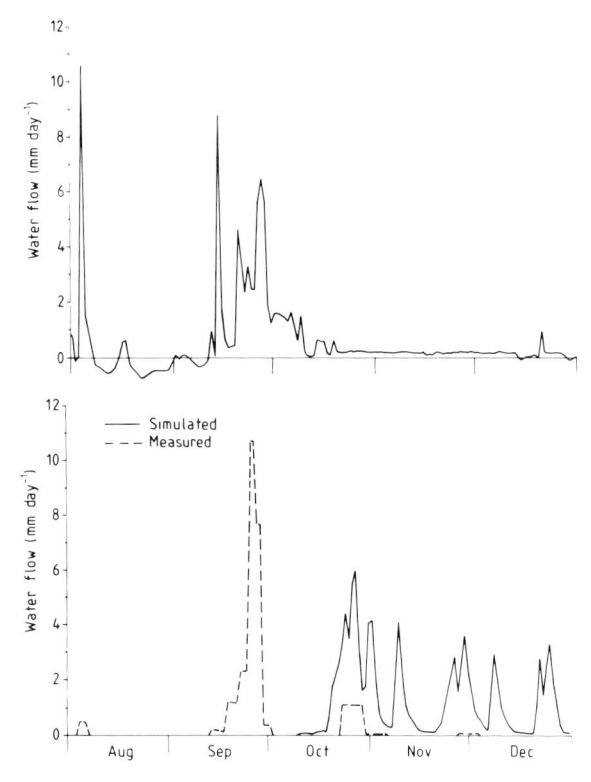

Fig. 7.15. Simulated water flow rate between the sand layer and the upper clay layer (top) and simulated and measured water flow rate through the drainage tiles (bottom) in GL, for the period Aug–Dec 1984. (From Bergström and Johnsson 1988).

Tab. 7.10. General characteristics of the sites used in the nitrogen model applications at locations in southern Sweden (from Jansson and Andersson 1988, Jansson et al. 1987). Information from Kjettslinge is included for comparison.

Site	Location	Mean annual temperature (°C)	Mean annual precipitation (mm)	Soil texture (top soil)	Loss on ignition (top soil) (%)	Crops in rotation	Simulation period (yr)
Lönnstorp	55°40' 13°06'	8.0	633	sandy loam	3.3	cereals, beets, horsebeans	5
Tönnersa	56°33' 12°58'	7.2	799	sand	2.2	cereals, potatoes, rapeseed	5
Ugerup	55°57' 14°08'	7.7	577	loamy sand	5.6	cereals, potatoes, beets	5
Råån	58°00' 12°45'	ca. 8	ca. 590	clay	ca. 5	mainly cereals	20
Kjettslinge	60°10' 17°38'	5.4	520	loam	5.4	barley, meadow fescue ley	3–4

Tab. 7.11. Crop rotation and fertilization rates (kg N ha^{-1} yr^{-1}) for the three field reseach stations in southern Sweden for the simulation time period. Nitrogen additions were as spring-applied Ca(NO$_3$)$_2$. (From Jansson et al. 1987).

Research site	1979	1980	1981	1982	1983
Lönnstorp					
Crop	Barley	Sugarbeet	Spring wheat	Horsebean	Barley
Fertilization	84	140	98	0	112
Tönnersa					
Crop	Potatoes	Barley	Turnip rape	Barley	Barley
Fertilization	126	98	126	98	112
Ugerup					
Crop	Potatoes	Barley	Sugar-beet	Barley	Winter rye
Fertilization	152	90	140	90	90

Plant uptake had the greatest effect on soil mineral N changes. Uptake by cereals was rapid and concentrated over a relativley short period of time, while sugar beets, at Lönnstorp in 1980 and Ugerup in 1981, had a more extended period of uptake and a larger total amount over the growing season (Fig. 7.17). There was less uptake of mineral N by the N$_2$-fixing horse-beans (at Lönnstorp in 1982) which received no N fertilizer (Fig. 7.17).

Nitrate leaching was largely restricted to time periods outside of the growing season (Fig. 7.17). The major exception to this was during the spring of 1983 where precipitation was high during April and May at all three sites. The high nitrate losses at Ugerup during 1980 and 1981 were a consequence of high levels of residual nitrate in the subsoil following a potato crop in 1979 (Fig. 7.18).

Simulated and measured N profiles are shown in Fig. 7.18. The rotation with predominantly cereal crops at Tönnersa exhibited a regular pattern of rapid depletion of mineral N in the top soil during the first part of the growing season. In the autumn there was a partial recharge of inorganic N from mineralization, while increases due to mineralization in the spring were masked

Fig. 7.16. Simulated net N mineralization rate in the upper 1 m of soil at three field experiments (Lönnstorp, Tönnersa, Ugerup) in southern Sweden. Mineralization from litter (i.e. crop residues, microbial biomass and unstabilized decomposition products) (—) and from humus (···). See Tabs 7.10–7.11 for site information. (From Jansson et al. 1987).

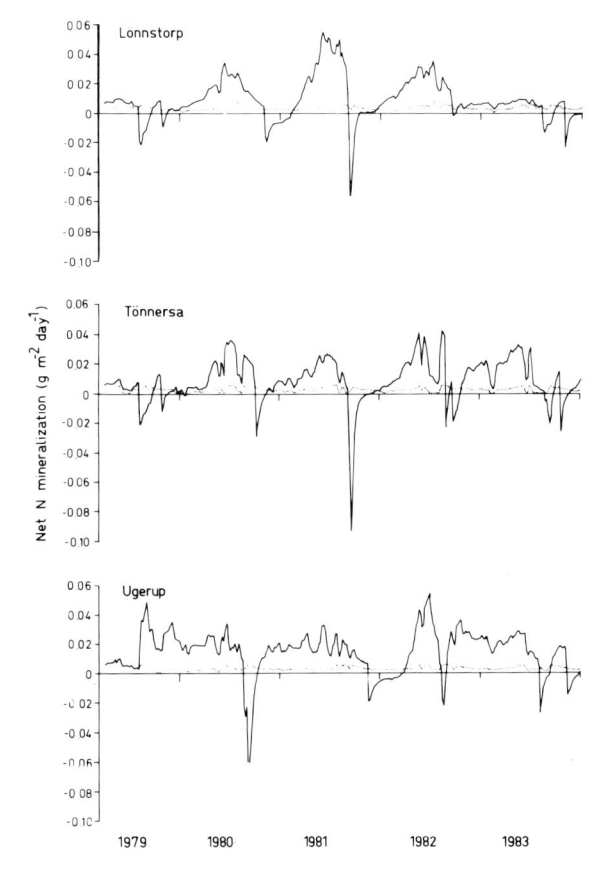

by the large fertilizer additions. There were greater year-to-year variations at Lönnstorp and Ugerup, which had more varied crop rotations. There was generally good agreement of simulated values with observations, particularly for the top 40 cm of soil. The greatest deviations between simulated and predicted values of subsoil nitrate were found at Tönnersa, where nitrate storage in the lower soil layer showed an irregular pattern. There was a buried A-horizon containing clay and organic material at a depth of around 1 m and the presence of this buried horizon may have contributed to the irregularity in measured nitrate amounts. In general, predictions of mineral N in the subsoil and nitrate leaching were complicated by uncertainties in drainage flow patterns, as was experienced in the simulations for Kjettslinge.

Long-term trends in nitrate leaching from an agricultural watershed

In the applications presented up to this point, the conceptional representation of the model has been of a single agricultural field, with a single crop species and management regime in any given year. The Råån application (Jansson and Andersson 1988), discussed below, differed in that the data base and simulation results represented an agricultural watershed of ca. 200 km^2 and a time period of 20 yr. The objective of the study was to examine the influence of changes in land use and agricultural practices on nitrate export from agricultural fields to the Råån river. Previous surveys showed that open, ploughed agricultural fields, with predominantly cereal cropping, were the main sources of nitrate transported to the river (Andersson 1986). Thus the model was used to represent an average cereal cropping field for the area and the nitrogen flows calculated for the watershed were scaled-up from the results for this average cropping system.

The water and heat model was run using climatic variables from two meteorological stations in the area. Soil properties such as water retention curves and unsaturated conductivities were estimated from data on typical soil profiles in the area. Drainage of the soil was assumed to occur both via drainage tiles at depths of 1 m and via the natural groundwater flow. Evapotranspiration controls in the model were adjusted to obtain a reasonable agreement between 20-yr totals of simulated and measured runoff. No adjustment was made for individual years.

Site-specific parameters in the nitrogen model, including average fertilizer and manure addition rates and crop information were derived from regional land use statistics for the province of Skåne, where Råån is located. Application rates of N-fertilizers increased from about 90 kg N ha^{-1} yr^{-1} during the 1960s to about 120 kg N ha^{-1} yr^{-1} after 1970. Yields increased from about 64 to 80 kg N ha^{-1} during the same period. These changes were incorporated in the model by altering the relevant model parameter for the time periods 1961–1970 and 1971–1982, respectively (Tab. 7.8).

Simulation results

One component determining nitrate leaching losses from the watershed was runoff as subsurface drainage and surface flow (Fig 7.19). Trends in simulated annual runoff generally corresponded with measurements, although between-year variation was more pronounced in the simulation than in the measurements. The most notable differences were for the period 1966–1972, where the model consistently underestimated runoff. However, when considering the entire twenty-year period the cumulative deviation between simulated and observed total runoff was small, less than 300 mm. This difference was acceptable when considering the accuracy in measurements of precipitation and runoff.

The increase in fertilizer use after 1970, not only increased the amounts of nitrogen added to the system but also increased crop production and crop residue input to the soil. To evaluate the influence of crop

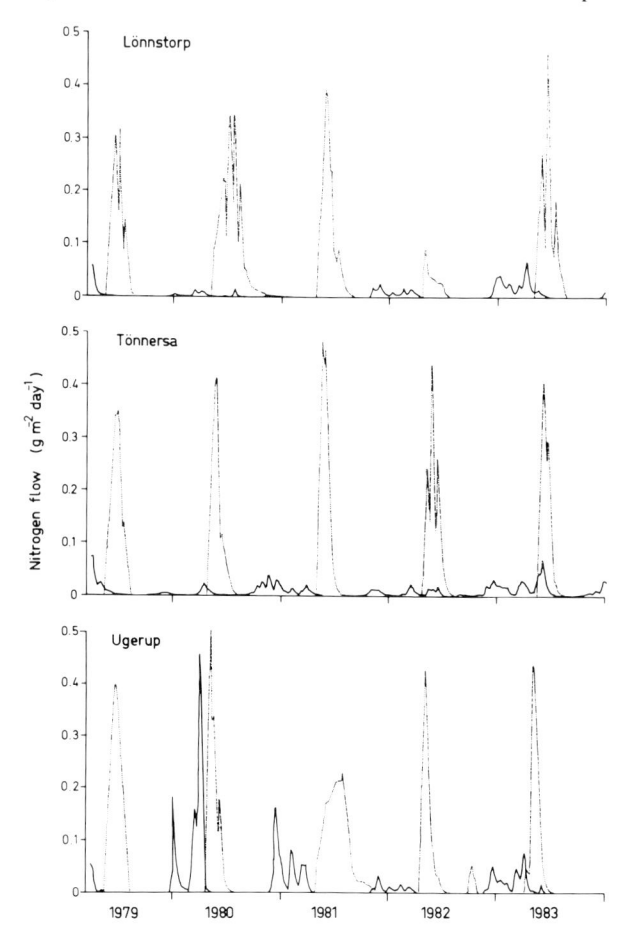

Fig. 7.17. Plant N uptake (···) and nitrate leaching (—) from the upper 1 m of soil at three field experiments (Lönnstorp, Tönnersa, Ugerup) in southern Sweden. See Tab. 7.11 for information on crop rotations. (From Jansson et al. 1987).

residues two model runs, representing "slow" vs "fast" turnover of nitrogen in the crop residues were compared. The parameters controlling humification rate and humus N mineralization were set so that in one instance most of the mineralized N was derived from the manure and litter (crop residues and unstabilized organic matter) pools ("fast" version). In the other instance, net N mineralization occurred only from the humus pool and litter and manure inputs caused net immobilization in these pools ("slow" version).

The two approaches used for simulating mineralization of organic matter yielded small differences in nitrate leaching during the first ten years (Fig. 7.20). Results from both approaches corresponded well with measurements except for the initial year simulated. However, substantial differences occurred during the second ten-year period. The higher application rates of N-fertilizer during the latter period did not increase leaching in the slow turnover approach, whereas the same change in N-fertilization gave a clear response in the fast turnover approach. Observation in the stream showed a pattern similar to the one simulated by the fast turnover approach (Fig. 7.20).

The effects of the different assumptions of organic matter turnover are illustrated by the mean values for net mineralization of litter and humus (Tab. 7.12). In-

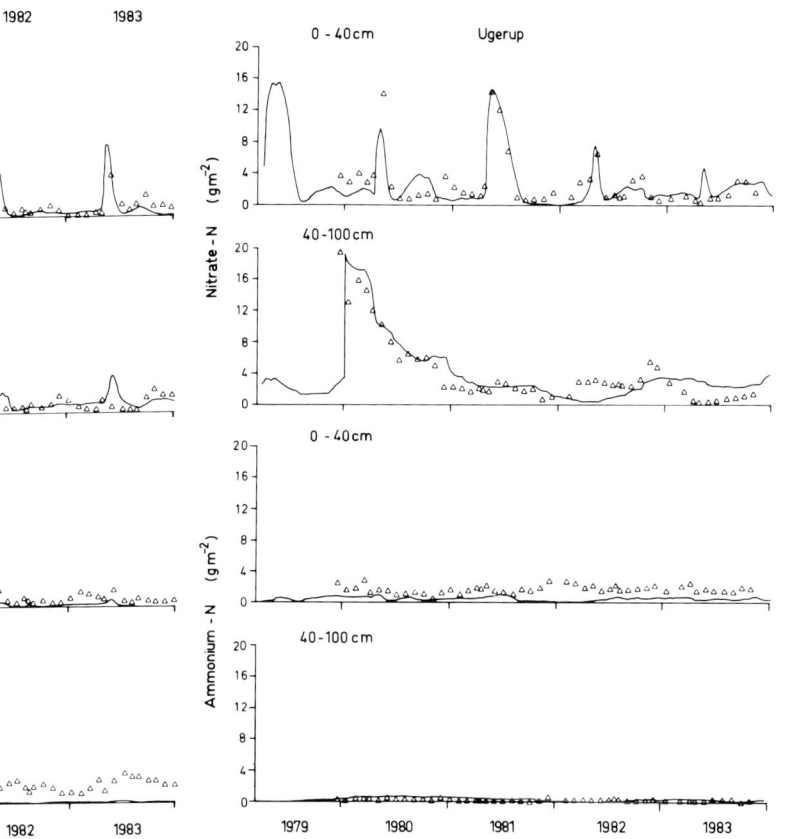

Fig. 7.18. Simulated (—) and measured (Δ) mineral N profiles in the upper 1 m of soil at three field experiments (Lönnstorp, Tönnersa, Ugerup) in southern Sweden. See Tabs 7.10–7.11 for site information. (From Jansson et al. 1987).

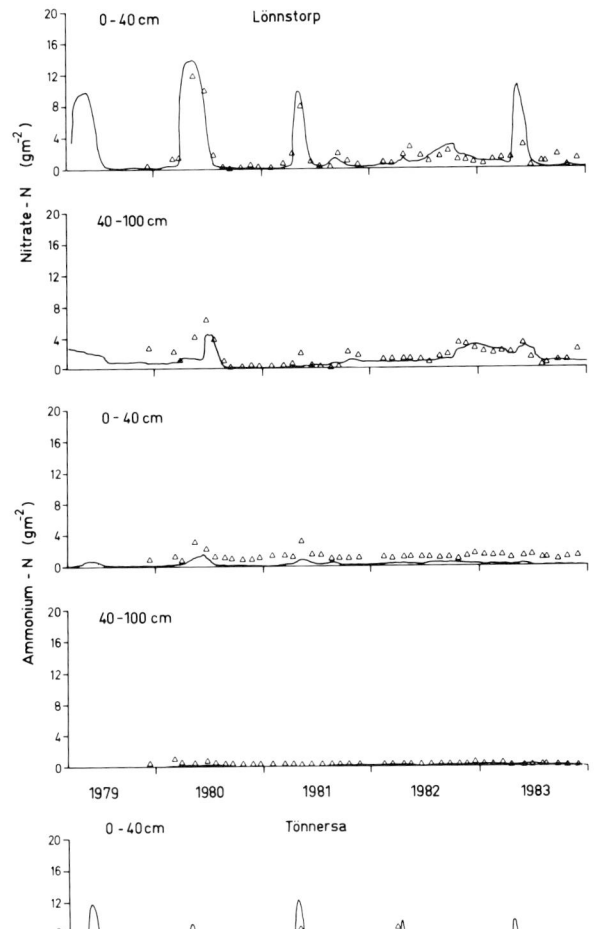

creased fertilization after 1971 resulted in higher crop production and higher residue inputs in the model. With a slow turnover of crop residue-N, there was a larger immobilization sink and net mineralization decreased compared with the previous ten-year period. In the fast turnover simulation the opposite occurred, i.e., with higher residue inputs, total net mineralization increased. In both cases the model predicted that the increased input of fertilizer N was mainly taken up by the crop or immobilized, and only a small fraction of fertilizer N was directly leached from the system.

The simulations for the Råån region illustrate that several assumptions concerning the flow of N between organic matter pools are crucial when investigating trends in leaching. The results suggested that direct losses of fertilizer N are low and therefore the behaviour of nitrogen in organic forms is of greatest importance. For example, increased N-fertilization need not lead to higher leaching rates over the short-term if immobilization increases due to greater crop residue inputs. However, the results suggest that a considerable amount of the N in plant residues is remineralized after an initial phase of immobilization, thus contributing to increased leaching later on.

Summary

The original intention of the experimental design at Kjettslinge, i.e., to obtain contrasting amounts and types of carbon and nitrogen inputs, was largely fulfilled. However, there was perhaps less difference than originally expected in the effective carbon input between the two barley crops, since the addition of fertilizer mainly affected the biomass produced aboveground and subsequently exported in harvest. Over the short-term, the effects on decomposer activity of the

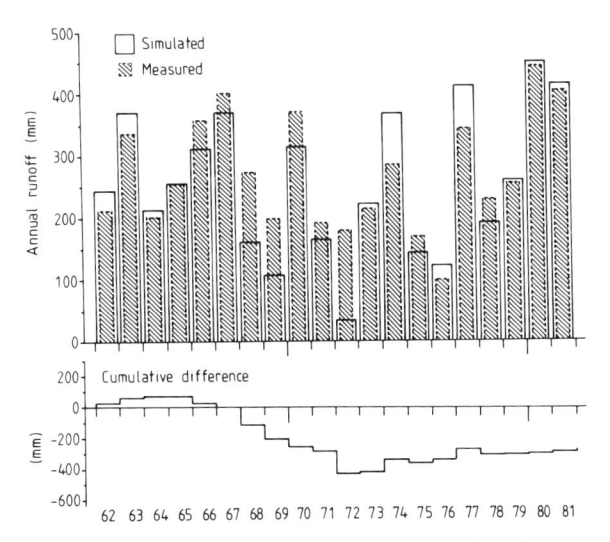

Fig. 7.19. Simulated and measured annual runoff and cumulative difference between simulated and measured values, for the Råån watershed. (From Jansson and Andersson 1988).

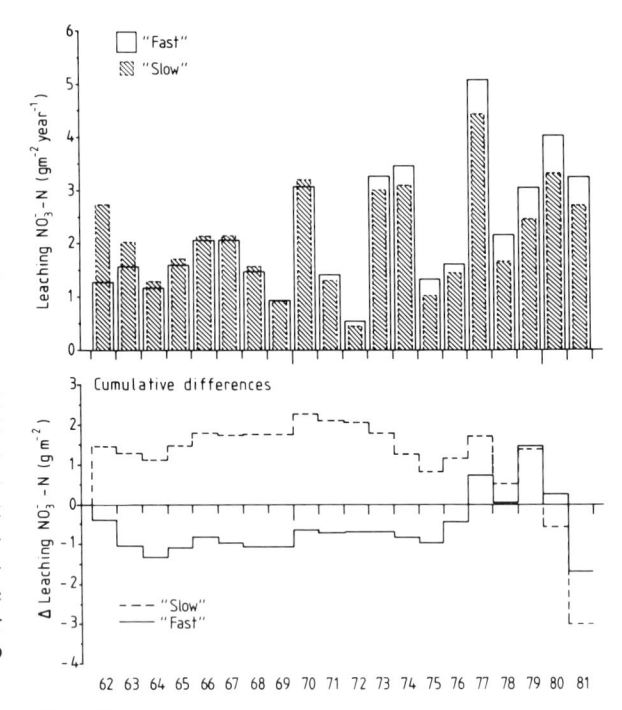

Fig. 7.20. Annual amounts of simulated nitrate leaching and cumulative differences between simulated and measured nitrate leaching from the Råån watershed. Two sets of assumptions concerning organic matter turnover were contrasted: "slow" turnover, with a high proportion of decomposing crop residues being stabilized in humus forms, and "fast" turnover of crop residues, with a low proportion becoming humified. (From Jansson and Andersson 1988).

slightly higher C inputs in B120 were largely counterbalanced by the drier conditions due to the higher evapotranspiration losses in B120. However, over a longer period of time soil organic matter would probably decline to a lower level in the unfertilized barley and consequently decreases in soil organism activity, compared with fertilized barley, would be expected (Schnürer et al. 1985, Jenkinson et al. 1987).

The intentions that the ley crops should represent high C input systems was realized. Organic matter inputs to the soil in the leys were about 60% greater than in barley and a net buildup of carbon occurred during the period the leys were maintained.

The hypothesis that microbial biomass levels would increase significantly in the high C input systems was not found to be true. While this may in part reflect the low resolution of methods available for measuring microbial biomass in soil (see Chapter 4), it is indicative of the relative stability of the microbial populations in soil. Results similar to ours were obtained by McGill et al. (1986) who found that the average microbial biomass in cereal rotations in Canada was affected by the amounts of organic matter added only after 5 yr or more. Thus, while microbial biomass has been found to be a more sensitive measure of changes in site fertility than total

Tab. 7.12. Annual means of simulated nitrogen flows (g N m^{-2} yr^{-1}) for simulations of the Råån watershed, with model versions assuming either high stabilization rates and slow turnover of crop residues ("slow") or low stabilization and rapid turnover of crop residues ("fast"). The simulation was divided into two time periods according to changes in fertilizer application practices, with higher rates of N fertilization occurring in the later period. Partial flows are given in parentheses. From Jansson and Andersson (1988).

Flow	1961–1970		1971–1980	
	Slow	Fast	Slow	Fast
Fertilization and deposition	9.9	9.9	12.9	12.9
Applied manure	2.0	2.0	2.0	2.0
Mineralization	9.1	7.8	8.1	9.4
Net litter mineralization	(−4.2)	(5.0)	(−5.4)	(6.5)
Humus mineralization	(13.3)	(2.8)	(13.5)	(2.9)
Humification	14.0	3.9	17.1	5.1
Plant uptake	15.3	14.2	17.8	17.9
Crop residues to litter	(9.2)	(8.5)	(10.7)	(10.7)
Harvest	(6.1)	(5.7)	(7.1)	(7.2)
Leaching	1.9	1.6	2.3	2.7
Denitrification	2.5	2.2	2.2	2.2

soil organic matter (Powlson et al. 1987) a fairly long treatment period may still be necessary for significant changes to occur.

Although differences between cropping systems in microbial biomass were not found, the activity of the decomposer communities and most soil C and N process rates were substantially higher in the leys. Faunal abundance and activity were good indicators of decomposer activity. The more rapid turnover of soil N and C in the leys was associated with higher organic matter inputs and a persistent plant cover. The apparent greater N mineralization relative to C mineralization in the grass ley was an unexpected result. While our results may reflect a transitional state in the development of the ley, the findings emphasize the importance of interactions between the plant and the microflora and fauna in controlling soil N turnover and N mineralization.

With respect to the retention of nitrogen in the different cropping systems, we had hypothesized that the barley crops would lose more nitrogen than the leys. For the two-year period for which the budgets were calculated this was not found to be the case – the higher leaching losses in the barley were offset by higher gaseous losses in the leys. Regarding the barley treatments, it should be added that the period for which the budgets were calculated (1982–83) had relatively low leaching losses and the highest nitrate leaching in barley was recorded in 1984, 3.6 g m^{-2} (from the large lysimeters; Chapter 6). Also the soil physical properties at Kjettslinge were not as conducive to high leaching losses as those in more coarse-textured soils (Chapter 8). Moreover, losses are highly dependent on weather patterns, i.e., precipitation amount and duration and whether the soil is frozen. Thus while annual crops may be as efficient in retaining nitrogen as perennial crops during a given year, they are more susceptible to major leaching losses occurring at irregular intervals. Contrary to expectations, the highest denitrification was found in LL and the lowest in the B120 (no intensive measurements were made in B0). In general, total N losses were low, ranging between 3–6 % of total nitrogen inputs.

The simulation model provided an additional means to analyze the various processes controlling nitrogen behaviour in soil. Budgeting and simulation proved to be complementary in comparing the cropping systems. The fact that the model, with an identical set of parameters, gave a satisfactory description of mineral N contents, leaching and overall N balances in both barley treatments, suggested that the factors controlling nitrogen dynamics operated in an identical fashion in both systems. Conversely, the results from the simulation of the grass ley supported the hypothesis put forward in the budget analysis that mineralization and immobilization controls differed markedly in the grass ley compared with barley. It was necessary to reject the assumption, implicit in the model, that soil organic N mineralization is a characteristic for a given soil and independent of the crop.

Simulation models also provide a means of isolating component parts of a set of processes, which is difficult or impossible to do with conventional data analysis techniques. This technique was utilized in the comparisons of measured and simulated time series of nitrate leaching and in the analysis of mineral nitrogen dynamics following ploughing of the grass ley. The application to the Råån watershed case study also demonstrated the utility of simulation models in evaluating alternative hypotheses, in this case, concerning the role of crop residues and nitrogen mineralization on nitrate leaching potential. The model application for the field sites in southern Sweden were more descriptive in nature. However, they served as an informal sensitivity analysis of the model by identifying parameters which had to be treated as site-specific.

The relationship between nitrogen supply (through mineralization and fertilization) and water flow in determining nitrate transport and distribution in the soil profile was a central focus of the modelling work. Under most circumstances the model provided an adequate description of the influence of abiotic controls, organic matter and fertilizer addition, and soil water flow on mineral N dynamics. The most serious failures of the model were related to apparent changes in drainage flow patterns. Macropore flow appeared to have been substantial on some occasions and model modifications to explicity treat macropore flow are presently being implemented.

There was a lack of generality in the nitrogen mineralization-immobilization formulation. The humus mineralization and humification rate constants were not site independent and had to be adjusted for some of the cropping systems, on the basis of comparisons with measured data. This is a weakness shared by other soil nitrogen models, which generally define one or more pools or fractions of soil organic matter for which rate constants and/or initial state variables are determined through model fitting to data. At present there are no proven methods for fractionating soil organic matter into functional components or for determining gradients in soil organic matter decomposibility and thus the independent estimation of model parameters or pool sizes is problematic.

While recognizing such limitations, further steps towards improving the generality of the model can be made. At present a crop growth submodel is being developed to replace the use of measured crop production data as driving variables. This will allow questions involving plant-nutrient interactions to be dealt with more explicitly and it is a necessary addition if the model is to be used for forecasting. Finally, the use of the model in new research projects has helped fulfill an important goal of the Ecology of Arable Land, that of stimulating work in new areas of agricultural ecology research in Sweden.

8. Agricultural outlook

Eliel Steen

8. Agricultural outlook. E. Steen

Introduction	183
Transitions in Swedish agriculture	183
Overview of the Kjettslinge field experiment in relation to practical agriculture	184
General	184
Components of the soil	184
Production and harvest	185
The balance of organic matter in arable soil	185
The role of nitrogen	186
Input through fixation and fertilizer	186
Mineralization	187
Leaching	187
Denitrification	188
Swedish agriculture – problems and challenges	189
Regional differences in Sweden	189
Swedish agriculture – the present situation	189
Future changes in Swedish agriculture	189
The first step – modifying present techniques	189
A somewhat speculative longer step	190
Step number three – a future scenario	191

Introduction

Transitions in Swedish agriculture

The characteristics of present-day Swedish agriculture began to emerge in the 19th century when radical changes in animal and plant husbandry were introduced. In the first half of the century there was a country-wide reallotment of land (*laga skifte* in Swedish; Hoppe 1983, Hoppe and Laughton 1986, Sporrong 1988). The village system, where 20–30 families worked a large number of extremely small fields (e.g., width 5–10 m, length 25–30 m) around the village, was abandoned. These fields were combined into larger farms which were more evenly spread over the countryside, each with its own area of ploughland, meadows, pastures and forest. This made it possible for the farmer to develop a cereal-fallow system on the larger arable fields, where the fallow was important for controlling weeds. Some farmers used the fallow for green forage production of oats and vetch, and later hay seed was introduced (*höfrö* in Swedish) containing red clover, timothy and fescues. From this, the ley (mainly red clover and timothy) was developed as a two-three year crop rotated with cereals in central Sweden and as one-two year crop rotated with wheat, barley and sugar beet (The Norfolk rotation) in southern Sweden (Skåne) (Osvald 1959, 1962).

Traditionally, meadows had served as the primary source of animal fodder, where grass, forbs, tree foliage and twigs were harvested and fed to livestock during winter. The livestock manure, containing nitrogen, phosphorus, potash and other plant nutrients, was then applied to the croplands, resulting in a nutrient export from the meadows to the fields. For this reason it was said that "the meadow is the mother of the field". Ley cultivation gradually replaced this earlier system and by 1866 leys were grown on 27% of the arable land in Sweden. Meadows were ploughed and incorporated in an arable system with cereal, forage crops and fallow. Crop rotations with 2–3 yr leys (clover/grass mixture) became typical since most farms had dairy cattle (Berglund 1956, Jansson 1961, 1966).

Scientific plant breeding led to the development of better varieties of cereals, red clover and ley grasses. Seed control was also improved, reducing the number of weeds in the seed material. Stronger horses (e.g., the Ardennes) and better plows made it possible to cultivate heavy clay soils. Many wetlands were drained and converted into arable land. Fields with low pH levels were treated with burnt lime (CaO) and later with ground limestone ($CaCO_3$) (Perman 1956). Plant nutrients were added to the soil in the form of farmyard manure and commercial fertilizers, such as basic slag, superphospate, and chloride of potash, which were gradually becoming available.

All these steps led to a considerable rise in productivity. It also was a near optimal system for soil conservation. This period continued through World War II, when the land area covered by clover-grass leys (45% of cropland) and the population of dairy cows were at their greatest.

After 1945, Sweden underwent another period of rapid change in agricultural techniques and management. Mechanization and extensive use of fertilizers and pesticides as well as further improvements in plant breeding, seed control, and tile drainage of arable land led to higher production, larger farms and specialization. Agricultural policy, including a complex system of farm subsidies, caused many farmers to abandon the production of perennial forage grass crops and cattle in favour of cereals, especially barley, and pigs. Many farmers on the south and central plains ceased raising dairy cows. They became cash-crop growers of wheat, barley and rapeseed. Some farmers specialized in large-scale swine production, with several thousand animals in a single operation. This development increased production in plant as well as animal husbandry. The average yields of barley increased from 2.4 to 3.4 t ha^{-1} during 1950–1980 (Granstedt 1986). During the same period the milk yield increased from 2.8 to 5.4 t cow^{-1} yr^{-1} (Statistical Yearbook 1951 and 1981).

These changes in Swedish agriculture were similar to those that occurred in all of northwestern Europe (Allison 1973, Russell 1977, 1978), and in other cold temperate climates with industrialized agriculture. Today, agricultural production in industrialized countries is highly specialized. Remarkably few crops are cultivated within each climatic zone, and cropping systems tend to be simple. In central Sweden, large areas of arable land are dominated by only three crops: barley, wheat and rapeseed. These simple systems are characterised by the intensive use of agro-chemicals, e.g., pesticides, fertilizers and growth regulators. Modern agricultural practices make it necessary to use heavy tractors and other machinery.

The post-war development caused a tremendous increase in agricultural productivity, as illustrated by current yield statistics. In 1987 average yields of major crops were 6.0 and 4.3 t ha yr^{-1} (15% water content) for winter wheat and barley, respectively, 32 t ha^{-1} yr^{-1} for potatoes (fresh weight) and 48 t ha^{-1} yr^{-1} for sugar beet (fresh weight). The mean production of milk per dairy cow was 6500 l yr^{-1} (Yearbook of Agricultural Statistics 1987). Thus, Sweden belongs to an elite group of countries in terms of agricultural productivity. This production is dependent on a constant flow of subsidiary energy in the form of large quantities of fertilizers, pesticides, animal feed concentrates and machinery. These production methods have also caused problems concerning soil productivity and the environment, which will be discussed later in this chapter.

Overview of the Kjettslinge field experiment in relation to practical agriculture

General

Agricultural land, and arable soil in particular, has traditionally been of limited interest to ecologists, perhaps due to the extreme degree of human perturbation and the overall simplicity of agroecosystems. On the other hand, this same simplicity facilitates studies of carbon and nitrogen flows, which are difficult to elucidate in more complex ecosystems. It is our belief that ecological investigations in agricultural systems can increase our understanding of important functional relationships at the ecosystem level and can contribute significantly to traditional agricultural knowledge and the solution of environmental problems related to modern agriculture.

In agricultural systems, energy fluxes, nutrient dynamics, and the hydrologic cycle are subject to human manipulation (Briggs and Courtney 1985). The arable soil – with its associated crops, soil organisms and abiotic soil components – constitutes a man-made system that eventually reaches a state of equilibrium as long as the same techniques, crops and crop rotations are applied. Equilibrium state is characterized by a large import of nitrogen and large export of carbon and nitrogen, few species of primary producers, and a system stability maintained by agricultural practices.

Most field trials on arable soil are of a short-term nature and normally have clearly defined production and management-oriented objectives, e.g., maximizing crop yield, controlling weeds and pests, increasing the efficiency of applied fertilizers or improving the quality of harvested crops. A second category of field trials is usually of a longer duration, i.e., over several crop rotations, and these studies usually concern the impact of agricultural practices on soil properties, including C- and N-balances.

The "Ecology of Arable Land" project was a relatively short-term study focussing on C- and N-dynamics. A multitude of methods were used to obtain as accurate a picture as possible of nitrogen and carbon flows, in the arable soil system, in relation to various organism groups. Results from the project included both information of a basic ecological nature as well as more traditional agronomic data (e.g., harvest yields). The project results serve as a background for a discussion of present problems within agriculture in industrialized countries.

Components of the soil

The field at Kjettslinge has been used as arable land for about 130 yr, before which it existed as grassland for many centuries (Chapter 2). During the study the content of carbon and nitrogen in the top soil (0–27 cm) was, on an average, 90 t of C and 9 t of N ha^{-1}. The soil carbon pool was dominated by non-living soil organic matter (Tab. 8.1). The soil organisms comprised about 3% of the pool, and there were no apparent differences between the barley plots and the perennial ley plots after the experimental period. However, there was a difference in the amount of roots, which represented 0.6–0.7 and 3.0–3.7% of the total soil carbon in the annual and perennial crops, respectively. The soil organism biomass was dominated by fungi and bacteria, with protozoa and other soil fauna comprising less than 10% of the total biomass (Tab. 8.2). The relative role of the heterotrophic soil organisms in comparison to that of roots was reflected by their relative respiration (Tab. 8.3).

Tab. 8.1. Organic carbon partitioning (%) in the top soil at Kjettslinge. Values from 1982–83 (Paustian et al. 1990).

Component	Cropping system			
	B0	B120	GL	LL
Dead organic matter	96	96	94	93
Roots	0.6	0.7	3.0	3.7
Other organisms	3.3	3.2	2.9	3.3

Tab. 8.2. Carbon partitioning (%) in different groups of soil organisms at Kjettslinge. Values from 1982–83 (Schnürer et al. 1986a, Lagerlöf 1987, Boström 1988a, Paustian et al. 1990).

Organism group	Cropping system			
	B0	B120	GL	LL
Fungi	63	71	60	64
Bacteria	32	27	36	32
Protozoa	5	1	3	2
Macro- and mesofauna	1	1	1	2

Tab. 8.3. Partitioning of total soil respiration (%) at Kjettslinge. Values from 1982–83 (Paustian et al. 1990).

Organism group	Cropping system			
	B0	B120	GL	LL
Microorganisms	67	66	64	65
Fauna	7	7	7	8
Roots	25	27	29	27

Production and harvest

The range in mean annual net primary production, i.e., from 5 t in the unfertilized barley to 19 t dry matter (dm) ha^{-1} yr^{-1} in the lucerne ley (Pettersson et al. 1986, Pettersson 1987) reflected the wide difference between an annual crop with a low nitrogen availability and a perennial crop with a high nitrogen availability. Application of 120 kg N ha^{-1} yr^{-1} in the annual cereal almost doubled production. The grass and lucerne leys yielded a maximum production in the second and third years of 18 and 21 t dm ha^{-1} yr^{-1}, respectively. The estimates of harvested biomass are somewhat higher than the means for central Sweden (Tab. 2.7) suggesting that this level of production and nitrogen fixation may only be representative for intensively managed farms in central Sweden.

The balance of organic matter in arable soil

High net primary production means an increased harvested biomass and a large input of organic matter to the soil. In the leys, at least half of the year's production (excluding rhizodeposition) was left in the field. There was a large difference in the proportion left in the field in unfertilized versus fertilized barley crops, 48% and 33%, respectively (Tab. 8.4). However, in absolute terms there was more organic matter added to the soil in barley given N-fertilizer than in barley given no N-fertilizer, about 3 versus 2.3 t dm ha^{-1} yr^{-1} (Fig. 5.5). It was more difficult to estimate organic matter inputs in the leys. Of the non-harvested production, a part remained in overwintering biomass and another part was incorporated into soil organic matter. Above-ground inputs could be estimated with reasonable accuracy (Pettersson et al. 1986, Pettersson 1987) whereas belowground inputs were less certain (Hansson and Andrén 1986, Pettersson et al. 1986). Organic matter inputs (excluding rhizodeposition) to the soil were 4–5 t dm ha^{-1} yr^{-1} in the leys (Fig. 5.5). The root biomass increased with the age of the perennial ley. Ploughing the leys after four years added an additional 8–12 t dm ha^{-1} organic matter to the soil from the standing biomass

Tab. 8.4. Partitioning (%) of total net primary production (NPP) into harvested and non-harvested components at Kjettslinge (Hansson 1987, Pettersson 1987).

Cropping system	Harvested Grain	Straw/ herbage	Non-harvested Above-ground[a]	Below-ground	Sum
B0	36	16	25	23	48
B120	43	26	17	16	33
GL	0	54	15	31	46
LL	0	49	21	30	51

[a] In the annual crops the percent of NPP not harvested aboveground equalled stubble and spillage, whereas in the perennial crops it equalled litterfall and spillage plus autumn production.

Tab. 8.5. Annual carbon budgets (kg C ha^{-1}) for the cropping systems at Kjettslinge. Mean of 1982 and 1983 (Paustian et al. 1990).

	Cropping system			
	B0	B120	GL	LL
Input from				
Net canopy carbon assimilation	3190	5430	8410	9030
Output to				
Harvest	1150	2980	3570	3510
Root respiration	570	610	1020	1070
Microbial respiration	1510	1510	2220	2620
Faunal respiration	160	170	240	310
Soil organic matter (by balance)	−200	+160	+1350	+1520
Sum of outputs	3190	5430	8410	9030

(Pettersson et al. 1986, Andrén et al. 1987). Total annual organic matter additions including estimations of rhizodeposition are given in Fig. 7.4.

Thus, the value of a perennial grass or legume ley as a producer of organic matter to the soil was well documented. On the other hand, if the barley straw had been left in the field and ploughed under, the annual input would have almost equalled that in the leys. The only difference would have been the extra input of organic matter at the breaking of the leys after four years (Andrén et al. 1987).

The rather complete estimates of carbon input and output to the soil at Kjettslinge made it possible to estimate the carbon economy of the four cropping systems during two years (Tab. 8.5). In a steady state, net carbon assimilation is equal to the sum of harvested C in biomass and total soil respiration. There was a small deficit in total carbon in barley without N-fertilizer, little change in the fertilized barley, and a substantial gain in the two leys. Most of the carbon increment was in the standing crops of roots and litter, part of which became incorporated in the soil organic matter after the leys were ploughed. Even though these results are based only on two years, they give a fairly representative picture of the great difference in carbon budgets between perennial leys and cereals where straw is removed.

The Kjettslinge field was an old meadow on somewhat wet land before being subjected to ploughing about 130 yr ago and tile draining in 1980. The carbon content in the top soil (ca. 90 t C ha^{-1}) is therefore comparatively high and is representative of young soils on lowland sediments. Lower values (ca. 60 t C ha^{-1}) are found in old cultivated soils in central Sweden cropped with cereals, rapeseed, and short-term leys (Persson 1974, 1978b, 1981, Parton et al. 1983). Similar amounts are typical for arable soils in Denmark (Kofoed and Nemming 1976), Holland (van Dijk 1982), and Germany (Sauerbeck 1982). Arable soils in the northeastern USA often contain about 60 t C ha^{-1} but due to the climate gradient this value gradually declines towards the southwest and varies between 5–60 t C ha^{-1} in the Great Plains (Cole et al. 1989, Parton et al. 1989).

The difference in carbon content between arable soil and old grassland soil was well demonstrated by long-term experiments at Rothamsted (Johnston 1976, 1982, Jenkinson and Johnston 1977). After 50 yr, the Hoosfield experiment on a clay soil had 80 t C ha^{-1} in the old grassland plots vs 55 t in the arable rotation. The site at Fosters, on a clay loam, had 60 and 35 t C ha^{-1} in grassland and arable rotation, respectively, and the Woburn site on sandy loam had 40 and 28 t C ha^{-1} in a grassland/ley and arable rotation, respectively.

The effects of fertilizer and manure on soil carbon contents have been investigated in a number of long-term experiments throughout Europe as well as in Sweden. Well known sites are Askov in Denmark (Kofoed and Nemming 1976, Kofoed 1982), Halle in East Germany (Körschens and Kundler 1984) and Dikopshof in West Germany (Söchtig and Sauerbeck 1982, Sauerbeck 1982). In Sweden there are 12 long-term field trials of this kind (Jansson 1975, 1978, 1983), in addition to a frame-plot trial which has also helped to elucidate the carbon balance in arable soil (Persson 1981, Parton et al. 1983, Mattsson 1987).

In the experiments at Askov (Kofoed and Nemming 1976), there was a clear difference in the carbon content in top soil between sites fertilized with NPK versus farmyard manure on loamy soil (1.44 and 1.54% C) and on sandy soil (0.77 and 0.88% C) respectively. On sandy soil at Halle, soil carbon content was 1.26% C on sites fertilized with NPK and 1.79% C where farmyard manure was applied (Körchens and Kundler 1984).

The experiments at Rothamsted revealed great differences between the effects of organic manures and fertilizers, but there was considerable variation between sites. Positive effects on C and N balance have been obtained using short-term leys over a 25-yr period. The resulting nitrogen concentration in the soil was considerably higher in the farmyard manure treatment (0.254% total N), and slightly higher in the NPK treatment (0.103%) than in the control (0.097%) (Johnston 1987).

In the Swedish long-term trials, a decline in soil carbon content, caused by the absence of perennial leys and farmyard manure, was counterbalanced by annual cropping and NPK-fertilizers only if all crop residues, including straw, were incorporated into the soil (Tab. 8.6).

Annual cropping can lead to a decline of soil organic carbon if farmyard manure is not applied (Kohlenbrander 1974). However, if NPK fertilization is high and crop residues are thoroughly restored in the order of 4–5 t ha^{-1} yr^{-1} to the soil, the soil C content can be conserved. If the cropping system contains perennial ley crops the input of new plant organic matter results in an annual gain of C to the soil pool and there is a slow build-up of soil organic matter. The experiment at Kjettslinge demonstrated this rather well. However the experiment did not elucidate long-term changes in organic matter contents or the composition of different organic matter fractions, as discussed by many others (Jansson 1958, 1963, van Veen and Paul 1981, Bauer and Black 1981, Juma and Paul 1981, Newbould 1982, Bakken 1983, Parton et al. 1983, Bjarnason 1988).

Tab. 8.6. Soil organic carbon (%) in Swedish long-term trials started in 1956 (Jansson 1986b).

Crop rotation	1956	1977
Ley and farmyard manure, 30 ton ha^{-1} each 4th year	2.04	2.23
Annual crops only, crop residues left[a]	1.99	2.15
Annual crops only, crop residues removed[a]	1.97	1.89

[a] 15 kg P, 40 kg P, 100 kg N ha^{-1} yr^{-1} in NPK.

The role of nitrogen

Input through fixation and fertilizer

For optimal harvests, annual crops, such as cereals, rapeseed, potatoes and root crops, need 100–200 kg N ha^{-1} yr^{-1}. Perennial grass crops cut two or more times each year need at least 200 kg N ha^{-1}. Production of high quality grass in northwestern Europe, involving three to five cuts, requires 80–100 kg N ha^{-1} for each regrowth, i.e., up to 500 kg N ha^{-1} yr^{-1}.

The high uptake of nitrogen in the lucerne crop at Kjettslinge (500 kg ha^{-1} yr^{-1}) is of great significance, since most of the nitrogen (380 kg ha^{-1} yr^{-1} during the two full harvest years) was fixed by *Rhizobium meliloti* (Tab. 8.7). It should be emphasized that this amount represented total fixed-nitrogen allocated to above- as well as below-ground parts. Literature data on nitrogen fixation are usually limited to above-ground estimates, since determination of nitrogen incorporated in root production is generally considered too tedious to perform. In the lucerne crop, 268 kg N ha^{-1} yr^{-1} from nitrogen fixation was recovered from the above-ground crop alone (Wivstad et al. 1987).

The allocation of nitrogen to various parts of the three crops is shown in Tab. 8.8. With fertilization, relatively more nitrogen ended up in the barley grain and less in stubble and roots. The total amount of nitrogen at harvest in the plant biomass in the leys was

Tab. 8.7. Plant uptake of biologically fixed and mineralized N (kg N ha^{-1} yr^{-1}) in LL at Kjettslinge. Values from 1982–83 (Wivstad et al. 1987, Paustian et al. 1990).

	N$_2$-fixation	Mineralization	Sum
Above ground	268	110	378[a]
Below ground	112	36	148
Sum	380	146	526

[a] 246 kg N was harvested.

Tab. 8.8. Partitioning of nitrogen (%) into harvested and non-harvested components at Kjettslinge (Hansson 1987, Pettersson 1987).

Cropping system	Harvested Grain	Harvested Straw/herbage	Non-harvested Above-ground	Non-harvested Below-ground
B0	43	14	18	25
B120	56	14	11	19
GL	0	63	16	21
LL	0	48	26	26

high. In addition to nitrogen supplied by fertilizer or symbiotic nitrogen fixation, mineralization contributed 150–210 kg N ha^{-1} yr^{-1} in the leys (Chapter 7).

A higher percentage of N was recovered in the harvest in the grass ley, compared with the lucerne ley, which retained a large percentage of nitrogen in stubble and roots.

Mineralization

Nitrogen mineralization is the transformation of N from the organic state into mineral forms performed by heterotrophic organisms in the soil. If there is enough N in the organic matter to satisfy the growth requirements of organism biomass there will be a net release of mineral N to the soil (Bartholomew 1965, Broadbent 1965, Tinker 1983, Clarholm 1985b). If growth requirements are not satisfied, immobilization occurs, i.e., free mineral N is taken up and transformed into organic forms. The two processes go on simultaneously throughout the growing season, and the combined processes are referred to as mineralization-immobilization turnover (Jansson and Persson 1982). However, a direct determination of the amounts of mineral nitrogen released and/or taken up by the decomposer organisms is, however, virtually impossible. For practical purposes agronomists are interested in the net release of mineral nitrogen and its availability to plants.

The mineralization rates in the four cropping systems at Kjettslinge should be viewed in relation to the high organic matter content of the soil and the methods used to estimate the rates. Net mineralization of N calculated for 1982–83 was 210 kg ha^{-1} yr^{-1} in the grass ley, 150 kg in the lucerne ley and 80–90 kg in the two barley crops (Chapter 7, Fig. 7.5). Our estimates of net mineralization are based on a balance of the mineral N pool, taking into account total plant N uptake and other inputs and outputs of mineral N (Chapter 7). Therefore, they are higher than those typically cited in practical agronomic situations. For example, 50 kg N ha^{-1} is often used as a "rule-of-thumb" measure in Sweden for the nitrogen delivered from the soil to the crop (Jansson and Siman 1978, Jansson 1982). This is calculated from the N recovered in above-ground biomass in unfertilized plots plus an additional 1/3 of that amount to account for N in roots. Using this approach, the unfertilized barley yielded close to this rule-of-thumb amount

where 46 kg N ha^{-1} yr^{-1} were recovered in above-ground biomass (grain and straw). This can be contrasted with the 80 kg N ha^{-1} yr^{-1} obtained from the more complete data base.

Leaching

During the last 25 yr many leaching experiments have been conducted in field plots or field lysimeters in northwestern Europe, and in humid areas of USA and Canada. Data are summarized or reviewed in recent reports by Kohlenbrander (1980, 1982), Frissel and van Veen (1981), Stevenson (1982), Hansen and Aslyng (1984) and Andersson (1986). These studies show that leaching of nitrogen mainly depends on four factors: soil type, crop, nitrogen input and climate.

Soil types vary in their capacity to store mineral nitrogen, e.g., clay soils have a higher storage capacity than sandy soils. The ability of the soil to retain mineral nitrogen in spring, when plant growth starts, and after harvest is of specific interest, and has been evaluated in Denmark by Poulsen and Hansen (1962) and Östergaard et al. (1983) and in Sweden by Brink (1982), Lindén (1983a, b), Lindén and Mattsson (1987) and Mattsson and Lindén (1988). These studies found storage capacities of 5–20 kg ha^{-1} in sandy soils and 25–50 kg in clay soils. In cereal crops, fertilized at the recommended rate, there is normally 20–30 kg ha^{-1} of mineral N remaining after harvest in clay and loam soils, and somewhat less in sandy soils (Hansen and Aslyng 1984). Due to their lower storage capacity, sandy soils are more susceptible to nitrogen losses through leaching than clay soils.

The importance of a crop with a long period of growth, such as sugarbeet and perennial grass, in diminishing N leaching has been well documented, especially in Denmark (Lindhard 1975, Simmelsgaard 1980, Hansen 1983, Kjellerup 1983, Hansen and Aslyng 1984) and in Sweden (Gustafson 1987, Bergström 1987a, c, Steen and Lindén 1987).

According to Swedish N leaching studies, the following annual leaching losses appear representative: barley with a high fertilizer level, 25 kg; barley without N fertilizer as well as standing grass and lucerne leys, 5 kg; ploughed ley, 45 kg (Andersson 1986, Bergström 1987a, c). These estimates agree well with results obtained in Norway with a lysimeter technique similar to one of the techniques used at Kjettslinge (Chapter 2), as reported by Uhlen (1978) and as concerns breaking of leys (Cameron and Wild 1984).

N-leaching is less of a problem in northern Scandinavia than in most of Europe, because the ground remains frozen much of the year and spring precipitation is low, limiting opportunities for leaching. The high proportion of perennial grass leys also reduces total leaching. In addition, since a short growing season limits the usefulness of N-fertilizers, much less is used in northern Scandinavia than in, for example, central Sweden. Consequently, typical values of N leaching are in the order of

5–10 kg ha^{-1} annually in northern Scandinavia, compared with 25–50 kg N ha^{-1} in southern Scandinavia as well as in the rest of northwestern Europe (Kohlenbrander 1980, 1982, Hansen and Aslyng 1984, Andersson 1986).

The losses of N through leaching vary from year to year in all soils and crops depending on the amounts of rain during the frost-free period and the amount of melting snow in spring. This was shown at Kjettslinge by the low leaching rate during the dry year of 1983 and the high leaching rates during the wet years of 1981 and 1984. It has also been shown in other studies in Sweden (Gustafson 1987), and Great Britain (Burns and Greenwood 1982). Losses of N through leaching from the four cropping systems at Kjettslinge (Tab. 8.9) were somewhat lower than those obtained in other experiments in central Sweden (Brink 1982, Gustafson 1983, Bergström and Brink 1986). This was probably due, in part, to relatively low precipitation in 1982 and 1983.

The experiment at Kjettslinge, as well as many other studies in northwestern Europe illustrates that modern agricultural practices with normal N-fertilization rates gradually increases mineral-N in the lower part of the soil (1–2 m depth) (Lindén 1985, Bergström 1986, Mattsson and Lindén 1988). Since only part of the nitrogen can be stored, nitrate is lost in drainage water during certain periods of the year, especially during autumn after the crop is harvested and before the ground freezes.

Denitrification
Nitrogen gas fluxes from arable soil to the atmosphere may occur as N_2, N_2O, NO, NO_2 and NH_3. All of these compounds except NH_3 are mainly produced through denitrification and/or nitrification (Firestone 1982). The major products emitted to the atmosphere from arable land are N_2 (denitrification only) and N_2O. Factors controlling the oxygen tension in the soil, i.e., soil moisture, organic matter content and soil texture, determine which process will dominate as the source of N_2O-fluxes (Knowles 1982).

When sufficient NO_3^- and an energy source is present, any factor decreasing oxygen in the soil promotes denitrification and influences the proportion of the main compounds produced. At high nitrate concentrations or at low oxygen levels, N_2O may dominate as a product from denitrification. A balanced proportion between energy supply to the denitrifiers and nitrate availability will, especially in completely oxygen depleted soil, result in mainly N_2. The availability of NO_3^- is dependent on deposition, fertilization, mineralization-nitrification, root uptake, leaching and the pH status of the soil. The nitrifiers produce N_2O and NO mainly under microaerophilic conditions. The gaseous nitrogen oxide formation by this group of bacteria is thus favoured by factors giving rise to low oxygen tensions.

From an economical and environmental point of view denitrification should be minimized. N_2O may affect the ozone layer and contribute to the greenhouse effect. Since NO_3^- may accumulate during denitrification it may leach into and poison groundwater supplies.

The results from Kjettslinge show higher denitrification losses in perennial leys than in annual barley (Svensson et al. 1990). The lower leaching in the barley during 1983 than in 1982 was accompanied by higher denitrification rates. The investigations did not indicate to what proportions N_2 and N_2O were emitted, however, the relatively dry conditions implied a dominance of N_2O. Laboratory experiments to determine the source of N_2O indicated that N_2 was probably the main product in the vicinity of plants and that N_2O dominated in soil less affected by roots, i.e., between plant rows (Svensson et al. 1985, 1986, Klemedtsson et al. 1988b).

From the results of the Kjettslinge study the following conclusions regarding environmental problems caused by denitrification, i.e., mainly the release of N_2O to the atmosphere, can be made:

1. Annual crops seem to denitrify less and leach more than perennial crops. Thus perennial crops may reduce leaching problems, while annual crops may reduce denitrification.

2. Catch crops can be used to minimize the leaching of N. However, denitrification may be stimulated by such crops. The catch crops should be effective in taking up nitrate, especially during autumn, when moist conditions may increase leaching. Denitrification is probably low during autumn because of low temperatures, but the main product is probably N_2O. Thus, if a catch crop stimulates denitrification, N_2O-emissions may be increased when compared with leaving the soil in bare fallow.

Leaching of nitrate from agricultural areas has contributed to the eutrophication of the Kattegat Sea in southern Sweden (Andersson 1986, Fleischer et al. 1987, Ryding 1988). Catch crops would decrease the leaching

Tab. 8.9. Annual nitrogen budgets (kg N ha^{-1}) for the four cropping systems at Kjettslinge. Mean of 1982–1983 (Paustian et al. 1990). ND = Not determined.

	Cropping system			
	B0	B120	GL	LL
Input from				
Fertilizer	0	120	200	0
N_2-fixation	ND	ND	ND	380
Seed	5	5	negligible	negligible
Deposition	5	5	5	5
Sum	10	130	205	385
Output to				
Harvest	36	127	241	246
Leaching	5	10	1	1
Gas loss	4	5	10	20
Standing plant biomass increment	0	0	(51)	(90)
Sum	45	142	252	267
Difference	−35	−12	−47	118

in these areas but it is not known if an increase in N_2O-emission would occur simultaneously.

Swedish agriculture – problems and challenges
Regional differences in Sweden

Geographically, Sweden is a long, narrow country with a large north to south gradient. Because of the cold climate, the only agricultural crops of importance in northern Sweden are perennial grasses. In southern Sweden there is a greater diversity, with winter wheat, winter rapeseed, and sugar beet as characteristic crops. There are also important differences between the eastern and western parts of Sweden, since the eastern regions have a drier climate and more calcareous soils than the western regions.

A great proportion of agricultural production takes place in the southern plains, where there is also a concentration of large farms producing pork, poultry and beef. About 60% of the arable land in this region is used for cereals, and only 10% for leys (grass and legume crops grown 1–2 yr). In the coastal areas in the north (the Bothnic coast), croplands consist of 30% cereals and 70% leys (Yearbook of Agricultural Statistics 1987).

Swedish agriculture – the present situation

The intensive system of agricultural production has negative effects on the environment. The most serious and acute problem is the leaching of plant nutrients, especially nitrogen. It has contributed to the pollution of the south-western coastal areas. A number of wells used for drinking water in southern Sweden have high nitrate concentrations (Brink et al. 1986). Much of this leached nitrate originates from large applications of pig and cow manure (Brink and Jernlås 1982). Many farmers also apply N-fertilizers at high rates. In the rest of the country the situation is less serious but can worsen if appropriate changes are not made (Gustafson 1982).

Intensive crop production requires the use of pesticides. Normally, modern agrochemicals are decomposed quickly in the soil. However, during the last two years, traces of the herbicides MCPA and dichlorprop, the fungicide metalaxyl, and the insecticide fenitrothion, have been detected in rivers in agricultural areas (Kreuger and Brink 1988).

Besides problems associated with pesticides and other additives, nitrogen is lost through ammonia volatilization and denitrification. In farms which keep livestock, losses of nitrogen through ammonia volatilization can be large (Beauchamp et al. 1982, Kohlenbrander 1982, Kirchmann 1985). Atmospheric transport and subsequent deposition of ammonia to non-agricultural areas can contribute to soil and water acidification.

Since the early 1950s, there has been a decline in the numbers of dairy farms and, because of this, there has been a decrease in perennial forage leys and supplies of farmyard manure. Replacing leys with annual crops caused a decline in soil carbon and a deterioration of soil structure, resulting in problems with puddling and compaction in some areas. To investigate the influence of various agricultural practices on soil organic matter content, a series of long-term field trials were initiated throughout Sweden. The results from these investigations have been used to make recommendations designed to maintain adequate soil organic levels (Jansson 1975, Ebbersten 1980, Mattsson 1985). Today the situation seems to have been stabilized since farmers are more aware of the importance of incorporating the crop residues. Addition of sewage sludge to arable land also has contributed to rebuilding soil carbon levels (Persson 1978b, 1981).

Landscapes in Sweden have become less heterogeneous in recent years. The decrease in grazing livestock has resulted in the re-colonization, by homogeneous stands of shrubs and native trees, of the diversified and fragmented habitats in the old open-woodland and marshy meadow pastures. A further reduction of habitat diversity has occurred on arable lands which now have fewer hedges, open ditches, small waterways, and wooded stony-moraine hills than before. The use of herbicides has reduced the diversity of the arable land flora, and the invertebrate and vertebrate fauna dependent on this flora (Larsson 1986, 1987, Svensson and Wigren 1986a, b, c).

There are other problems in present-day Swedish agriculture, such as accumulations of heavy metals in the soil. A new threat is the accelerating acidification of water and soil, due to external factors such as fossil fuel combustion and use of acidifying fertilizers.

Future changes in Swedish agriculture

The seriousness of the present economic and environmental problems in Swedish agriculture is debatable. However, some modifications of the present production techniques, which have resulted in over-production of food as well as nitrogen and pesticide leaching, are necessary. Some people favour alternative farming while others argue that there is no need for change or that there is a need for only small adjustments of present practices.

The first step – modifying present techniques
A recent problem has been the over-production of cereals. The urgent need to reduce grain production led to policy recommendations in 1987 that black fallow should be practiced. Consequently, farmers received economic compensation for each hectare that they did not sow. Since black fallow increases nitrogen leaching this system was replaced after one year by green fallow, i.e., the fallow is sown with a catch crop such as rapeseed, white mustard, Italian rye grass or some other fast-growing annual. A recommended alternative is to

sow winter wheat or other overwintering crops, if the growing period during autumn is more than 30 d. It is also possible to undersow catch crops such as red clover, white clover, Italian rye grass, and rapeseed in spring sown cereals, which also reduces the yield of cereals by 10–15%. The catch crop could even be used as an energy source, i.e., as fuel. Another strategy is to plant shrubs and trees (fast-growing energy forest) or timber-producing trees (pine and spruce).

A perennial crop is the best solution for reducing N leaching, but the only one presently available is the ley with grasses and/or legumes. Recently it has been decided to give economic compensation to farmers if they convert cereal areas into perennial leys. However, a substantial increase in ley farming is not advisable, since a decrease in cattle production is also desirable. Leys are normally highly effective at retaining nitrogen. However, considerable amounts of organic matter and nitrogen accumulates during the ley years. When the ley is broken, nitrogen is quickly mineralized and may be subjected to leaching losses. Such a situation can lead to more leaching as well as higher gaseous losses compared with an annual crop unless a new crop is sown immediately. The experiment at Kjettslinge illustrated this, suggesting that denitrification and other gaseous losses to the atmosphere occur to a greater extent in leys, especially legume leys, than in annual crops, and that substantial N leaching can occur when the ley is broken.

Losses of nitrogen through leaching are generally considered to be a more acute pollutant than gaseous losses. However considerable losses can occur as ammonia volatilization after application of farmyard manure and pig slurry (Torello et al. 1983, Kirchmann 1985, Christensen 1986, Nilsson 1986, Ryding 1988, Steineck 1988). Thus farmyard manure may be a large contributor to air pollution whereas commercial fertilizers are only marginal contributors except when urea (<1% of N-fertilizers in Sweden) is applied (Ferm 1983).

Intensive soil cultivation during periods when the soil is bare can increase mineralization and leaching losses. The loss of organic matter can be decreased using techniques similar to the stubble mulch or direct drill methods widely adopted in the USA (Doran et al. 1987). It is also important to minimize heavy-vehicle traffic on fields. Such traffic causes soil compaction, decreases water holding capacity, and creates more anaerobic conditions leading to increased nitrogen loss by surface runoff, denitrification and reduced earthworm populations (Håkansson and Danfors 1981, Boström 1986b).

Decreased tillage would positively affect the dynamics of soil-nitrogen utilization. More nitrogen would be immobilized in plant litter, increasing the short-term need for fertilizer nitrogen. It would also affect soil organisms and their role in decomposition. As shown by Doran (1980) nitrifiers as well as denitrifiers would increase considerably. Earthworms would also be more essential, as would other faunal groups (Stinner and Crossley 1980, Coleman et al. 1984).

However, decreased tillage would create new problems, unrelated to nitrogen, in a cold temperate climate. Infestation of couch grass and other weeds would become more troublesome, and problems with fungal diseases and insects could also increase. Thus, increased use of agrochemicals would probably be necessary.

Further improvements in fertilizer application technology are needed to increase nutrient utilization efficiency. In addition to improving the timing of application, other novel approaches to fertilization need to be developed, i.e., slow-release fertilizers, such as sulphur-coated urea and plastic-coated N-fertilizer pellets, foliar application of ammonia and urea, carbon supplements, and nitrification inhibitors (Goring 1962, Keeney and Huber 1980, Nelson 1982). A new technique for applying nitrogen and other plant nutrients tailored to the actual requirements of the plant during each stage of development, has been advocated by Ingestad (1982). In this procedure, now practiced to some extent in northwestern Europe, N-fertilizer can be spread from long application rods attached to tractors driven in permanent tracks or the fertilizer can be dissolved in irrigation water and spread via a system of pipes. These techniques permit frequent applications of nitrogen in small quantities well adapted to the needs of the crop at the time.

The use of pesticides should be reduced as much as possible, but there is no easy and rapid way to bring about this reduction. An increase in the cultivation of perennial leys and the introduction of catch crops should decrease the need for herbicides directed against annual weeds, while problems with perennial weeds would certainly increase. The replacement of ploughing by surface tillage would probably also exacerbate the problem with biennial and perennial weeds. A reduction of the acreage in cereals should diminish the need for pesticides, but the remaining cereal fields would require treatment as would potatoes, sugar beet and rapeseed.

A somewhat speculative longer step
Clues to the future of Swedish agriculture can be obtained by examining agricultural systems in other countries. Many problems in Sweden are shared by the rest of northwestern Europe and the northeastern USA and Canada. It is possible to compare and evaluate situations in these areas, especially if the analysis is restricted to nitrogen and carbon. It is much more difficult to predict or even to speculate about the character of agriculture in the future.

Presently in Sweden there are intense discussions about agricultural surpluses, alarming reports about nutrient pollution of the Kattegat Sea and the Sound (Öresund), and increasing levels of nitrate in drinking water (Andersson 1986). Changes to remedy this situation will have consequences for producers as well as consumers. Reduction of production intensity including a decrease in the use of nitrogen fertilizers and pesticides has been eagerly discussed as one change. A special tax on fertilizers and pesticides has been introduced to bring about such a reduction.

From an exclusively biological point of view it is comparatively easy to convert to a ley-farming system within a ten-year period. It would be based on leys for livestock production and perennial crops for food, livestock and bioenergy production. Crops such as cereals, potatoes, sugar beet and rapeseed would be grown in reduced acreages due to partial replacement of feed grains with forage. Catch crops and minimum-tillage would be used extensively. Crops for fibre, phytochemical purposes, and energy would play an important role. The number of beef cattle and sheep would increase somewhat while the number of pigs would decrease. There would be a decrease in pigs by about 300 000 head and a corresponding increase in cattle. Sheep production could also increase (Tab. 8.10).

These changes could result in a decrease of cereals in the order of 300 000 ha and a corresponding increase in leys. Crops for fibre, energy, and for use by the chemical industry would cover 200 000 ha of arable land. Marginally arable and semi-natural grasslands (ca. 500 000 ha) could be used for energy forestry or more long-term timber production (Finansdepartementet 1987).

A farming system of this kind would probably reduce nitrogen leaching from 25–30 kg to approximately 5–10 kg ha^{-1} yr^{-1}. The need for herbicides (on an areal basis) would probably decrease considerably, and the use of fungicides and insecticides would also decrease. The use of fertilizers would decrease by about 25% since there would be an increase in the use of farmyard manure. The new system would also contribute to a more diversified landscape.

The over-production of cereals (grain) would cease, but a smaller surplus of beef and even milk could occur instead. It would also be necessary to make new investments in buildings and machines, to develop new cropping techniques, and to modify techniques for livestock production and handling of farmyard manure. Finally, an increased number of farmers would become involved in livestock management. Therefore, this model for the near future is probably not the best one.

A modification of this model would be to reduce the role of livestock – and consequently of leys – in favour of fibre crops, phytochemical crops, and energy crops which can be called the food-feed-fibre-energy system (Tab. 8.11). This would avoid some of the old problems resulting from surpluses of meat, butter and milk as well as from the handling and use of manure from large cattle, pig and poultry units. It would probably also be better for farmers, from a socio-economic point of view, to be occupied more with crops and less with livestock management.

Step number three – a future scenario
There have been many poor predictions made about the future of agriculture. It is very difficult to put forth scenarios with reasonable certainty; they often give a ridiculous impression and often only reflect wishful thinking. Nevertheless, it is urgent to continue addressing the problems associated with agriculture in the industrialized world. There is a food surplus that we cannot handle properly; there are also problems of food costs and distribution – i.e., there are still undernourished people in the world even though there is a food surplus in some areas; the environment is being polluted; the quality of some agricultural products has decreased; and there are unnecessary reductions in food quality resulting from industrial processing, transport and storage.

However, it is doubtful that there is any realistic chance of developing a better agricultural system, let us call it "alternative agriculture", until we alter our myopic perspective of the problem. We must perceive agriculture as one part of industrialized society. All parts of the system are interdependent. We must design a system that efficiently recycles nutrients through minimizing nutrient losses from the system. On arable soil there should not only be cash crops and forage crops but

Tab. 8.10. Proposed use of acreage and livestock resulting from an increase of a ley-farming system.

Crop/livestock	Before (1987)	After (1997)
Crop acreages (10^3 ha)		
Cut leys	700	1000
Cereals, feed	1220	900
Cereals, food	390	350
Rapeseed	170	160
Potatoes	40	40
Sugar beet	50	50
Grazed ley	180	300
Energy crops	0	200
Grassland	300	400
Livestock (10^3 head)		
Dairy cows	655	600
Dairy cattle used for beef	62	300
Beef cattle	1160	1400
Sheep	177	300
Pigs	2600	2300

Tab. 8.11. Proposed use of arable land in a future food-feed-fibre-energy system.

Crops	Acreage (10^3 ha)
Cereals, food	300
Cereals, feed	600
Rapeseed	180
Potatoes	40
Sugar beet	50
Leys (legume/grass)	1000[a]
Fibre crops	300
Phytochemical crops	100
Energy crops	300
Grazed ley	250
Sum	3120

[a]Part of the ley can be fibre crops (lucerne, canary grass).

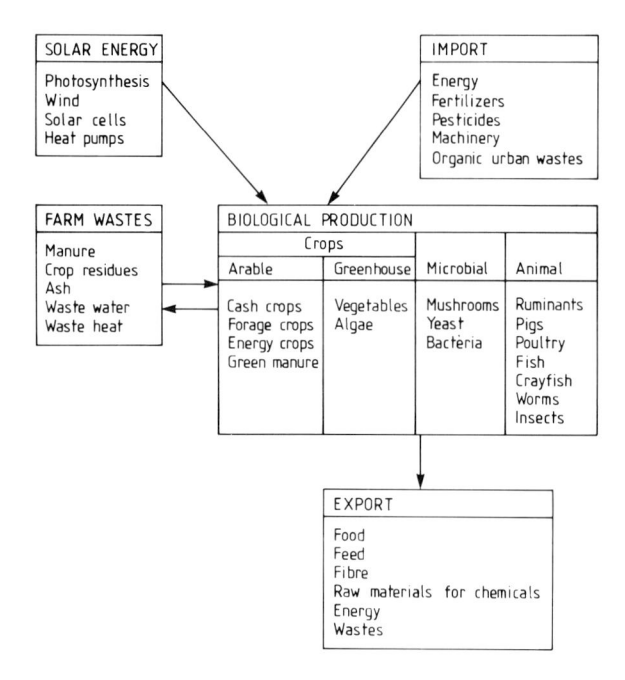

SOLAR ENERGY	IMPORT
Photosynthesis	Energy
Wind	Fertilizers
Solar cells	Pesticides
Heat pumps	Machinery
	Organic urban wastes

FARM WASTES	BIOLOGICAL PRODUCTION			
Manure	Crops			
Crop residues	Arable	Greenhouse	Microbial	Animal
Ash	Cash crops	Vegetables	Mushrooms	Ruminants
Waste water	Forage crops	Algae	Yeast	Pigs
Waste heat	Energy crops		Bacteria	Poultry
	Green manure			Fish
				Crayfish
				Worms
				Insects

EXPORT
Food
Feed
Fibre
Raw materials for chemicals
Energy
Wastes

Fig. 8.1. A future agricultural system.

also crops providing leaf protein, fibre, green manure and energy. This includes trees, shrubs and vegetables. The harvesting of greenhouse products and algal crops should be expanded. Animal production must be regarded in its entirety; e.g., cattle, sheep, poultry, fish, crayfish, mussels, earthworms and insects. Mushrooms and other microorganisms should also be exploited (Anderson 1979, Steen 1988a, b). All of these components must be interlinked and responsive to changes in other parts of the system. However, this is still not enough.

There will be a need for recirculation of surpluses, residues and wastes from urban areas as well as from the primary, secondary and tertiary steps of production. The list of material to be recycled includes organic matter, nitrogen, phosphorus, calcium, methane, carbon dioxide, waste water and waste heat. This is not an easy task to accomplish since it deeply affects people's way of life. It will be necessary to separate different fractions of garbage and wastes at the consumer source. New sewage systems, capable of separating organic materials and plant nutrients from other chemicals, metals, plastics, etc. will be needed.

The production of food, feed and fibre should be coordinated within geographically restricted areas. The components of such a system have been fitted into a conceptual model (Fig 8.1). Some major flows between them are indicated. Others are left open. This is not a system for tomorrow, not even for the near future. However, it is time to begin planning for such alternatives, since the time may come when we will be left with no choice but to change.

9. Epilogue

Thomas Rosswall, Olof Andrén, Torbjörn Lindberg and Keith Paustian

9. Epilogue. T. Rosswall, O. Andrén, T. Lindberg and K. Paustian
Project history 195
Project administration 197
Post-graduate education................... 199

Project history

In December 1975, researchers at several departments at the Swedish University of Agricultural Sciences (SLU) initiated discussions about developing a multidisciplinary research programme to study Swedish agriculture from an ecological and resource conservation perspective. In January 1977, Professors Börje Norén at the Department of Microbiology and Eliel Steen at the Department of Ecology and Environmental Research submitted a grant proposal for an agroecosystem project with the title "Ecological Systems of Arable Land. How are They Influenced by Swedish Agriculture?" to several research councils. The proposal was shelved at the Natural Science Research Council (NFR) and the Council for Forestry and Agricultural Research (SJFR). The proposal was also sent to the Swedish Council for Planning and Coordination of Research (FRN). At the initiative of FRN, a committee was established by the research councils to serve as a reference group for further development of a new project proposal. FRN awarded a small planning grant to the committee to develop the project proposal. The members of this ad hoc committee were Bengt Lundholm (FRN), Nils Malmer (NFR) and Olle Johansson (SJFR). In cooperation with researchers at SLU, a detailed project proposal was submitted to the research councils in January 1979. The committee also developed a budget outline for the project, and this was to serve as a financial guideline for the allocation of funds and for the relative contributions from the different research councils. This plan was very valuable, although it did not represent any binding commitment from the research councils.

As part of the preparatory activities, two workshops were arranged for further planning and a hearing was organized with participation from the research councils, the scientists involved in the preparation of the revised grant proposal and researchers not involved in the preparatory work. These were important steps in the development of a revised version of the grant proposal. Four sub-project proposals were sent to SJFR and NFR in January 1978, and two of those received financial support starting on 1 July, 1978. A revised proposal for an integrated ecosystems project was submitted to the research councils in January 1979.

Historically, ecology had not been a major research area at the agricultural faculty of SLU, probably because agricultural research was often carried out by persons with a management-oriented agronomic background. Basic ecosystem research was carried out at university departments of botany and zoology. Ecologists had not yet discovered that agroecosystems are ideal for studying many aspects of ecology, especially ecosystem perturbations.

Swedish agricultural research has traditionally been almost exclusively production-oriented. Most ecological studies in agricultural settings were of an autecological nature and concerned the relationship of a single crop, weed or pathogen with its abiotic and biotic environment. The results were evaluated in economic terms with short-term profitability as the dominant value-judgement factor. The initiators of the "Ecology of Arable Land" project felt that increased attention should be given to research aimed at a thorough understanding of agroecosystems with an ultimate aim of designing agricultural systems capable of long-term sustainable production without negative environmental consequences. It was felt that such a study could provide a knowledge base, which could be utilized at SLU and the extension services in the development of new cropping systems for Swedish agriculture. The intention was not to produce results which would be of direct use to an individual farmer's choice of cropping or management systems.

It was decided that the major attention of the project would be devoted to soil processes. Although soil is a major component of the agricultural production system, little attention had been devoted to detailed studies of biological components. Soil had often been considered as a black box in studies of nutrient cycling and energy flows. In the 1979 proposal it was stated that a focus on soil animals and microorganisms was necessary, since they are the biotic components affecting the actual and potential fertility of arable land. Until recently, there had been no research at SLU on soil animals apart from studies of plant pathogens, and even the soil microbiology research had been very limited and mainly devoted to studies of symbiotic nitrogen fixation by bacteria/legumes.

The objective of the project in this first grant proposal was as follows:

– To investigate the role of soil fauna and soil microorganisms, especially their importance for the circulation of carbon and nitrogen, to illuminate important biological processes related to soil fertility.

Special attention was to be given to cropping systems of current use and three crop rotations were selected. Four subordinate objectives were also formulated:

- to quantify the nitrogen cycle in three cropping systems (monoculture of a grain crop, a grass ley and a lucerne ley)
- to describe the dynamic processes regulating the nitrogen cycle within a systems-analysis framework
- to describe the effect of management practices on soil organic matter and to develop a simulation model to clarify which factors were important in determining the effects of such practices in a long-term perspective
- to determine the effects of cropping systems (additions of organic and inorganic fertilizers, crop residues and the use of certain pesticides) on important groups of soil organisms.

The grant proposal divided the project into two major components: (1) nitrogen cycling and the long-term effects on soil organic matter and (2) studies on the effects of management practices on important groups of soil organisms.

The first part of the grant proposal was centered around nitrogen flows and transformations and a number of sub-projects were proposed, that dealt with the various transformations in detail. The second part of the proposal contained sub-projects on the effects of cropping systems on the soil fauna, soil fauna as primary consumers, the role of soil fauna in the turnover of crop residues and manure, the effect of soil fauna on soil physical characteristics, studies on microbial diversity, and the role of fauna in the above-ground crop.

It was suggested that the major parts of the project would focus on three cropping systems: (1) monoculture of barley with removal of all crop residues after harvest, (2) a four-year lucerne ley, and (3) a four-year grass ley. The barley systems would receive three levels of nitrogen fertilizers and the grass ley two, giving a total of six treatments. The field plan called for three parallel plots for each treatment giving a total of 18 plots. It was also stated that for certain studies, especially those concerning the effect of management practices on soil organic matter content, long-term fertility experimental plots already established at SLU should be used.

Although the grant had been approved in 1979, the research site had not yet been selected, but certain site selection criteria had been specified; (1) a relatively light soil to facilitate soil sampling and especially the root studies, (2) an underlying clay layer to facilitate leaching measurements, (3) close proximity to SLU (Ultuna, Uppsala). It had been decided that most efforts would be concentrated at one site since the financial and personnel resources would only be adequate for detailed investigations at one location. The rationale behind this decision was that if we could understand the selected cropping systems at one site in detail, this knowledge could then be extrapolated to other soil types, climatic regions and cropping systems.

None of the four sites originally selected and listed in the 1979 grant proposal were later considered suitable. The research site finally selected, located at Kjettslinge, Örbyhus estate, was leased for a seven-year period (see Chapter 2).

The 1979 project proposal was successful and grants of SEK 1,359,000 were received. This was considerably less than the planned budget of SEK 2,708,000, but it was enough to initiate the project. Börje Norén and Eliel Steen accepted the responsibility of principal investigators and Thomas Rosswall was appointed scientific coordinator.

In preparing for the next grant proposal, major revisions were made and the title of the project was changed to "Ecology of Arable Land. The Role of Organisms in Nitrogen Cycling", which better reflected the contents of the revised project. The original design with six croppings with three replicates was modified to four cropping systems with four replicates.

The main objective was reformulated, i.e., "to investigate the functions of soil microorganisms and soil fauna, with particular attention to their importance for the circulation of nitrogen in four cropping systems with differing nitrogen input, primary production above- and below-ground, and organic matter incorporation in the soil."

The subordinate objectives were:

- to quantify the nitrogen budget in four cropping systems, including an analysis of the dynamic processes regulating nitrogen circulation
- to evaluate the influence of the cropping systems on soil microorganisms and soil fauna and their function in relation to nitrogen transformations and decomposition.

These objectives did not change during the project (except for minor details, see Chapter 1), but the emphasis gradually shifted from individual, although coordinated, investigations to a more integrated approach. This was in accordance with the original long-term plans, which comprised extensive field investigations during the 1981–1984 field seasons and supplementary experiments and synthesis work thereafter. The total grant received during the project amounted to over 20 M SEK. Annual funding was greatest in 1983/84 (Fig. 9.1). Additional funding was obtained in the form of salaries for permanent staff and doctoral student fellowships, especially during the later parts of the project. The project involved 25 scientists, including doctoral students, during the most intensive periods (Andrén 1988). Most researchers were affiliated with the Departments of Ecology and Environmental Research; Microbiology; and Soil Sciences at the Swedish University of Agricultural Sciences. In addition, scientists from the Departments of Meteorology and Zoology at the University of Stockholm, Department of Water and Environmental Research at the University of Linköping and the Swedish Air and Water Pollution Laboratory, Gothenburg participated. Project results to date have been

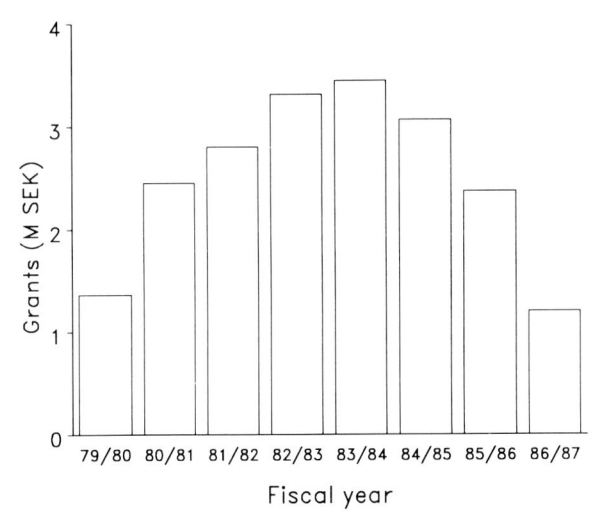

Fig. 9.1. Grants received from funding agencies during 1979–1987 (Andrén 1988).

presented in over 150 scientific publications (See Appendix 3).

Project administration

The research councils appointed two committees. One committee, the Project Committee, consisted of representatives from the research councils, SLU and the Swedish Farmers' Association (LRF) (Tab. 9.1), and served as the contact between the project and the research councils and between the councils and the Scientific Advisory Committee (SAC). The second committee, SAC, which was appointed by the councils in October 1979, was set up to "assist in planning and evaluating the project and also to keep the project in contact with the international research front". The members of the SAC are listed in Tab. 9.1. This committee wrote a yearly evaluation report to the research councils. These reports by the SAC provided a constructive outside review of the project and influenced the development of the project in many ways. At their first meeting in January of 1980, the committee stated their: "... unanimous approval of the need for a project of this type and in the general principles of experimentation described. We believe it will be a significant contribution to both fundamental science and to future methods of agricultural resource management judged on both a European and a world basis; only a few

Tab. 9.1. Members of the Project Management Committee, the Project Synthesis Committee and Editorial Committee and Synthesis Volume Editors, the Project Committee and the Scientific Advisory Committee (SAC) during 1979–1989.

Project Management Committee 1979/80–1986/87

Principal Investigators
Eliel Steen 1979/80–1982/83
Börje Norén 1979/80–1982/83
Thomas Rosswall 1983/84–1988/89

Scientific Coordinators
Thomas Rosswall 1979/80–1982/83
Bo Svensson 1983/84–1984/85
Olof Andrén 1983/84–1986/87
Torbjörn Lindberg 1985/86–1986/87
Ann-Charlotte Hansson 1987/88

Site Coordinator
Ruben Johansson 1979/80–1985/86

Other members
Olof Andrén 1980/81, 1982/83
Lars Bergström 1983/84–1986/87
Solveig Geidnert 1980/81–1982/83
Ann-Charlotte Hansson 1981/82–1986/87
Leif Klemedtsson 1986/87
Jan Lagerlöf 1979/80
Keith Paustian 1983/84–1986/87
Jan Persson 1979/80–1983/84
Johan Schnürer 1984/85–1985/86
Bo Svensson 1980/81–1982/83

Project Synthesis Committee 1983–1985

Olof Andrén (Chairman)
Keith Paustian
Thomas Rosswall
Eliel Steen
Bo Svensson

Project Editorial Committee 1986–1987

Olof Andrén (Chairman)
Marianne Clarholm
Ann-Charlotte Hansson
Per-Erik Jansson
Torbjörn Lindberg
Keith Paustian
Thomas Rosswall
Eliel Steen

Synthesis Volume Editors 1986–1989

Olof Andrén
Torbjörn Lindberg
Keith Paustian
Thomas Rosswall

Project Committee 1979/80–1986/87

Bert Bolin (NFR) Chairman 1979–1987
Rune Andersson (SNV) 1981–1987
August Håkansson (SLU) 1979–1982
Olle Johansson (SJFR) 1979–1982
Waldemar Johansson (SLU) 1982–1987
Bo Lindborg (NFR) 1982
Bengt Lundholm (FRN) 1979–1982
Nils Malmer (NFR) 1983–1987
Ingmar Månsson (SJFR) 1982–1987
Anders Nilsson (LRF) 1979–1987
Birgitta Norkrans (NFR) 1982–1983
Tryggve Troedsson (FRN) 1982–1987
(SNV = the Swedish Environment Protection Board)

Scientific Advisory Committee 1979/80–1986/87

James P. Curry (Ireland)
Klaus H. Domsch (FRG)
Martin J. Frissel (the Netherlands)
Peter Newbould (UK)
James M. Tiedje (USA)
Robert G. Woodmansee (USA)

integrated projects of this nature are known to us ... We admire the desire to bring together soil zoologists, microbiologists, meteorologists, botanists, agronomists and soil chemists to tackle four of the main problems central to the development of studies in this part of the biogeochemical cycle:

– the application of a systems approach to agriculture
– the collection of data on mineralization and immobilization of nitrogen, which is urgently needed in studies of the nitrogen cycle
– the linkage of studies on carbon and nitrogen cycles in nature
– the attempt to integrate work on soil fauna with the processes of primary production and decomposition of organic matter".

The committee also strongly recommended that only four cropping systems with four replicates should be investigated and not the six systems with three replicates suggested in the grant proposal. In general, the committee commended both the overall aims of the project and the contents of individual sub-projects.

The project was also evaluated in an international review of research in the environmental sciences at SLU. The evaluation report, which was made available in 1984, was very complimentary to the project: "In all, I am convinced that this programme is of outstanding scientific value. The intensity by which the problems are studied and the completeness of the objectives are unique in this field of research. The results will become the base for further applied research or, at least, will be the key for understanding research already obtained" (K. Baeumer, FRG). "This is a splendid project of very great importance to all ecological studies involving nitrogen. It is excellently planned, the work is very broad in scientific scope and many types of difficult measurements are being made" (G. W. Cooke, UK).

The late Börje Norén and Eliel Steen, the prime movers behind the conception of the project and its initiation, acted as co-principal investigators during the first years. A Project Management Committee consisting of project scientists was also set up, which was responsible for assuring that the project was developed according to the original plans and the grant proposals. This management committee had a rotating membership (Tab. 9.1) and it was later elected by the project participants among those involved in the project, including the technical staff. It was felt that a democratically elected management committee would increase the sense of involvement of all project personnel in the scientific and administrative management and that there would be a mutual trust between the governing body and all project participants. The principal investigator (s) and the scientific coordinator(s) as well as the site coordinator were ex officio members of the project management committee.

Decisions about the division of available funds between the different sub-projects were made by the project management committee. Sub-project leaders were given the responsibility for their own grants, and the money could be spent as they saw fit – as long as they kept within the budget. This transfer of responsibilities to sub-project leaders, who were mostly graduate students, created an efficient and decentralized project structure, and forced the students to take responsibility both for their own work and for the work of collaborators and technical personnel. The project management committee, however, always reserved funds, in the order of SEK 100 000 per year, to be used for unforeseen expenses. Central funds were also set aside for site management, meteorological data collection, travel, both within Sweden and internationally, as well as for project seminars and visits by SAC. Through a local administrative computer system, each sub-project leader received a monthly economic statement. The responsibility for appointments of the individuals rested with the department where the persons worked. The project paid for the salaries but the responsibility for employment security did not rest with the project.

The project management committee was discontinued in 1987, when the experimental phase of the project was finished. The responsibility of directing the final synthesis of project results was changed to a Project Synthesis Committee, selected from members of the project management committee. When the senior authors of each chapter in the synthesis volume had been selected, these were, together with the editors of the proceedings of the 'Ecology of Arable Land – The Role of Organisms in Nitrogen Cycling – Perspectives and Challenges' (see Preface) included to form a Project Editorial Committee.

The decentralized and democratic way in which the project was run has created interest outside the project. The National Accounting and Audit Bureau, in a study of research within the agricultural sector, showed a keen interest to learn more about the project management and considered our organization as a possible model for other research projects/programmes involving more than one department at the university (Anonymous 1988).

Those of us who worked on the project feel that the ambitious integrated approach was appropriate but that the synthesis, in spite of this volume, is still incomplete. There are many reasons for this. The amount of time required to prepare primary publications, from several intensive years of field work, delays the coordination and synthesis needed to progress from individual sub-project papers to integrated project publications. In addition, synthesis of primary results from a project of this size is both time and labour demanding. We estimate that about four full-time scientists would be needed during two years for an optimal level of synthesis after the completion of the core project.

Post-graduate education

One of the aims of the project was to serve as a training programme in cooperative research for graduate students. This caused some initial problems, since the project was understaffed with senior scientists having a thorough knowledge of scientific research, proper methodological skills and experience in scientific writing. On the other hand, the younger scientists' enthusiasm and willingness to adapt compensated for their lack of experience.

In order for the graduate students to learn from past experiences and to make personal contacts with the international scientific community, it was the policy of the project to ensure that the students could visit appropriate research groups abroad. Once they started to get results from their research, active participation in international meetings was also encouraged and supported with project funds. The project management committee also decided that all publications from the project should be in English, something which is not that common at SLU. Yearly progress reports were also written in English. In order for the students to get training in scientific writing, they were asked to write in the form of short "scientific papers" for the annual progress reports (Appendix 2). Students also presented their work in English at numerous required seminars, e.g., during the yearly visits of the SAC.

Many Ph. D. theses were fully or partly written as part of the project, and several are still to be completed:

Andrén, O. 1984. Soil Mesofauna of Arable Land and its Significance for Decomposition of Organic Matter. – Dept of Ecology and Environmental Research, Report 16. Swedish Univ. of Agr. Sci., Uppsala.

Berg, P. 1986. Nitrifier Populations and Nitrification Rates in Agricultural Soil. – Dept of Microbiology, Report 34. Swedish Univ. of Agr. Sci., Uppsala.

Bergström, L. 1987. Transport and Transformation of Nitrogen in an Arable Soil – Ekohydrologi 23. Dept of Soil Science, Swedish Univ. of Agr. Sci., Uppsala.

Bonde, T. 1990. In preparation. – Dept of Water and Environmental Research, Univ. of Linköping, Linköping.

Boström, S. 1988. Morphological and Systematic Studies of the Family Cephalobidae (Nematoda: Rhabditida). – Dept of Zoology, Univ. of Stockholm, Stockholm.

Boström, U. 1988. Ecology of Earthworms in Arable Land. Population Dynamics and Activity in Four Cropping Systems. – Dept of Ecology and Environmental Research, Report 34. Swedish Univ. of Agri. Sci., Uppsala.

Clarholm, M. 1983. Dynamics of Soil Bacteria in Relation to Plants, Protozoa and Inorganic Nitrogen. – Dept of Microbiology, Report 17. Swedish Univ. of Agr. Sci., Uppsala.

Ferm, M. 1986. Concentration Measurements and Equilibrium Studies of Ammonium. Nitrate and Sulphur Species in Air and Precipitation. – Dept of Inorganic Chemistry, Univ. of Göteborg, Göteborg.

Hansson, A.-C. 1987. Roots of Arable Crops: Production, Growth Dynamics and Nitrogen Content. – Dept of Ecology and Environmental Research, Report 28. Swedish Univ. of Agr. Sci., Uppsala.

Johansson, C. 1988. Exchange of Nitrogen Oxides between the Atmosphere and Terrestrial Surfaces. – Dept of Meteorology, Univ. of Stockholm, Stockholm.

Johansson, G. In preparation. – Dept of Soil Science, Swedish Univ. of Agr. Sci., Uppsala.

Johnsson, H. In preparation. Nitrogen Losses from Agricultural Soils – A Model Approach. – Dept of Soil Science, Swedish Univ. of Agr. Sci., Uppsala.

Klemedtsson, L. 1986. Denitrification in Arable Soil with Special Emphasis on the Influence of Plant Roots. – Dept of Microbiology, Report 32. Swedish Univ. of Agr. Sci., Uppsala.

Lagerlöf, J. 1987. Ecology of Soil Fauna in Arable Land. Dynamics and Activity of Microarthropods and Enchytraeids in Four Cropping Systems. – Dept of Ecology and Environmental Research, Report 29. Swedish Univ. of Agr. Sci., Uppsala.

Lofs-Holmin, A. 1983. Earthworms in Arable Land. Methods for Studying Agricultural Impact, Especially Pesticides. – Dept of Ecology and Environmental Research. Swedish Univ. of Agr. Sci., Uppsala.

Lundin, L.-C. 1989. Water and Heat Flows in Frozen Soils. – Acta Univ. Ups. 186. Dept of Hydrology, Uppsala Univ.

Mårtensson, A. 1985. Ecology of *Rhizobium*. Methods for Strain Identification and Quantification of Nitrogen Fixation. – Dept of Microbiology, Report 30. Swedish Univ. of Agr. Sci., Uppsala.

Paustian, K. 1987. Theoretical Analyses of C and N Cycling in Soil. – Dept of Ecology and Environmental Research, Report 30. Swedish Univ. of Agr. Sci., Uppsala.

Pettersson, R. 1987. Primary Production in Arable Crops: Above-ground Growth Dynamics, Net Production and Nitrogen uptake. – Dept of Ecology and Environmental Research, Report 31. Swedish Univ. of Agr. Sci., Uppsala.

Robertson, K. In preparation. Nitrogen Cycling in a Regional Context with Special Emphasis on Denitrification. – Dept of Water and Environmental Research, Univ. of Linköping, Linköping.

Schnürer, J. 1985. Fungi in Arable Soil. Their Role in Carbon and Nitrogen Cycling. – Dept of Microbiology, Report 29. Swedish Univ. of Agr. Sci., Uppsala.

Wessén, B. 1983. Decomposition of Some Forest Leaf Litters and Barley Straw – Some Rate-Determining Factors. – Dept of Microbiology, Report 19. Swedish Univ. of Agr. Sci., Uppsala.

References

Ågren, G. and Axelsson, B. 1980. Population respiration: a theoretical approach. – Ecol. Model. 11: 39–54.
– and Ingestad, T. 1987. Root: shoot ratio as a balance between nitrogen productivity and photosynthesis. – Plant Cell Environ. 10: 579–586.
Ångström, A., Liljeqvist, G. H. and Wallén, C. C. 1974. Sveriges Klimat. – Generalstabens litografiska anstalt, Stockholm.
Abdel Wahab, A. M. 1980. Diurnal variations of $N_2(C_2H_2)$-fixing activity in *Medicago sativa* under water stress. – Nat. Monspel. Ser. Bot. 35: 1–7.
Adams, T. McM. and Laughlin, R. J. 1981. The effects of agronomy on the carbon and nitrogen in the soil biomass. – J. Agric. Sci., Cambridge 97: 319–327.
Allison, F. E. 1955. The enigma of soil nitrogen balance sheets. – Advan. Agron. 7: 213–250.
– 1973. Soil organic matter and its role in crop production. – Developments in Soil Science 3. Elsevier, Amsterdam-London-New York.
Alvenäs, G., Johnsson, H. and Jansson, P.-E. 1986. Meteorological conditions and soil climate of four cropping systems. Measurements and simulations from the project "Ecology of Arable Land". – Dep. of Ecology and Environmental Research, Report 24. Swedish Univ. of Agricultural Sciences, Uppsala.
Ambrosiani, B. 1983. Background to the boat-graves of the Mälaren valley. – In: Lamm, J. P. and Nordström, H.-A. (eds), Vendel Period Studies. The Museum of National Antiquities, Stockholm, Studies 2, pp. 17–22.
Andersen, A. 1986. Rodvækst i forskellige jordtyper. – Tids. Planteavl Specialserie. Beretning nr. S 1827: 1–90.
Andersen, H. J. 1979a. Migratory nematodes in Danish barley fields. I. The qualitative and quantitative composition of the fauna. – Tids. Planteavl 83: 1–8.
– 1979b. Migratory nematodes in Danish barley fields. II. Population dynamics in relation to continuous barley growing. – Tids. Planteavl 83: 9–27.
Andersen, N. C. 1983. Nitrogen turnover by earthworms in arable plots treated with farmyard manure and slurry. – In: Satchell, J. E. (ed.), Earthworm ecology from Darwin to vermiculture. Chapman and Hall, London and New York, pp. 139–150.
– 1987. Investigations of the ecology of earthworms (Lumbricidae) in arable soils. – Tids. Planteavl Specialserie. Beretning nr. S 1871. Copenhagen, pp. 195.
Anderson, I. C. and Levine, J. S. 1987. Simultaneous field measurements of nitric oxide and nitrous oxide. – J. Geophys. Res. 92: 965–976.
Anderson, J. M. and Ineson, P. 1984. Interactions between microorganisms and soil invertebrates in nutrient flux pathways of forest ecosystems. – In: Anderson, J. M., Rayner, A. D. M. and Watson, D. W. S. (eds), Invertebrate-microbial interactions. Cambridge Univ. Press, Cambridge, pp. 89–132.
Anderson, R. E. 1979. Biological paths to self-reliance. A guide to biological solar energy conversion. – Van Nostrand Reinhold, New York.
Anderson, R. V., Coleman D. C., and Cole, C. V. 1981. Effects of saprotrophic grazing on net mineralization. – In: Clark, F. E. and Rosswall, T. (eds), Terrestrial nitrogen cycles. Ecol. Bull. (Stockholm) 33: 201–216.
Andersson, R. 1986. Förluster av kväve och fosfor från åkermark i Sverige. – Ph. D. Thesis, Dep. of Soil Sciences, Swedish Univ. of Agricultural Sciences, Uppsala (English summary: Losses of nitrogen and phosphorus from arable land in Sweden).
Andrássy, I. 1956. Die Rauminhalts- und Gewichtsbestimmung der Fadenwürmer (Nematoden). – Acta Zool. Acad. Sci. Hung. 2: 1–15.
Andrén, O. 1984. Soil mesofauna of arable land and its significance for decomposition of organic matter. – Ph. D. Thesis, Dep. of Ecology and Environmental Research, Report 16. Swedish Univ. of Agricultural Sciences, Uppsala.
– 1985. Microcomputer-controlled extractor for soil microarthropods. – Soil Biol. Biochem. 17: 27–30.
– 1987. Decomposition of shoot and root litter of barley, lucerne and meadow fescue under field conditions. – Swed. J. agric. Res. 17: 113–122.
– 1988. Ecology of Arable Land – an integrated project. – Ecol. Bull. (Copenhagen) 39: 131–133.
– and Lagerlöf, J. 1980. The abundance of soil animals (Microarthropoda, Enchytraeidae, Nematoda) in a crop rotation dominated by ley and in a rotation with varied crops. – In: Dindal, D. (ed.), Soil biology related to land use practices. Environment Protection Agency, Washington D. C., pp. 274–279.
– and Lagerlöf, J. 1983. Soil fauna (microarthropods, enchytraeids, nematodes) in Swedish agricultural cropping systems. – Acta Agric. Scand. 33: 33–52.
– and Schnürer, J. 1985. Barley straw decomposition with varied levels of microbial grazing by *Folsomia fimetaria* (L.) (Collembola, Isotomidae). – Oecologia (Berl.) 68: 57–62.
– and Paustian, K. 1987. Barley straw decomposition in the field: a comparison of models. – Ecology 68: 1190–1200.
– Hansson, A.-C. and Pettersson, R. 1987. Contributions to soil organic matter from four arable crops. – INTECOL Bull. 15: 41–47.
– Paustian, K. and Rosswall, T. 1988. Soil biotic interactions in the functioning of agroecosystems. – Agric. Ecosys. Environ. 24: 57–68.
Andrzejewska, L. 1976. The influence of mineral fertilization on the meadow phytophagous fauna. – Pol. Ecol. Stud. 2: 93–109.
Anonymous, 1988. Går det att förändra forskning? Exemplet Sveriges lantbruksuniversitet. Revisionsrapport. – Riksrevisionsverket (National Audit Bureau), Stockholm.
Aulakh, M. S., Rennie, D. A. and Paul, E. A. 1983. Field studies on gaseous nitrogen losses from soils under continuous wheat versus a wheat-fallow rotation. – Pl. Soil 75: 15–27.
– Rennie, D. A. and Paul, E. A. 1984. Gaseous nitrogen losses from soil under zero-till as compared with conventional-till management systems. – J. Environ. Qual. 13: 130–136.
Axelsson, B., Lohm, U. and Persson, T. 1984. Enchytraeids, lumbricids, and soil arthropods in a northern deciduous woodland – a quantitative study. – Holarct. Ecol. 7: 91–103.
Bååth, E. and Söderström, B. 1979. The significance of hyphal diameter in calculation of fungal biovolume. – Oikos 33: 11–14.
– and Söderström, B. 1980. Comparison of the agar film and membrane-filter methods for the estimation of hyphal lengths in soil, with particular reference to the effects of magnification. – Soil Biol. Biochem. 12: 385–387.
– and Söderström, B. 1982. Seasonal and spatial variation in fungal biomass in a forest soil. – Soil Biol. Biochem. 14: 353–358.
– , Lohm, U., Lundgren, B., Rosswall, T., Söderström, B. and Sohlenius, B. 1981. Impact of microbial-feeding animals on total soil activity and nitrogen dynamics: a soil microcosm experiment. – Oikos 37: 257–264.
Bailey, L. D. 1976. Effects of temperature and roots on denitrification in a soil. – Can. J. Soil Sci. 56: 79–87.
Baker, J. L. and Johnson, H. P. 1981. Nitrate-nitrogen in tile drainage as affected by fertilization. – J. Environ. Qual. 10: 519–522.
Bakken, L. R. 1983. The turnover of C and N in cultivated soil at different fertilizer levels. – Ph. D. Thesis, Dep. of Microbiology, Agricultural Univ. of Norway, Ås.

Balandreau, J. and Dommergues, Y. 1973. Assaying nitrogenase (C_2H_2) activity in the field. – In: Rosswall, T. (ed.), Modern methods in the study of microbial ecology. Bull. Ecol. Res. Comm. (Stockholm) 17: 247–254.

Balderston, W. L., Sherr, B. and Payne, W. J. 1976. Blockage by acetylene of nitrous oxide reduction in *Pseudomonas perfectomarinus*. – Appl. Environ. Microbiol. 31: 504–508.

Balsberg, A.-M. 1982. Plant biomass, primary production and litter disappearance in a *Filipendula ulmaria* meadow ecosystem, and the effects of cadmium. – Oikos 38: 72–90.

Barber, S. A. 1977. Nutrients in soil and their flow to plant roots. – In: Marshall, J. K. (ed.), The belowground ecosystem: A synthesis of plant-associated processes. Range Sci. Dept Sci. Ser. No 26, Colorado State Univ., Fort Collins, Co, pp. 161–170.

Barley, K. P. 1959. The influence of earthworms on soil fertility. I. Earthworm populations found in agricultural land near Adelaide. – Austr. J. agric. Res. 10: 171–178.

– and Jennings, A. C. 1959. Earthworms and soil fertility. III. The influence of earthworms on the availability of nitrogen. – Austr. J. agric. Res. 10: 364–370.

Bartholomew, W. V. 1965. Mineralization and immobilization of nitrogen in the decomposition of plant and animal residues. – In: Bartholomew, W. V. and Clark, F. E. (eds), Soil nitrogen. Agronomy 10: 285–306.

Bauer, A. and Black, A. L. 1981. Soil carbon, nitrogen and bulk density comparison in two cropland tillage systems after 25 years and in virgin grassland. – Soil Sci. Soc. Am. J. 45: 1166–1170.

Bavel, C. H. M. van and Hillel, D. 1976. Calculating potential and actual evaporation from a bare soil surface by simulation of concurrent flow of water and heat. – Agric. Meteorol. 17: 453–476.

Beadle, C. L. 1985. Plant growth analysis. – In: Coombs, J., Hall, D. O., Long, S. P. and Scarlock, J. M. O. (eds), Techniques in bioproductivity and photosynthesis. 2nd edition. Pergamon Press, Oxford, pp. 20–25.

– , Long, S. P., Imbamba, S. K., Hall, D. O. and Olembo, R. J. 1985. Photosynthesis in relation to plant production in terrestrial environments. Light responce curve. – Tycooly Publ. Ltd., Oxford, pp. 60–61.

Beauchamp, E. G., Kidd, G. E. and Thurtell, G. 1982. Ammonia volatilization from liquid dairy cattle manure in the field. – J. Soil Sci. 62: 11–19.

Belford, R. K., Klepper, B. and Rickman, R. W. 1987. Studies of intact shoot-root systems of field-grown winter wheat. II. Root and shoot developmental patterns as related to nitrogen fertilizer. – Agron. J. 79: 310–318.

Belser, L. W. 1979. Population ecology of nitrifying bacteria. – Ann. Rev. Microbiol. 33: 309–333.

– and Mays, E. L. 1982. Use of nitrifier activity measurements to estimate the efficiency of viable nitrifier counts in soils and sediments. – Appl. Environ. Microbiol. 43: 945–948.

Benckiser, G., Haider, K. and Sauerbeck, D. 1986. Field measurements of gaseous nitrogen losses from an alfisol planted with sugar-beets. – Z. Pflanzenernaehr. Bodenk. 149: 249–261.

Bengtsson, A. 1985. Stråsäd. Trindsäd. Oljeväxter. Sortval 1986. – Aktuellt från Lantbruksuniversitetet 348. Sveriges Lantbruksuniversitet, Uppsala (in Swedish).

Bengtsson, G., Erlandsson, A. and Rundgren, S. 1988. Fungal odour attracts soil Collembola. – Soil Biol. Biochem. 20: 25–30.

Bengtsson, S.-A., Nilsson, A., Nordström, S. and Rundgren, S. 1979. Short-term colonization success of lumbricid founder populations. – Oikos 33: 308–315.

Berg, P. 1986. Nitrifier populations and nitrification rates in agricultural soil. – Ph. D. Thesis, Dept of Microbiology, Report 34, Swedish Univ. of Agricultural Sciences, Uppsala.

– and Rosswall, T. 1985. Ammonium oxidizer numbers, potential and actual oxidation rates in two Swedish arable soils. – Biol. Fert. Soils 1: 131–140.

– and Rosswall, T. 1987. Seasonal variations in abundance and activity of nitrifiers in four arable cropping systems. – Microb. Ecol. 13: 75–87.

– and Rosswall, T. 1989. Abiotic factors regulating nitrification in arable soil. – Biol. Fertil. Soils. In press.

– , Klemedtsson, L. and Rosswall, T. 1982. Inhibitory effect of low partial pressures of acetylene on nitrification. – Soil Biol. Biochem. 14: 301–303.

Bergersen, F. J. 1970. The quantitative relationship between nitrogen fixation and the acetylene reduction assay. – Austr. J. Biol. Sci. 23: 1015–1025.

Berglund, E. 1956. Gödslingens och kalkningens historik. Handelsgödseln i vårt land – en historisk återblick. – Handbok om Växtnäring I. Tidskriftsaktiebolaget Växtnärings-Nytt, Stockholm, pp. 1–24 (in Swedish).

Bergström, L. 1986. Distribution and temporal changes of mineral nitrogen in soils supporting annual and perennial crops. – Swed. J. agric. Res. 16: 105–112.

– 1987a. Nitrate leaching and drainage from annual and perennial crops in tile-drained plots and lysimeters. – J. Environ. Qual. 16: 11–18.

– 1987b. Leaching of 15-N labeled nitrate fertilizer applied to barley and a grass ley. – Acta Agric. Scand. 37: 199–206.

– 1987c. Transport and transformation of nitrogen in an arable soil. – Ph. D. Thesis, Ekohydrologi 23. Div. of Water Management, Swedish Univ. of Agricultural Sciences, Uppsala.

– and Brink, N. 1986. Effects of differential applications of fertilizer N on leaching losses and distribution of inorganic N in the soil. – Pl. Soil 93: 333–345.

– and Johnsson, H. 1988. Simulated nitrogen dynamics and nitrate leaching in a perennial grass ley. – Pl. Soil 105: 273–281.

Billès, G., Gandais-Riollet, N. and Bottner, P. 1986. Effet d'une culture de graminées sur la décomposition d'une litière végétale, marquée au 14C et 15N, dans le sol, en conditions contrôlées. – Acta Oecologia/Oecologia Plant. 7: 273–286.

Birch, H. F. 1960. Nitrification in soils after different periods of dryness. – Pl. Soil 12: 81–96.

Biscoe, P. V. and Gallagher, J. N. 1978. A physiological analysis of cereal yield. 1. Production of dry matter. – Agric. Prog. 53: 34–50.

Bjarnason, S. 1988. Turnover of organic nitrogen in agricultural soils and the effects of management practices on soil fertility. – Ph. D. Thesis, Dept of Soil Sciences, Swedish Univ. of Agricultural Sciences, Uppsala.

Bleakley, B. H. and Tiedje, J. M. 1982. Nitrous oxide production by organisms other than nitrifiers and denitrifiers. – Appl. Environ. Microbiol. 44: 1342–1348.

Bock, E., Koops, H.-P. and Harms, H. 1987. Cell biology of nitrifying bacteria. – In: Prosser, J. I. (ed.), Nitrification. Spec. Publ. Soc. Gen. Microbiol. IRL Press, Oxford, pp. 17–38.

Bolton, E. F., Aylesworth, J. W. and More, F. R. 1970. Nutrient losses through tile drains under three cropping systems and two fertility levels on a Brookston clay soil. – Can. J. Soil Sci. 50: 275–279.

Bolton, J. L. 1962. Alfalfa. – Leonard Hill, London, pp. 97–115.

Bolton, P. J. and Phillipson, J. 1976. Burrowing, feeding, egestion and energy budgets of *Allolobophora rosea* (Savigny) (Lumbricidae). – Oecologia (Berl.) 23: 225–245.

Bonde, T. and Rosswall, T. 1987. Seasonal variation of potentially mineralizable nitrogen in four cropping systems. – Soil Sci. Soc. Am. J. 51: 1508–1514.

– , Schnürer, J. and Rosswall, T. 1988. Microbial biomass as a fraction of potentially mineralizable nitrogen in soils from long-term field experiments. – Soil Biol. Biochem. 20: 447–452.

Booth, R. G. and Anderson, J. M. 1979. The influence of fungal food quality on the growth and fecundity of *Folsomia candida* (Collembola Isotomidae). – Oecologia (Berl.) 38: 317–324.

Borg, Å. 1971. Några undersökningar över Kollemboler (Collembola) i åkerjord och deras påverkan av pesticider. – Statens växtskyddsanstalt. Meddelanden 15 (137): 53–85 (in Swedish).

Bormann, F. H., Likens, G. E. and Melillo, J. M. 1977. Nitrogen budget for an aggrading northern hardwood forest ecosystem. – Science 196: 981–983.

Boström, S. 1985. Description and morphological variability of *Chiloplacus minimus* (Thorne, 1925) Andrássy, 1959 (Nematoda: Cephalobidae). – Nematologica 30: 151–160.

– 1986a. Description of *Acrobeloides emarginatus* (de Man, 1880) Thorne, 1937 and proposal of *Acrolobus* n. gen. (Nematoda: Cephalobidae). – Rev. Nématologie 8: 335–340.

– 1988c. Morphological and systematic studies of the family Cephalobidae (Nematoda: Rhabditida). – Ph. D. Thesis, Dept of Zoology, Univ. of Stockholm, Stockholm.

– and Sohlenius, B. 1986. Short-term dynamics of nematode communities in arable soil. Influence of a perennial and an annual cropping system. – Pedobiologia 29: 345–357.

Boström, U. 1986b. The effect of soil compaction on earthworms (Lumbricidae) in heavy clay soil. – Swed. J. agric Res. 16: 137–141.

– 1987. Growth of earthworms *(Allolobophora caliginosa)* in soil mixed with either barley, lucerne or meadow fescue at various stages of decomposition. – Pedobiologia 30: 311–321.

– 1988a. Ecology of earthworms in arable land. Population dynamics and activity in four cropping systems. – Ph. D. Thesis, Dept of Ecology and Environmental Research, Report 34. Swedish Univ. of Agricultural Sciences, Uppsala.

– 1988b. Growth and cocoon production by the earthworm *Aporrectodea caliginosa* in soil mixed with various plant materials. – Pedobiologia 32: 77–80.

– and Lofs-Holmin, A. 1986. Growth of earthworms *(Allolobophora caliginosa)* fed shoots and roots of barley, meadow fescue and lucerne. Studies in relation to particle size, protein, crude fiber content and toxicity. – Pedobiologia 29: 1–12.

Bouché, M. B. 1977. Stratégies lombriciennes. – In: Lohm, U. and Persson, T. (eds), Soil organisms as components of ecosystems. Ecol. Bull. (Stockholm) 25: 122–132.

Boye Jensen, M. 1985. Interactions between soil invertebrates and straw in arable land soil. – Pedobiologia 28: 59–69.

Breitenbeck, G. A., Blackmer, A. M. and Bremner, J. M. 1980. Effects of different nitrogen fertilizers on emission of nitrous oxide from soil. – Geophys. Res. Lett. 7: 85–88.

Bremner, J. M. and Blackmer, A. M. 1978. Nitrous oxide: Emission from soils during nitrification of fertilizer nitrogen. – Science 199: 295–296.

Breymeyer, A. I. and Van Dyne, G. M. (eds), 1980. Grasslands, systems analysis and man. IBP 19. – Cambridge Univ. Press, Cambridge.

Briggs, D. and Courtney, F. 1985. Agriculture and environment. The physical geography of temperate agricultural systems. – Longman, London – New York.

Brink, N. 1982. Measurement of mass transport from arable land in Sweden. – Ekohydrologi 12: 29–36. Div. of Water Management, Swedish Univ. of Agricultural Sciences, Uppsala.

– and Jernlås, R. 1982. Utlakning vid spridning höst och vår av flytgödsel. Ekohydrologi 12: 3 14. Div. of Water Management, Swedish Univ. of Agricultural Sciences, Uppsala (in Swedish).

– , Gustafsson, A. and Torstensson, G. 1986. Odlingsåtgärdernas inverkan på kvalitet hos yt- och grundvatten. – Ekohydrologi 21: 24–31. Div. of Water Management, Swedish Univ. of Agricultural Sciences, Uppsala (in Swedish).

Bristow, A. W., Ryden, J. C. and Whitehead, D. C. 1987. The fate at several time intervals of ^{15}N-labelled ammonium nitrate applied to an established grass sward. – J. Soil Sci. 38: 245–254.

Broadbent, F. E. 1965. Effect of fertilizer nitrogen on the release of soil nitrogen. – Soil Sci. Soc. Am. Proc. 29: 692–695.

Brookes, P. C., Powlson, D. S. and Jenkinson, D. S. 1982. Measurement of microbial biomass phosphorus in soil. – Soil Biol. Biochem. 14: 319–329.

– , Landman, A., Pruden, G. and Jenkinson, D. S. 1985. Chloroform fumigation and the release of soil nitrogen: A rapid direct extraction method to measure microbial biomass nitrogen in soil. – Soil Biol. Biochem. 17: 837–842.

Brooks, R. H. and Corey, A. T. 1964. Hydraulic properties of porous media. – Hydrology Paper No 3, Colorado State Univ., Fort Collins, CO.

Brown, P. L. and Dickey, D. D. 1970. Losses of wheat straw residue under simulated field conditions. – Soil Sci. Soc. Am. Proc. 34: 118–121.

Bruce-Mitford, R. 1979. The Sutton Hoo Ship Burial: A Handbook. – British Museum Publ. Ltd., London.

Brussaard, L., van Veen, J. A., Kooistra, M. J. and Lebbink, G. 1988. The Dutch programme on soil ecology of arable farming systems. I. Objectives, approach and some preliminary results. – Ecol. Bull. (Copenhagen) 39: 35–40.

Burns, I. G. and Greenwood, D. J. 1982. Estimation of the year-to-year variation in nitrate leaching in different soils and regions of England and Wales. – Agric. Environ. 7: 35–45.

Buyanovsky, G. A., Kucera, C. L. and Wagner, G. H. 1987. Comparative analyses of carbon dynamics in native and cultivated ecosystems. – Ecology 68: 2023–2031.

Cameron, K. C. and Wild, A. 1984. Potential aquifer pollution from nitrate leaching following the plowing of temporary grassland. – J. Environ. Qual. 13: 274–278.

Campbell, C. A., Jame, Y. W. and Winkleman, G. E. 1984. Mineralization rate constants and their use for estimating nitrogen mineralization in some Canadian prairie soils. – Can. J. Soil Sci. 64: 333–343.

Cannell, R. Q., Belford, R. K., Gales, K. and Colin, W. D. 1980. A lysimeter system used to study the effect of transient waterlogging in crop growth and yield. – J. Sci. Food Agric. 31: 105–116.

Carter, A., Lagerlöf, J. and Steen, E. 1985. Effects of major disturbances in different agricultural cropping systems on soil macroarthropods. – Acta Agric. Scand. 35: 67–77.

Carter, M. R. and Rennie, D. A. 1982. Changes in soil quality under zero tillage farming systems: Distribution of microbial biomass and mineralizable C and N potentials. – Can. J. Soil Sci. 62: 587–597.

Carter, P. R. and Sheaffer, C. C. 1983. Alfalfa response to soil water deficits. III. Nodulation and N_2 fixation. – Crop Sci. 23: 985–990.

Cassell, D. K., Kreuger, T. H., Schroer, F. W. and Norum, E. B. 1974. Solute movement through disturbed and undisturbed soil cores. – Soil Sci. Soc. Am. Proc. 38: 36–40.

Catchpoole, V. R., Harper, L. A. and Myers, R. J. K. 1983. Annual losses of ammonia from grazed pasture fertilized with urea. – Proc. 14th Int. Grassland Congress, Lexington, Kentucky, pp. 344–347.

Christensen, B. 1961. Studies on cyto-taxonomy and reproduction in the Enchytraeidae. – Hereditas 47: 387–450.

Christensen, B. T. 1986. Ammonia volatilization loss from surface applied manure. – In: Dam Kofoed, A., Williams, J. H. and L'Hermite, P. (eds), Efficient land use of sludge and manure. Elsevier, London, pp. 193–203.

Christensen, S. and Bonde, G. J. 1985. Seasonal variation in

Clarholm, M. 1981. Protozoan grazing of bacteria in soil – impact and importance. – Microb. Ecol. 7: 343–350.
— 1984. Heterotrophic, free-living protozoa: Neglected microorganisms with an important task in regulating bacterial populations. – In: Klug, M. J. and Reddy, C. A. (eds), Current perspectives in microbial ecology. Am. Soc. Microbiol., Washington D.C., pp. 321–326.
— 1985a. Interactions of bacteria, protozoa and plants leading to mineralization of soil nitrogen. – Soil Biol. Biochem. 17: 181–187.
— 1985b. Possible roles for roots, bacteria, protozoa and fungi in supplying nitrogen to the plants. – In: Fitter, A. H., Atkinson, D., Read, D. J. and Usher, M. B. (eds), Ecological interactions in soil: plants, microbes and animals. Brit. Ecol. Soc. Spec. Publ. 4. Blackwell Sci. Publ., Oxford, pp. 355–365.
— in press. Effects of plant-bacterial-amoebal interactions on plant uptake of nitrogen under field conditions. – Biol. Fertil. Soils.
— and Rosswall, T. 1980. Biomass and turnover of bacteria in a forest soil and a peat. – Soil Biol. Biochem. 12: 49–57.
— and Bergström L. (eds) 1989. Ecology of Arable Land. Perspectives and Challenges. – Developments in Plant and Soil Sciences 39. Kluwer, Dordrecht.
Clark, F. E. and Paul, E. A. 1970. The microflora of grassland. – Adv. Agron. 23: 375–435.
Clement, C. R. and Williams, T. E. 1967. Leys and soil organic matter. II. The accumulation of nitrogen in soils under different leys. – J. agric. Sci., Camb. 69: 133–138.
Cochran, W. G. 1977. Sampling techniques. – Wiley, New York.
Colburn, P. and Dowdell, R. J. 1984. Denitrification in field soils. – Pl. Soil 76: 213–226.
— , Ryden, J. C. and Dollard, G. J. 1987. Emission of NO_x from urine-treated pasture. – Environ. Pollut. 46: 253–261.
Cole, C. V., Stewart, J. W. B., Ojima, D. S., Parton, W. J. and Schimel, D. 1989. Modelling land use effects on soil organic matter dynamics in the North American Great Plains. – In: Clarholm, M. and Bergström, L. (eds), Ecology of Arable Land – Perspectives and Challenges. Developments in Plant and Soil Sciences 39, pp. 89–98. Kluwer, Dordrecht 39.
Coleman, D. C. 1973. Soil carbon balance in a successional grassland. – Oikos 24: 195–199.
— 1985. Through a ped darkly: an ecological assessment of root-microbial-faunal-interactions. – In: Fitter, A. H., Atkinson, D., Read, D. J. and Usher, M. B. (eds), Ecological interactions in soil: plants, microbes and animals. Brit. Ecol. Soc. Spec. Publ. 4. Blackwell Sci. Publ., Oxford, pp. 1–21.
— , Cole, C. V. and Elliott, E. T. 1984. Decomposition, organic matter turnover, and nutrient dynamics in agroecosystems. – In: Lowrance, R., Stinner, B. R. and House, G. J. (eds), Agricultural ecosystems. Unifying concepts. Wiley, New York, pp. 83–104.
Corbett, D. C. M. 1972. The effect of *Pratylenchus fallax* on wheat, barley and sugar beet roots. – Nematologica 18: 303–308.
Cox, W. J. and Reisenauer, H. M. 1973. Growth and ion uptake by wheat supplied nitrogen as nitrate or ammonium or both. – Pl. Soil 38: 363–380.
Crawley, M. J. 1983. Herbivory. The dynamics of animal-plant interactions. – Studies in Ecology. Vol. 10. Blackwell, Oxford.
Currie, J. A. 1984. Gas diffusion through soil crumbs: the effect of compaction and wetting. – J. Soil Sci. 35: 1–10.
Curry, J. P. 1969. The decomposition of organic matter in soil. Part 1. The role of the fauna in decaying grassland herbage. – Soil Biol. Biochem. 1: 253–258.
— 1986. Above-ground arthropod fauna of four Swedish cropping systems and its role in carbon and nitrogen cycling. – J. Appl. Ecol. 23: 853–870.
— and O'Neill, N. 1978. A comparative study of the arthropod communities of various swards using the D-vac suction sampling technique. – Proc. Roy. Irish Acad. 79(B): 247–258.
Cutler, D. W., Lettice, M. A., Crump, M. and Sandon, H. 1923. A quantitative investigation of the bacterial and protozoan population of the soil, with an account of the protozoan flora. – Phil. Trans. Roy. Soc., London B 211: 317–347.
Czarnecki, A. 1983. Participation of Collembola in the decomposition of crop residues (In Polish with English summary). – Rocz. Glebozn. 34: 75–83.
Darbyshire, J. F., Wheatley, R. F., Greaves, M. P. and Inkson, R. H. E. 1974. A rapid micromethod for estimating bacterial and protozoan populations in soil. – Rev. Ecol. Biol. Sol 11: 465–475.
Davidson, R. H. 1976. The viking road to Byzantium. – George Allen and Unwin Ltd., London.
Decker, H. 1961. Die Bedeutung wurzelparasitischer Nematoden für den Anbau von Gramineen. – Wiss. z. Univ. Halle 10, Math. – Nat. Reihe, H 2/3: 297–302.
Delany, A. C., Fitzgerald, D. L., Lenschow, D. H., Pearson, Jr., R., Wendel, G. J. and Woodruff, B. 1986. Direct measurements of nitrogen oxides and ozone fluxes over grasslands. – J. Atmos. Chem. 4: 429–444.
Diekmahns, E. C. 1972. Morphology. – In: Spedding, C. R. W. and Diekmahns, E. C. (eds), Grasses and legumes in British agriculture. Commonwealth Bureau of Pastures and Field Crops. Bulletin 49: 283–286. Alden Press, Oxford.
Dijk, H. van 1982. Survey of Dutch soil organic matter research with regard to humification and degradation rates in arable land. – In: Boels, D., Davies, D. B. and Johnston, A. E. (eds), Soil Degradation. A. A. Balkema, Rotterdam, pp. 133–144.
Domsch, K. H., Beck, Th., Anderson, J. P. E., Söderström, B., Parkinson, D. and Trolldenier, G. 1979. A comparison of methods for soil microbial population and biomass studies. – Z. Pflanzenernaehr. Bodenkd. 142: 520–533.
Domurat, K. 1970. Nematode communities occurring in spring barley crops. – Ekol. Pol. Ser. A 18: 682–740.
Doran, J. W. 1980. Soil microbial and biochemical changes associated with reduced tillage. – Soil Sci. Soc. Am. J. 44: 765–771.
— , Mielke, L. N. and Power, J. F. 1987. Tillage/residue management interactions with soil environment, organic matter, and nutrient cycling. – INTECOL Bull. 15: 33–39.
Dowdell, R. J. 1982. Fate of nitrogen applied to agricultural crops with particular reference to denitrification. – Phil. Trans. Roy. Soc., London B 296: 363–373.
Dowson, C. G., Rayner, A. D. M. and Boddy, L. 1986. Outgrowth patterns of mycelial cord-forming basidiomycetes from and between woody resource units in soil. – J. Gen. Microbiol. 132: 203–211.
Duxbury, J. M. and McConnaughey, P. K. 1986. Effect of fertilizer source on denitrification and nitrous oxide emissions in a maize-field. – Soil Sci. Soc. Am. J. 50: 644–648.
Ebbersten, S. 1980. Optimerad växtodling III. Ett försök till analys av dagens diskussion om olika växtodlingssystem. – K. Skogs- och Lantbr.-akad tidskr. 119: 367–385 (in Swedish).
Edwards, C. A. 1984. Changes in agricultural practices and their impact on soil organisms. – In: Jenkins, D. (ed.), Agriculture and the environment. Inst. of Terrestrial Ecology, Great Britain, pp. 56–65.
— and Lofty, J. R. 1969. The influence of agricultural practice on soil micro-arthropod populations. – In: Sheals, J. G. (ed.), The soil ecosystem. Systematics Association Publ. No. 8, pp. 237–247.
— and Thompson, A. R. 1973. Pesticides and the soil fauna. –

Residue Rev. 45: 1–80.
- and Lofty, J.R. 1975. The invertebrate fauna of the Park Grass Plots. 1. Soil fauna. Rothamsted Exp. Stn Rep. 1974, Part 2, pp. 133–154.
- and Lofty, J. R. 1977. The influence of invertebrates on root growth of crops with minimal or zero cultivation. – Ecol. Bull. (Stockholm) 25: 348–356.
Efron, B. 1979. Computers and the theory of statistics: Thinking the unthinkable. – SIAM Rev. 21: 460–480.
- and Tibshirani, R. 1986. Bootstrap methods for standard errors, confidence intervals and other measures of statistical accuracy. – Statist. Sci. 1: 54–77.
Eitminaviciuté, I., Bagdancviciené, Z, Kadyté, B, and Lazauskiené, L. 1976. Characteristic successions of microorganisms and soil invertebrates in the decomposition process of straw and lupine. – Pedobiologia 16: 106–115.
Ekman, R. 1982. Distribution av kväve i ovanjordisk biomassa hos korn. – Sv. Utsädesför. Tidskr. 92: 241–245 (in Swedish).
El-Haris, M. K., Cochran, V. L., Elliott, L. F. and Bezdicek, D. F. 1983. Effect of tillage, cropping, and fertilizer management on soil nitrogen mineralization potential. – Soil Sci. Soc. Am. J. 47: 1157–1161.
Elliott, E. T. and Coleman, D. C. 1977. Soil protozoan dynamics in a shortgrass prairie. – Soil Biol. Biochem. 9: 113–118.
- , Anderson, R. V., Coleman, D. C. and Cole, C. V. 1980. Habitable pore space and microbial trophic interactions. – Oikos 35: 327–335.
- , Horton, K., Moore, J. C. and Coleman, D. C. 1984. Mineralization dynamics in fallow dryland wheat plots, Colorado. – Pl. Soil 76: 149–155.
Emmanuel, N., Curry, J. P. and Evans, G. O. 1985. The soil Acari of barley plots with different cultural treatments. – Exp. Appl. Acarology 1: 101–113.
Epps, H. M. R. and Gale, E. F. 1942. The influence of the presence of glucose during growth on the enzymatic activities of *Escherichia coli*: comparison of the effect with that produced by fermentation acids. – Biochem. J. 36: 619–623.
Eriksson, B. 1981. The potential evapotranspiration in Sweden. – Swedish Meteorological and Hydrological Institute. Report No. RMK 28. SMHI, Norrköping.
- 1983. Data rörande Sveriges nederbördsklimat. Normalvärden för perioden 1951–80. – Sveriges Meteorologiska och Hydrologiska Institut. Klimatsektionen. Rapp. 1983: 28. SMHI, Norrköping (in Swedish).
Evans, A. C. and Guild, W. J. McL. 1948a. Studies on the relationships between earthworms and soil fertility. IV. On the life cycles of some British Lumbricidae. – Ann. Appl. Biol. 35: 471–484.
- and Guild, W. J. McL. 1948b. Studies on the relationships between earthworms and soil fertility. V. Field populations. – Ann. Appl. Biol. 35: 485–493.
Fairley, R. I. and Alexander, I. J. 1985. Methods of calculating fine root production in forests. – In: Fitter, A. H., Atkinson, D., Read, D. J. and Usher, M. B. (eds), Ecological interactions in soil: Plants, microbes and animals. Brit. Ecol. Soc. Spec. Publ. 4. Blackwell Sci. Publ., Oxford, pp. 37–42.
FAO 1974. Soil map of the world, Vol. 1. – UNESCO, Rome.
- 1977. Guidelines for soil profile description. – Soil resources development and conservation service, Land and water development division, FAO, Rome.
Fenchel, T. 1982. Ecology of heterotrophic microflagellates. II. Bioenergetics and growth. – Mar. Ecol. Progr. Ser. 8: 225–231.
Ferm, M. 1983. Ammonia volatilization from arable land – An evaluation of the chamber technique. – In: Observation and measurement of atmospheric contaminants. WHO Spec. Environ. Report 16. WHO, Geneva, pp. 145–172.
- 1986. Concentration measurements and equilibrium studies of ammonium, nitrate and sulphur species in air and precipitation. – Ph. D. Thesis, Dept of Inorganic Chemistry, Univ. of Gothenburg, Gothenburg.
- and Christensen, B. T. 1987. Determination of NH_3 volatilization from surface-applied cattle slurry using passive flux samplers. – In: Proc. EURASAP Symposium on Ammonia and acidification, Bilthoven, (the Netherlands) 13–15 April 1987, pp. 28–41.
Ferris, V. R. and Ferris, J. M. 1974. Inter-relationships between nematode and plant communities in agricultural ecosystems. – Agro-ecosystems 1: 275–299.
Finansdepartementet 1987. Vägar ut ur jordbruksprisregleringen – några ideskisser. – Ds Fi 1987: 4. Regeringskansliets offsetcentral, Stockholm (in Swedish).
Firestone, M. K. 1982. Biological denitrification. – In: Stevenson, F. J. (ed.), Nitrogen in agricultural soils. Agronomy 22: 289–326.
- , Smith, M. S., Firestone, R. B. and Tiedje, J. M. 1979. The influence of nitrate, nitrite and oxygen on the composition of gaseous products of denitrification in soil. – Soil Sci. Soc. Am. J. 43: 1140–1144.
Fleischer, S., Hamrin, S. and Kindt, T. 1987. Coastal eutrophication in Sweden: Reducing nitrogen in land runoff. – Ambio 16: 246–251.
Flühler, H., Ardakani, M. S., Szyszkiewicz, T. E. and Stolsy, L. H. 1976. Field-measured nitrous oxide concentrations, redoxpotentials, oxygen diffusion rates, and oxygen partial pressures in relation to denitrification. – Soil Sci. 122: 107–114.
Focht, D. D. and Verstraete, W. 1977. Biochemical ecology of nitrification and denitrification. – Adv. Microb. Ecol. 1: 135–214.
Fogel, R. and Hunt, G. 1979. Fungal and arboreal biomass in a western Oregon Douglas-fir ecosystem: distribution pattern and turnover. – Can. J. For. Res. 9: 245–256.
- and Hunt, G. 1983. Contribution of mykorrhizae and soil fungi to nutrient cycling in a Douglas-fir ecosystem. – Can. J. For. Res. 13: 219–232.
Folorunso, O. A. and Rolston, D. E. 1984. Spatial variability of field-measured denitrification gas fluxes. – Soil Sci. Soc. Am. J. 48: 1214–1219.
Frankland, J. C. 1975. Estimation of live fungal biomass. – Soil Biol. Biochem. 7: 339–340.
- 1982. Biomass and nutrient cycling by decomposer basidiomycetes. – In: Frankland, J. C., Hedger, J. N. and Swift, M. J. (eds), Decomposer basidiomycetes. Their biology and ecology. Brit. Mycol. Soc. Symp. 4. Cambridge Univ. Press, Cambridge, pp. 241–262.
Freckman, D. W., Duncan, D. A. and Larson, J. R. 1979. Nematode density and biomass in an annual grassland ecosystem. – J. Range Manage. 32: 418–422.
Fried, M. and Broeshart, H. 1975. An independent measurement of the amount of nitrogen fixed by a legume crop. – Pl. Soil 43: 707–711.
Frissel, M. 1977. Cycling of mineral nutrients in agricultural ecosystems. – Agroecosystems 4: 1–354.
- and van Veen, J. A. (eds) 1981. Simulation of nitrogen behaviour of soil-plant systems. – Pudoc, Wageningen.
Galbally, I. E. 1985. The emission of nitrogen to the remote atmosphere: Background paper. – In: Galloway, J. N., Charlson, R. J., Andreae, M. O. and Rodhe, H. (eds), The biogeochemical cycling of sulfur and nitrogen in the remote atmosphere. D. Riedel, Dordrecht, pp. 27–53.
- and Roy, C. R. 1978. Loss of fixed nitrogen by exhalation of nitric oxide. – Nature, Lond. 275: 734–735.
- and Johansson, C. 1989. A model relating laboratory measurements of rates of nitric oxide production and field measurements of nitric oxide emissions from soils. – J. Geophys. Res. In press.
- Roy, C. R., Elsworth, C. M. and Rabich, H. A. H. 1985. The measurement of nitrogen oxide (NO, NO_2) exchange

over plant/soil surfaces. – CSIRO, Div. of Atmospheric Research, Technical Paper No. 8, pp. 1–23. Aspendale, Australia.
– , Freney, J. R., Muirhead, W. A., Simpson, J. R., Trevitt, A. C. and Chalk, P. M. 1987. Emission of nitrogen oxides (NO$_x$) from a flooded soil fertilized with urea: relation to other nitrogen loss processes. – J. Atmos. Chem. 5: 343–365.
Gallagher, J. N. and Biscoe, P. V. 1978. A physiological analysis of cereal yield. I. Production of dry matter. – Agric. Prog. 53: 34–49.
Gamble, T. N., Betlach, H. R. and Tiedje, J. M. 1977. Numerically dominant denitrifying bacteria from world soils. – Appl. Environ. Microbiol. 33: 926–939.
Gerard, B. M. 1967. Factors affecting earthworms in pastures. – J. Anim. Ecol. 36: 235–252.
Gill, R. W. 1969. Soil microarthropod abundance following old field manipulation. – Ecology 50: 805–816.
Goreau, T. J., Kaplan, W. A., Wofsy, S. C., McElroy, M. B., Valois, F. W. and Watson, S. W. 1980. Production of NO and N$_2$O by nitrifying bacteria at reduced concentrations of oxygen. – Appl. Environ. Microbiol. 40: 526–532.
Goring, C. A. I. 1962. Control of nitrification by 2-chloro-6-(trichloromethyl) pyridine. – Soil Sci. 93: 211–218.
Goudrian, J. 1977. Crop micrometeorology: A simulation study. – Pudoc, Wageningen.
Graff, O. 1953a. Bodenzoologische Untersuchungen mit besonderer Berücksichtigung der terrikolen Oligochaeten. – Z. Pflern. Düng. Bodenk. 61: 72–77.
– 1953b. Die Regenwürmen Deutschlands. – Schriftenreihe der Forschungsanstalt für Lantwirtschaft Braunschweig-Volkenrode. Heft 7.
– 1971. Stickstoff, Phosphor und Kalium in der Regenwurmlösung auf der Wiesenversuchsfläche des Sollingprojektes. – 4. Colloq. Pedobiologie. (Ann. Zool. – Ecol. Anim.) Hors Serie, pp. 503–511.
Granstedt, A. 1986. Växtodlingens förutsättningar. – In: Granstedt, A., Högborg, E., Johansson, L. and Weidow, B. (eds), Växtodlingens grunder. LT:s förlag, Stockholm, pp. 13–29 (in Swedish).
Gregory, P. J., Marshall, B. and Biscoe, P. V. 1981. Nutrient relations of winter wheat 3. Nitrogen uptake, photosynthesis of flag leaves and translocation of nitrogen to grain. – J. Agric. Sci. 96: 539–547.
Griffin, D. M. 1981. Water and microbial stress. – In: Alexander, M. (ed.), Advances in microbial ecology, Vol. 5, Plenum Press, New York, pp. 91–136.
Groat, R. G. and Vance, C. P. 1981. Root nodule enzymes of ammonia assimilation in alfalfa (*Medicago sativa* (L.)). – Plant Physiol. 67: 1198–1203.
Groffman, P. M. 1985. Nitrification and denitrification in conventional and no-tillage soil. – Soil Sci. Soc. Am. J. 49: 329–334.
Groth, R. 1987. Messungen zum Wurzelwachstum von Winterweizen auf Tieflehm – Fahlerde. – Arch. Acker – Pflanzenbau Bodenk., Berlin. 31: 197–204.
Guild, W. J. McL. 1948. Studies on the relationship between earthworms and soil fertility. III. The effect of soil type on the structure of earthworm populations. – Ann. appl. Biol. 35: 181–192.
– 1955. Earthworms and soil structure. – In: Kevan, D. K. McE. (ed.), Soil zoology. Butterworths, London. pp. 83–98.
Gustafson, A. 1982. Växtnäringsförluster från åkermark i Sverige. – Ekohydrologi 11: 19–27. Div. of Water Management, Swedish Univ. of Agricultural Sciences, Uppsala (in Swedish).
– 1983. Leaching of nitrate from arable land into groundwater in Sweden. – Environ. Geol. 5: 65–71.
– 1987. Water discharge and leaching of nitrate. – Ph. D. Thesis, Ekohydrologi 22. Div. of Water Management, Swedish Univ. of Agricultural Sciences, Uppsala.

Håkansson, I. and Danfors, B. 1981. Effects of heavy traffic on soil conditions and crop growth. – Proc. 7th Int. Conf. Int. Soc. for Terrain-Vehicle Systems, Vol. 1, pp. 239–253.
Haarløv, N. 1955. Vertical distribution of mites and Collembola in relation to soil structure. – In: Kevan, D. K. McE. (ed.), Soil zoology. Butterworths, London. pp. 167–179.
Hadas, A., Molina, J. A. E., Feigenbaum, S. and Clapp, C. E. 1987. Simulation of nitrogen-15 immobilization by the model NCSOIL. – Soil Sci. Soc. Am. J. 51: 102–106.
Haider, K., Mosier, A. and Heinemeyer, O. 1985. Phytotron experiments to evaluate the effect of growing plants on denitrification. – Soil Sci. Soc. Am. J. 49: 636–641.
– , Mosier, A. and Heinemeyer, O. 1987. The effect of growing plants on denitrification at high soil nitrate concentrations. – Soil Sci. Soc. Am. J. 51: 97–102.
Hallmark, S. L. and Terry, R. E. 1985. Field measurement of denitrification in irrigated soil. – Soil Sci. 140: 35–44.
Hanlon, R. D. G. and Anderson, J. M. 1979. The effects of Collembola grazing on microbial activity in decomposing leaf litter. – Oecologia (Berl.) 38: 93–100.
Hansen, L. 1983. Kvaelstoftab til draen- og grundvand. Bilag Statens Planteavlsmöde. – Statens Planteavlsforsög: 18–22, Köpenhamn (in Danish).
Hansen, S. and Aslyng, H. C. 1984. Nitrogen balance in crop production. – Hydrotechnical Laboratory, The Royal Veterinary and Agricultural Univ., Copenhagen.
Hansson, A.-C. 1987. Roots of arable crops: Production, growth dynamics and nitrogen content. – Ph. D. Thesis, Dept of Ecology and Environmental Research, Report 28. Swedish Univ. of Agricultural Sciences, Uppsala.
– and Steen, E. 1984. Methods of calculating root production and nitrogen uptake in an annual crop. – Swed. J. agric. Res. 14: 191–200.
– and Andrén, O. 1986. Below-ground plant production in a perennial grass ley (*Festuca pratensis*) assessed with different methods. – J. Appl. Ecol. 23: 657–666.
– and Andrén, O. 1987. Root dynamics in barley, lucerne and meadow fescue investigated with a mini-rhizotron technique. – Pl. Soil. 103: 33–38.
– and Pettersson, R. 1989. Uptake and above- and below-ground allocation of soil mineral-N and fertilizer-^{15}N in a perennial grass ley (*Festuca pratensis* Huds.). – J. Appl. Ecol. 26: 259–271.
– , Pettersson, R. and Paustian, K. 1987. Shoot and root production and nitrogen uptake in barley, with and without nitrogen fertilization. – J. Agron. Crop Sci. 158: 163–171.
– , Andrén, O. and Steen, E. In press. Root production of four arable crops in Sweden and its effect on abundance of soil organisms. – In: Atkinson, D. (ed.), Plant root systems: Their effect on ecosystem composition and structure. Blackwell, Oxford.
Hart, P. B. S., Goh, K. M. and Ludecke, T. E. 1979. Nitrogen mineralisation in fallow and wheat soils under field and laboratory conditions. – N. Z. J. Agric. Res. 22: 115–125.
Hauk, R. D. 1971. Quantitative estimates of nitrogen-cycle processes. Concepts and review. – In: Nitrogen-15 in soil plant studies. IAEA, Vienna, pp. 65–80.
– and Bremner, J. M. 1976. Use of tracers for soil and fertilizer nitrogen research. – Adv. Agron. 28: 219–266.
Heal, O. W. and MacLean, S. F. 1975. Comparative productivity in ecosystems-secondary productivity. – In: van Dobben, W. H. and Lowe-McConnell, R.H. (eds), Unifying concepts in ecology. Junk, Wageningen, pp. 89–108.
– and Dighton, J. 1985. Resource quality and trophic structure in the soil system. – In: Fitter, A. H., Atkinson, D., Read, D. J. and Usher, M. B. (eds), Ecological interactions in soil: Plants, microbes and animals. Brit. Ecol. Soc. Spec. Publ. 4. Blackwell Sci. Publ., Oxford, pp. 339–354.
Heath, G. W. 1962. The influence of ley management on earthworm populations. – J. Br. Grassl. Soc. 17: 237–244.
Helal, H. M. and Sauerbeck, D. 1981. Ein Verfahren zur

trennung von Bodenzon underschiedlichen Wurzelnähe. – Z. Pflanzenernähr. Bodenk. 144: 524–527.
– and Sauerbeck, D. 1987. Direct and indirect influences of plant roots on organic matter and phosphorus turnover in soil. – INTECOL Bull. 15: 49–58.
Hendrix, P. F., Parmelee, R. W., Crossley, D. A. Jr., Coleman, D. C., Odum, E. P. and Groffman, P. M. 1986. Detritus food webs in conventional and no-tillage agroecosystems. – BioScience 36: 374–380.
– , Crossley, D. A. Jr., Coleman, D. C., Parmelee, R. W. and Beare, M. H. 1987. Carbon dynamics in soil microbes and fauna in conventional and no-tillage agroecosystems. – INTECOL Bull. 15: 59–63.
Hénin, S., Monnier, G. and Turc, L. 1959. Un aspect de la dynamique des matières organiques du sol. – Comptes rendus de l'Académie de Sciences (Paris) 248: 138–141.
Hood, A. E. M. 1977. High fertilizer application on grassland. – Rep. Welsh Soils Discussion Group 18: 25–42.
Hoppe, G. 1983. Skiftesreformerna och agrarutvecklingen. – Bebyggelsehistorisk tidskrift 5: 32–44 (in Swedish).
– and Laughton, J. 1986. Time-geography and economic development: The changing structure of livelihood positions on arable farms in nineteenth century Sweden. – Geogr. annaler ser. B 1986: 2.
Hoyt, G. D., McLean, E. O., Reddy, G. Y. and Logan, T. J. 1977. Effect of soil, cover crop and nutrient source on movement of soil, water and nitrogen under simulated rainslope conditions. – J. Environ. Qual. 6: 285–290.
Hunt, G. A. and Fogel, R. 1983. Fungal hyphal dynamics in a western Oregon Douglas-fir stand. – Soil Biol. Biochem. 15: 641–649.
Hunt, H. W. 1977. A simulation model for decomposition in grasslands. – Ecology 58: 469–484.
– , Coleman, D. C., Ingham, E. R., Ingham, R. E., Elliott, E. T., Moore, J. C., Rose, S. L., Reid, C. P. P. and Morley, C. R. 1987. The detrital food web in a shortgrass prairie. – Biol. Fert. Soils 3: 57–68.
Hunt, R. 1978. Plant growth analysis. – Studies in Biology no. 96. Camelot Press. Southampton.
Hynes, R. K. and Knowles, R. 1978. Inhibition by acetylene of ammonia oxidation in *Nitrosomonas europaea*. – FEMS Microbiol. Lett. 4: 319–321.
Ingestad, T. 1982. Relative addition rate and external concentration; Driving variables used in plant nutrition research. – Plant Cell Environ. (1982): 443–453.
Ingham, E. B. and Klein, D. A. 1984. Soil fungi: measurement of hyphal length. – Soil Biol. Biochem. 16: 279–280.
Ingham, R. E. and Detling, J. K. 1984. Plant-herbivore interactions in a North American mixed-grass prairie. III. Soil nematode populations and root biomass on *Cynomys ludovicianus* colonies and adjacent uncolonized areas. – Oecologia (Berl.) 63: 307–313.
– , Trofymow, J. A., Ingham, E. R. and Coleman, D. C. 1985. Interactions of bacteria, fungi and their nematode grazers: Effects on nutrient cycling and plant growth. – Ecol. Monogr. 55: 119–140.
Jagnow, G. 1983. Relations between aerobic heterotrophic, nitrite reducing, denitrifying, nitrogen-fixing bacteria, nitrogenase activity, soil moisture and nitrogen fertilization in soil and rhizosphere of field-grown wheat and barley. – In: Lebrun, H. M., André, A., DeMedts, A., Gregoire-Wibo, C. and Wauthy, G. (eds), New trends in soil biology. Dieu-Brichart, Ottignies – Louvain-la-Neuve, pp. 225–235.
Jansson, P.-E. 1986a. The importance of soil properties when simulating water dynamics for an agricultural crop-system. – In: Haldorsen, S. and Perntsen, E. J. (eds), Water in the unsaturated zone. NHP rep. 15, Aas, pp. 219–232.
– and Halldin, S. 1979. Model for annual water and energy flow in layered soil. – In: Halldin, S. (ed.) Comparison of forest water and energy exchange models. Int. Soc. Ecol. Model., Copenhagen, pp. 145–163.
– and Halldin, S. 1980. Soil water and heat model. Technical description. – Swedish Coniferous Forest Project Tech. Report 26, Swedish Univ. of Agricultural Sciences, Uppsala.
– and Christoffersson, L. 1985. PGRA Version 1.0, User's Manual. – Dept of Soil Science, Swedish Univ. of Agricultural Sciences, Uppsala.
– and Thoms-Hjärpe, C. 1986. Simulated and measured water dynamics of fertilized and unfertilized barley. – Acta Agric. Scand. 36: 162–172.
– and Gustafsson, A. 1987. Simulation of surface runoff and pipe discharge from an agricultural soil in northern Sweden. – Nordic Hydrol. 18: 151–166.
– and Andersson, R. 1988. Simulation of runoff and nitrate leaching from an agricultural district in Sweden. – J. Hydrol. 99: 33–47.
– , Borg, G. C., Lundin, L.-G. and Lindén, B. 1987. Simulation of soil nitrogen storage and leaching – Applications to different Swedish agricultural systems. – Nat. Swed. Environ. Prot. Board, Report 3356, Solna.
Jansson, S. L. 1958. Tracer studies on nitrogen transformations in soil with special attention to mineralization – immobilization relationships. – Ann. Roy. Agr. Coll. Sweden 24: 101–361.
– 1961. Rationell gödsling förr, nu och i framtiden – en principdiskussion. – Växtnärings-Nytt 17: 1–7 (in Swedish).
– 1963. Balance sheet and residual effects of fertilizer nitrogen in a 6-year study with N15. – Soil Sci. 95: 31–37.
– 1966. Odlingsmarken, växtnäringen och markvården. – Skånes Natur 53: 7–36 (in Swedish).
– 1975. Bördighetsstudier för markvård. Försök i Malmöhus län 1957–74. – K. Skogs- och Lantbr.-akad. tidskr. Suppl. 10 (English summary: Long-term soil fertility studies. Experiments in Malmöhus county 1957–74).
– 1978. Bördighetsstudier för markvård. – K. Skogs- och Lantbr.-akad. tidskr. 117: 77–93 (in Swedish).
– 1982. Nitrogen balance in Swedish agriculture. – 2nd Nat. Symp. Biol. Nit. Fix. Helsinki, 8–10 June 1982. The Finnish National Fund for Research and Development, Helsinki, pp. 293–305.
– 1983. Kväve- och fosforbudget för två regioner i Sverige (Malmöhus och Värmlands län) med särskild hänsyn till jordbrukets roll. – K. Skogs- och Lantbr.-akad. tidskr. 122: 293–302 (in Swedish).
– 1986b. Markbiologi-växtproduktion-bördighetsstudier. Skörderesterna som bördighetsfaktor. – K. Skogs- och Lantbr.-akad. tidskr. Suppl. 18: 9–31 (in Swedish).
– and Siman, G. 1978. Kväveekonomi och energiutbyte i det svenska jordbruket, särskilt växtodlingen. – Tekniska Högskolans Energiarbetsgrupp, rapport nr. 3, Stockholm (in Swedish).
– and Persson, J. 1982. Mineralization and immobilization of soil nitrogen. – In: Stevenson, F. J. (ed.), Nitrogen in agricultural soils. Agronomy 22: 229–252.
Jenkinson, D. S. 1977. Studies on the decomposition of plant material in soil. V. The effects of plant cover and soil type on the loss of carbon from ^{14}C labelled ryegrass decomposing under field conditions. – J. Soil Sci. 28: 209–213.
– and Powlson, D. S. 1976. The effects of biocidal treatments on metabolism in soil – V. A method for measuring soil biomass. – Soil Biol. Biochem. 8: 203–213.
– and Johnston, A. E. 1977. Soil organic matter in the Hoosefield continuous barley experiment. – Rothamsted Exp. Stn. Report 1976. Harpenden, Herts. England, pp. 81–101.
– , Powlson, D. S. and Wedderburn, R. W. M. 1976. The effects of biocidal treatments on metabolism in soil. 3. The relationship between soil biovolume measured by optical microscopy and the flush of decomposition caused by fumigation. – Soil Biol. Biochem. 8: 189–202.
– , Ladd, J. N. and Rayner, J. H. 1981. Microbial biomass in soil – measurement and turnover. – In: Paul, E. A. and Ladd, J. N. (eds), Soil biochemistry, vol. 5. Marcel Dekker, New York, pp. 415–471.

- Hart, P. B. S., Rayner, J. H. and Parry, L. C. 1987. Modelling the turnover of organic matter in long-term experiments at Rothamsted. – INTECOL Bull. 15: 1–8.
Jensen, M. B. 1985. Interactions between soil invertebrates and straw in arable soil. – Pedobiologia 28: 59–69.
Jensen, S. E. 1979. Model ETFOREST for calculation of actual evapotranspiration. – In: Halldin S. (ed.), Comparison of forest water and energy exchange models. Int. Soc. Ecol. Model., Copenhagen, pp. 165–172.
Jönsson, N. and Nilsdotter-Linde N. 1987. Vallväxtsorter för Södra och Mellersta Sverige 1987. Avkastning i Försök. – Aktuellt från Lantbruksuniversitetet 355. Sveriges Lantbruksuniversitet, Uppsala (in Swedish).
Johansson, C. 1984. Field measurements of emission of nitric oxide from fertilized and unfertilized forest soils in Sweden. – J. Atmos. Chem. 1: 429–442.
– 1988. Exchange of nitrogen oxides between the atmosphere and terrestrial surfaces. – Ph. D. Thesis, Dept of Meteorology, Univ. of Stockholm, Stockholm.
– 1989. Fluxes of NO_x above soil and vegetation – In: Andreae, M. O. and Schimel, D. S. (eds), Exchange of trace gases between terrestrial ecosystems and the atmosphere. Wiley, Chichester (in press).
– and Galbally, I. E. 1984. Production of nitric oxide in loam under aerobic and anaerobic conditions. – Appl. Env. Microbiol. 47: 1284–1289.
– and Granat, L. 1984. Emission of nitric oxide from arable land. – Tellus 36: 25–37.
– and Sanhueza, E. 1989. Emission of NO from savanna soils during rainy season. – J. Geophys. Res. 94D: 6473–6480.
– , Rodhe, H. and Sanhueza, E. 1988. Emission of NO in a tropical savanna and a cloud forest during the dry season. – J. Geophys. Res. 93D: 7180–7193.
Johnen, B. G. and Sauerbeck, D. R. 1977. A tracer technique for measuring growth, mass and microbial breakdown of plant roots during vegetation. – In: Lohm, U. and Persson, T. (eds), Soil organisms as components of ecosystems. Ecol. Bull. (Stockholm) 25: 366–373.
Johnsson, H., Bergström, B., Jansson, P.-E. and Paustian, K. 1987. Simulated nitrogen dynamics and losses in a layered agricultural soil. – Agric. Ecosys. Environ. 18: 333–356.
Johnston, A. E. 1976. Additions and removals of nitrogen and phosphorus in long-term experiments at Rothamsted and Woburn and the effect of the residues on total soil nitrogen and phosphorus. – In: Agriculture and water quality. Ministry of Agriculture, Fisheries and Food. Tech. Bull. 32: 111–144.
– 1982. The effects of farming systems on the amount of soil organic matter and its effect on yield at Rothamsted and Woburn. – In: Boels, D., Davies, D. B. and Johnston, A. E. (eds), Soil degradation. Balkema, Rotterdam, pp. 187–202.
– 1987. Effects of soil organic matter on yields of crops in long term experiments at Rothamsted and Woburn. – INTECOL Bull. 15: 9–16.
Jones, M. B. and Woodmansee, R. G. 1979. Biogeochemical cycling in annual grassland ecosystems. – Bot. Rev. 45: 111–144.
Jones, P. C. T. and Mollison, J. E. 1948. A technique for the quantitative estimation of soil microorganisms. – J. Gen. Microbiol. 2: 54–69.
Juma, N. G. and Paul, E. A. 1981. Use of tracers and computer simulation techniques to assess mineralization and immobilization of soil nitrogen. – In: Frissel, M. J. and van Veen, J. A. (eds), Simulation of nitrogen behaviour of soil-plant systems. Pudoc, Wageningen, pp. 145–154.
Kajak, A. 1980. Invertebrate predator subsystem. – In: Breymeyer, A. J. and Van Dyne, G. M. (eds), Grasslands, systems analysis and Man. IBP 19, Cambridge Univ. Press, Cambridge, pp. 539–589.
Karg, W. 1967. Synökologische Untersuchungen von Bodenmilben aus forstwirtschaftlich und landwirtschaftlich genutzten Böden. – Pedobiologia 7: 198–214.
– 1971. Die freilebenden Gamasina (Gamasides), Raubmilben. – Tierwelt Dtl. 59: 1–475.
– 1983. Verbreitung und Bedeutung von Raubmilben der Cohors Gamasina als Antagonisten von Nematoden. – Pedobiologia 25: 419–432.
Kasprzak, K. 1982. Review of enchytraeid (Oligochaeta, Enchytraeidae) community structure and function in agricultural ecosystems – Pedobiologia 23: 217–232.
Kassim, G., Martin, J. P. and Haider, K. 1981. Incorporation of a wide variety of organic substrate carbons into soil biomass as estimated by the fumigation procedure. – Soil Sci. Soc. Am. J. 45: 1106–1112.
Keeney, D. R. and Huber, D. M. 1980. Performance of nitrification inhibitors in the Midwest (east). – In: Meisinger, J. J. (ed.), Nitrification inhibitors – potentials and limitations. ASA Spec. Publ. 38: 75–88.
Keith, H., Oades, J. M. and Martin, J. K. 1986. Input of carbon to soil from wheat plants. – Soil Biol. Biochem. 18: 445–449.
Kelly, J. M., Van Dyne, G. M. and Harris, W. F. 1974. Comparison of three methods of assessing grassland productivity and biomass dynamics. – Am. Midl. Nat. 92: 357–369.
Killham, K. 1986. Heterotrophic nitrification. – In: Prosser, J. I. (ed.), Nitrification. Spec. Publ. Soc. Gen. Microbiol. 20. IRL Press, Oxford, pp. 117–126.
Kirchmann, H. 1985. Losses, plant uptake and utilization of manure nitrogen during a production cycle. – Acta Agric. Scand. Suppl. 24.
Kjellerup, V. 1983. Kvaelstofsgødslningens indflytelse på draenvandets indhold af nitratkvaelstof 1973–81. – Meddelelse 1736, Statens Planteavlsforsøg 85 (in Danish).
Kjöller, A. and Struwe, S. 1982. Microfungi in ecosystems: fungal occurrence and activity in litter and soil. – Oikos 39: 389–422.
Klemedtsson, L. 1986. Denitrification in arable soil with special emphasis on the influence of plant roots. – Ph. D. Thesis, Dept of Microbiology, Report 32. Swedish Univ. of Agricultural Sciences, Uppsala.
– , Svensson, B. H., Lindberg, T. and Rosswall, T. 1977. The use of acetylene inhibition of nitrous oxide reductase in quantifying denitrification in soil. – Swed. J. Agric. Sci. 7: 179–185.
– , Svensson, B. H. and Rosswall, T. 1987a. Dinitrogen and nitrous oxide production by denitrification and nitrification in soil with and without roots. – Pl. Soil 99: 303–319.
– , Berg, P., Clarholm, M., Schnürer, J. and Rosswall, T. 1987b. Microbial nitrogen transformations in the root environment of barley. – Soil Biol. Biochem. 19: 551–558.
– , Svensson, B. H. and Rosswall, T. 1988a. A method of selective inhibition to distinguish between nitrification and denitrification as sources of nitrous oxide in soil. – Biol. Fert. Soils 6: 112–119.
– , Svensson, B. H. and Rosswall, T. 1988b. Relationships between soil moisture content and nitrous oxide production during nitrification and denitrification. – Biol. Fert. Soils. 6: 106–111.
– , Simkins, S., Svensson, B. H. and Rosswall, T. 1990. Soil denitrification in three cropping systems characterized by differences in carbon and nitrogen supply. II. Water and NO_3^- effects on the denitrification process. Submitted to Biogeochemistry.
Kmoch, H. G., Ramig, R. E., Fox, R. L. and Koehler, F. E. 1957. Root development of winter wheat as influenced by soil moisture and nitrogen fertilization. – Agron. J. 49: 20–25.
Knowles, R. 1982. Denitrification. – Microbiol. Rev. 46: 43–70.
Körschens, M. and Kundler, P. 1984. Dauerfeldversuche der

DDR. – Akad. der Landwirtschaftswiss, Berlin.
Kofoed, D. 1982. Humus in long term experiments in Denmark. – In: Boels, D., Davies, D. B. and Johnston, A. E. (eds), Soil degradation. Balkema, Rotterdam, pp. 241–258.
– and Nemming, O. 1976. Askov 1894: Fertilizers and manure on sandy and loamy soils. – Ann. agron. 27: 538–610.
Kohlenbrander, G. J. 1974. Efficiency of organic manure in increasing soil organic matter content. – Trans. 10th Int. Congr. Soil Sci. 2: 129–126.
– 1980. Nitrogen as a potential source of pollution. – Instituut Bodemvruchtbaarheid, Nota 83, Haren.
– 1981. Leaching of nitrogen in agriculture. – In: Brogan, J. C. (ed.), Nitrogen losses and surface run-off from land-spreading of manures. Developments in plant and soil sciences, Vol. 2. Martinus Nijhoff/Dr. W. Junk Publishers, The Hague, pp. 199–216.
– 1982. Fertilizers and pollution. – In: Kanwar, J. S. (ed.), Whither soil research. Panel Discussion Papers. Trans. 12th Int. Congr. Soil Sci., pp. 248–266.
Kotanska, M. 1975. Primary productivity in the meadow of the Hieracio-Nardetum association in the Gorce Mountains (Southern Poland). – Bull. Acad. Polon. Sci. Ser. Sci. Biol. Cl. II 23: 623–627.
Krantz, G. W. 1975. A manual of acarology. – Oregon State Univ. Book Stores Inc., Corvallis, OR.
Kreuger, J. K. and Brink, N. 1988. Losses of pesticides from agricultural land. – FAO/IAEA Int. Symp. on changing perspectives in agrochemical: isotopic techniques for the study of food and environmental implications. Neuherberg, FRG, 24–27 November, 1987.
Krogh, P. M. 1985. Toxisk perturbation af jordbundens microarthropodsamfund. – Specialerapport, Biologisk Inst., Odense Universitet, Odense (in Danish).
Kubiëna, W. L. 1955. Animal activity in soil as a decisive factor in establishment of humus forms. – In: Kevan, D. K. McE. (ed.), Soil zoology. Butterworths, London, pp. 73–82.
Kuenen, J. G. and Robertson, L. A. 1987. Ecology of nitrification and denitrification. – In: Cole, J. A. and Ferguson (eds), The nitrogen and sulphur cycles. Cambridge Univ. Press, Cambridge, pp. 161–218.
Kutschera, L. 1960. Wurzelatlas mitteleuropäischer Ackerunkräuter und Kulturpflanzen. – DLG-Verlags-Gmbh, Frankfurt am Main.
Lagerlöf, J. 1987. Ecology of soil fauna in arable land. Dynamics and activity of microarthropods and enchytraeides in four cropping systems. – Ph. D. Thesis, Dept of Ecology and Environmental Research, Report 29. Swedish Univ. of Agricultural Sciences, Uppsala.
– and Andrén, O. 1985. Succession and activity of microarthropods and enchytraeids during barley straw decomposition. – Pedobiologia 28: 343–357.
– and Andrén, O. 1988. Abundance and activity of soil mites (Acari) in four cropping systems. – Pedobiologia 32: 129–145.
– and Andrén, O. In press. Abundance and activity of Collembola, Protura and Diplura (Insecta, Apterygota) in four cropping systems. – Pedobiologia.
– and Scheller, U. 1989. Abundance and activity of Pauropoda and Symphyla (Myriapoda) in four cropping systems. – Pedobiologia 33: 315–321.
–, Andrén, O. and Paustian K. 1989. Dynamics and contribution to carbon flows of Enchytraeidae (Oligochaeta) under four cropping systems. – J. Appl. Ecol. 26: 183–199.
Langdon, K. R., Struble, F. B. and Young, H. C. 1961. Stunt of small grains, a new disease caused by the nematode *Tylenchorhynchus brevidens*. – Plant Dis. Reptr. 45: 248–252.
Larsson, B. M. P. 1986. Landskap, flora och fauna. Jordbrukets miljöeffekter. – Rapport från Svenska naturskyddsföreningens höstkonferens 1985, Stockholm, pp. 3–15 (in Swedish).

– 1987. Kulturlandskapets omvandling i ett historiskt perspektiv. Orsaker, omfattning och ekologiska konsekvenser. Naturvård i kulturmiljö. – Meddelande från Nåtö biologiska station 1: 22–32, Mariehamn (in Swedish).
Lawlor, D. W. 1987. Photosynthesis, plant production and yield. – In: Photosynthesis: metabolism, control and physiology. Longman Singapore Publishers (Pte) Ltd., Singapore, pp. 243–250.
Legg, J. O. and Meisinger, J. J. 1982. Soil nitrogen budgets. – In: Stevenson, F. J. (ed.), Nitrogen in agricultural soils. Agronomy 22: 503–566.
–, Day, W., Lawlor, D. W. and Parkinson, K. J. 1979. The effect of drought on barley growth: models and measurements showing the relative importance of leaf area and photosynthetic rate. – J. Agric. Sci., Cambridge 92: 703–716.
–, Cannon, K. R., Robertson, J. A. and Cook, F. D. 1986. Dynamics of soil microbial biomass and water-soluble organic C in Breton L after 50 years of cropping to two rotations. – Can. J. Soil Sci. 66: 1–19.
Lemeé G. 1975. Recherches sur les écosysteme des réserve biologique de sur les caractères et l'activité biologique du mull acide. – Rev. Ecol. Biol. Sol 12: 145–166.
Lemeur, R. and Rosenberg, N. J. 1979. Simulation of the quality and quantity of short-wave radiation within and above canopies. – In: Halldin, S. (ed.), Comparison of forest water and energy exchange models. Int. Soc. Ecol. Model., Copenhagen, pp. 77–100.
Letey, J., Blear, J.M., Devitt, D., Lund, L.J. and Nash, P. 1977. Nitrate-nitrogen in effluent from agricultural tile drains in California. – Hilgardia 45: 289–319.
Lethbridge, G. and Davidson, M. S. 1983. Root-associated nitrogen-fixing bacteria and their role in the nitrogen nutrition of wheat estimated by ^{15}N isotope dilution. – Soil Biol. Biochem. 15: 365–374.
–, Davidson, M. S. and Sparling, G. P. 1982. Critical evaluation of the acetylene reduction test for estimating the activity of nitrogen-fixing bacteria associated with the roots of wheat and barley. – Soil Biol. Biochem. 14: 27–35.
Lindberg, T., Bonde, T., Bergström, L., Pettersson, R., Rosswall, T. and Schnürer, J. 1989. Distribution of ^{15}N in the soil-plant system during a four-year field lysimeter study with barley (*Hordeum distichum* L.) and perennial meadow fescue (*Festuca pratensis* Huds.). – Pl. Soil 119: 23–37.
Lindén, B. 1983a. Movement, distribution and utilization of ammonium- and nitrate nitrogen in Swedish agricultural soils. – Ph. D. Thesis, Dept of Soil Sciences, Swedish Univ. of Agricultural Sciences, Uppsala.
– 1983b. De övervintrande mineralkväveförrådens storlek och variationer. – K. Skogs- o. Lantbr.-akad. rapp. 6: 4–21 (in Swedish).
– 1985. Mineral nitrogen present in the root zone in early spring and nitrogen mineralized during the growing season – their contribution to the nitrogen supply of crops. – In: Neeteson, J. J. and Dilz, K. (eds), Assessment of nitrogen fertilizer requirement. Inst. of Soil Fertility, Haren, pp. 37–49.
– and Mattson, L. 1987. Variationer i markens mineralkväveförråd. En undersökning på olika jordar i Uppland och Västergötland. – Dept of Soil Sciences, Report 167. Swedish Univ. of Agricultural Sciences, Uppsala (in Swedish).
Lindhard, J. 1975. Kvaelstof i afgröde og gennemsivningsvand efter tilförsel af nitrat- og ammoniumkvaelstof. Lysimeterforsög 1962–72. – Tidskr. Planteavl 79: 536–544 (in Danish).
Linn, D. M. and Doran, J. W. 1984. Effect of water-filled pore space on carbon dioxide and nitrous oxide production in tilled and nontilled soils. – Soil Sci. Soc. Am. J. 48: 1267–1272.
Lipschultz, F., Zafiriou, O. C., Wofsy, S. C., McElroy, M. B., Valois, F. W. and Watson, S. W. 1981. Production of NO

and N_2O by soil nitrifying bacteria. – Nature, Lond. 294: 641–643.
Lofs-Holmin, A. 1983. Reproduction and growth of common arable land and pasture species of earthworms (Lumbricidae) in laboratory cultures. – Swed. J. agric. Res. 13: 31–37.
Lohm, U. 1979. The present faunistic knowledge of terrestrial Enchytraeidae in Sweden. – Zoon 7: 63–66.
Long, S. P. 1986. Instrumentation for the measurement of CO_2 assimilation by crop leaves. – In: Gensler, W.G. (ed.), Advanced agricultural instrumentation, design and use. Dordrecht, Boston, pp. 39–91.
– and Hall, D. O. 1987. Nitrogen cycles in perspective. – Nature, Lond. 329: 584–585.
Lundgren, B. 1981. Flourescein diacetate as a stain of metabolically active bacteria in soil. – Oikos 36: 17–22.
Lundin, L.-C. 1989. Water and heat flows in frozen soils. – Acta Univ. Ups. 186. Dept of Hydrology, Uppsala Univ., Uppsala.
Lunt, H. A. and Jacobson, G. M. 1944. The chemical composition of earthworm casts. – Soil Sci. 58: 367–375.
Lynch, J. M. 1976. Products of soil microorganisms in relation to plant growth. – CRC Crit. Rev. Microbiol. 5: 67–107.
– and Panting, L. M. 1980. Variations in the size of the soil biomass. – Soil Biol. Biochem. 12: 547–550.
Mårtensson, A. M. and Ljunggren, H. D. 1984a. Nitrogen fixation in an establishing alfalfa (*Medicago sativa* L.) ley in Sweden, estimated by three different methods. – Appl. Env. Microbiol. 48: 702–707.
– and Ljunggren, H. D. 1984b. A comparison between the acetylene reduction method, the isotope dilution method and the total nitrogen difference method for measuring nitrogen fixation in lucerne (*Medicago sativa* L.). – Pl. Soil 81: 177–184.
Macfadyen, A. 1963. The contribution of the microfauna to total soil metabolism. – In: Doeksen, J. and van der Drift, J. (eds), Soil organisms. North Holland, Amsterdam, pp. 3–17.
Mallow, D., Snider, R. J. and Robertson, L. S. 1985. Effects of different management practices on Collembola and Acarina in corn production systems. – Pedobiologia 28: 115–131.
Mankau, R. 1980. Biological control of nematode pests by natural enemies. – Ann. Rev. Phytopathol. 18: 415–440.
Marshall, V. G. 1977. Effects of manures and fertilizers on soil fauna: a review. – Commonwealth Bureau of Soils, spec. publ. 3. Commonwealth Agricultural Bureaux.
Martin, J. K. 1977. Factors influencing the loss of organic carbon from wheat roots. – Soil Biol. Biochem. 9: 1–7.
– 1987. Carbon flow through the rhizosphere of cereal crops – A review. – INTECOL Bull. 15: 17–24.
– and Kemp, J. R. 1986. The measurement of C transfers within the rhizosphere of wheat grown in field plots. – Soil Biol. Biochem. 18: 103–107.
Mattson, L. 1985. Markbördighetsförsök i Norrland. – Dept of Soil Sciences, Report 164. Swedish Univ. of Agricultural Sciences, Uppsala (in Swedish).
– 1987. Long-term effects of N fertilizer on crops and soils. Ph. D. Thesis, Dept of Soil Sciences, Report 170. Swedish Univ. of Agricultural Sciences, Uppsala.
– and Lindén, B. 1988. Kväveförsök i potatis med bestämning av mineralkväve i marken. – Dept of Soil Sciences, Report 174. Swedish Univ. of Agricultural Sciences, Uppsala (in Swedish).
McDonald, R. M. 1979. Population dynamics of the nitrifying bacterium *Nitrosolobus* in soil. – J. Appl. Ecol. 16: 529–535.
McGill, W. B. and Cole, C. V. 1981. Comparative aspects of cycling of organic C, N, S and P through soil organic matter. – Geoderma 26: 267–286.
–, Hunt, H. W., Woodmansee, R. G. and Reuss, J. O. 1981. Phoenix – a model of the dynamics of carbon and nitrogen in grassland soils. – In: Clark, F. E. and Rosswall, T. (eds), Terrestrial nitrogen cycles. Ecol. Bull. (Stockholm) 33: 49–115.
Merckx, R., den Hartog, A. and van Veen, J. A. 1985. Turnover of root-derived material and related microbial biomass formation in soils of different texture. – Soil Biol. Biochem. 17: 565–569.
Milner, C. and Hughes, R. E. 1968. Methods for the measurement of the primary production of grassland. – IBP 6. Blackwell Sci. Publ., Oxford and Edinburgh.
Milthorpe, F. L. and Moorby, J. 1979. An introduction to crop physiology. – Cambridge Univ. Press, Cambridge.
Minderman, G. 1968. Addition, decomposition and accumulation of organic matter in forests. – J. Ecol. 56: 355–362.
Molina, J. A. E. and Rovira, A. D. 1964. The influence of plant roots on autotrophic nitrifying bacteria. – Can. J. Microbiol. 10: 249–257.
–, Gerard, G. and Mignolet, R. 1979. Asynchronous activity of ammonium oxidizer cluster in soil. – Soil Sci. Soc. Am. J. 43: 728–731.
Monteith, J. L. 1965. Evaporation and environment. – In: Fogg, G. E. (ed.), The state and movement of water in living organisms. 19th Symp. Soc. Exp. Biol. The Company of Biologists, Cambridge, pp 205–234.
– 1981. Evaporation and surface temperature. – Quart. J. R. Met. Soc. 107: 1–24.
Mosier, A. R., Guenzi, W. D. and Schwizer, E. E. 1986. Soil losses of dinitrogen and nitrous oxide from irrigated crops in northeastern Colorado. – Soil Sci. Soc. Am. J. 50: 344–348.
Mosier, A., Haider, K. and Heinemeyer, O. 1987. Effects of plants on organic carbon and nitrogen turnover in soil. – Agron. Abstr. 1987. Am. Soc. Microbiol., Madison, p. 188.
Mualem, Y. 1976. A new model for predicting the hydraulic conductivity of unsaturated porous media. – Water Resour. Res. 12: 513–522.
Müller, G. 1959. Untersuchung über das Nahrungswahlvermögen einiger in Ackerboden häufig vorkommender Collembolen und Milben. – Zool. Jb. (syst.) 87: 231–255.
Myers, R. J. K. and Paul, E. A. 1971. Plant uptake and immobilization of ^{15}N-labelled ammonium nitrate in a field experiment with wheat. – In: Nitrogen-15 in soil-plant studies. IAEA, Vienna, pp. 55–64.
Myrold, D. D. and Tiedje, J. M. 1985. Diffusional constraints on denitrification in soils. – Soil Sci. Soc. Am. J. 49: 651–657.
Naglitsch, F. 1966. Über veränderungen der zusammensetzung der mesofauna während der rotte organischer substanzen im boden. – Pedobiologia 6: 178–194.
Nannipieri, P., Johnson, R. L. and Paul, E. A. 1978. Criteria for measurement of microbial growth and activity in soil. – Soil Biol. Biochem. 10: 223–229.
Needham, A. E. 1957. Components of nitrogenous excreta in the earthworms *Lumbricus terrestris* L. and *Eisenia foetida* (Savigny). – J. Exp. Biol. 34: 425–446.
Nelson, D. W. 1982. Gaseous losses of nitrogen other than through denitrification. – In: Stevenson, F. J. (ed.), Nitrogen in agricultural soils. Agronomy 22: 327–364.
Newbould, P. 1982. Losses and accumulation of organic matter in soils. – In: Boels, D., Davies, D. B. and Johnston, A. E. (eds), Soil degradation. Balkema, Rotterdam, pp. 107–131.
Nicholas, W. L. 1984. The biology of free-living nematodes. 2nd ed. – Clarendon Press, Oxford.
Nielsen, C. O. 1949. Studies on the soil microfauna. II. The soil inhabiting nematodes. – Nat. Jutl. 2: 1–132.
– 1955. Studies of the Enchytraeidae 5. Factors causing seasonal fluctuation in numbers. – Oikos 6: 153–169.
Nilsson, J. (ed.) 1986. Ammoniakförluster från stallgödsel, åkermark och växter. Ammoniakutsläpp och dess effekter. – Statens naturvårdsverk Rapport nr. 3188, Stockholm (in Swedish).
Nömmik, H. 1956. Investigations on denitrification in soil. –

Acta Agric. Scand. 6: 195–228.
Nørlund, T., Gottschau, K., Thorhauge, N. E., Nielsen, N. E. and Jensen, H. E. 1985. The dynamics of nitrogen under field grown spring barley as affected by nitrogen application, irrigation and undersown catch crops. – Dept of Soil Fertility and Plant Nutrition and Hydrotechnical Laboratory, Report 1129. The Royal Veterinary and Agricultural Univ., Copenhagen.
Nordström, S. 1976. Growth and sexual development of lumbricids in southern Sweden. – Oikos 27: 476–482.
– and Rundgren, S. 1973. Associations of lumbricids in southern Sweden. – Pedobiologia 13: 301–326.
Norton, D. C. 1978. Ecology of plant-parasitic nematodes. – Wiley, New York.
Nosek, J. 1963. Zur Kenntnis der Apterygoten der kleinkarpatishen Wald – und Dauergrünlandböden. – Pedobiologia 2: 108–131.
– 1973. The European Protura. – Museum d'Histoire naturelle, Genève.
Nowak, E. 1975. Population density of earthworms and some elements of their production in several grassland environments. – Ekol. Pol. 23: 459–492.
Nye, P. H. and Tinker, P. B. 1977. Solute movement in the soil-root system. – Studies in Ecology 4. Blackwell, Oxford.
O'Connor, F. B. 1962. The extraction of Enchytraeidae from soil. – In: Murphy, P. W. (ed.), Progress in soil zoology. Butterworths, London, pp. 279–285.
– 1967. The Enchytraeidae. – In: Burges, A. and Raw, F. (eds), Soil biology. Academic Press, London, pp. 213–257.
Öborn, I. and Johnsson, H. 1990. Mollic Gleysol at Kjettslinge. – Dept Forest soils and Div. of Soil Science and Ecochemistry, Soil Description series, No 2, Swedish Univ. of Agricultural Sciences, Uppsala. In press.
Östergaard, H. S., Hvelplund, E. K. and Rasmussen, D. 1983. Kvaelstof-prognoser. Bestemmelse af optimalt kvaelstofbehov på grundlag af jordanalyser og klimamålinger för verkstsaesonen. – Landskontoret for Planteavl, Viby (in Danish).
O'Neill, R. V., DeAngelis, D. L., Waide, J. B. and Allen, T. F. H. 1986. A hierchical concept of ecosystems. – Monographs in Population Biology 23. Princeton Univ. Press, Princeton, NJ.
Odell, R. T., Melsted, S. W. and Walker, W. M. 1984. Changes in organic carbon and nitrogen of Morrow plot soils under different treatments, 1904–1973. – Soil Sci. 137: 160–171.
Odin, H., Eriksson, B. and Perttu, K. 1983. Temperature climate maps of Swedish forestry. – Dept of Forest Soils, Report 45. Swedish Univ. of Agricultural Sciences, Uppsala.
Odum, E. P. 1984. Properties of agroecosystems. – In: Lowrance, R., Stinner, B. R. and House, G. J. (eds), Agricultural ecosystems. Unifying Concepts. Wiley, New York, pp. 5–11.
Olive, O. J. and Clarc, R. B. 1978. Physiology of reproduction. – In: Mill, P. J. (ed.), Physiology of Annelids. Academic Press, London, pp. 271–368.
Osvald, H. 1959. Åkerns nyttoväxter. – AB Svensk Litteratur / Esselte AB, Stockholm (in Swedish).
– 1962. Vallodling och växtföljder. Uppkomst och utveckling i Sverige. – Natur och Kultur, Uppsala (in Swedish).
Page, F. C. 1988. A new key to freshwater and soil Gymnamoebae. – Freshwater Biol. Ass., Ambleside.
Paine, R. T. 1988. Food webs: Road maps of interactions or grist for theoretical development? – Ecology 69: 1648–1654.
Parkinson, D. 1983. Functional relationships between soil organisms. – In: Lebrun, P., André, H. M., De Medts, A., Grégoire-Wibo, C. and Wauthy, G. (eds), New trends in soil biology. Dieu-Brichart, Ottignies Louvain-la-Neuve, pp. 153–165.
Parle, J. N. 1963. A microbiological study of earthworm casts. – J. Gen. Microbiol. 31: 13–22.

Parton, W. J., Persson, J. and Anderson, D. W. 1983. Simulation of organic matter changes in Swedish soils. – In: Lauenroth, W. K., Skogerboe, G. V. and Flug, M. (eds), Analysis of ecological systems: State-of-the-art in ecological modelling. Elsevier, Amsterdam, pp. 511–516.
–, Schimel, D. S., Cole, C. V. and Ojima, D. S. 1987. Analysis of factors controlling soil organic matter levels in Great Plains grasslands. – Soil Sci. Soc. Am. J. 51: 1173–1179.
–, Cole, C. V., Stewart, J. W. B., Ojima, D. S. and Schimel, S. 1989. Simulating regional patterns of soil C, N, and P dynamics in the U.S. central grasslands regions. – In: Clarholm, M. and Bergström, L. (eds), Ecology of arable land – Perspectives and challenges. Developments in Plant and Soil Sciences 39. Kluwer, Dordrecht, pp. 99–108.
Patni, N. K. 1978. Physical quality and sediment transport in drainage water from a manured and fertilized cropping operation. – J. Environ. Sci. Health 13: 269–285.
Paul, E. A. and van Veen, J. A. 1978. The use of tracers to determine the dynamic nature of organic matter. – Trans. 11th Int. Congr. Soil Sci. 3: 61–102.
– and Voroney, R. P. 1980. Nutrient and energy flow through soil microbial biomass. – In: Ellwood, D. C., Hedger, J. N., Latham, M. J., Lynch, J. M. and Slater, J. H. (eds), Contemporary microbial ecology. Academic Press, London, pp. 215–237.
– and Juma, N. G. 1981. Mineralization and immobilization of soil nitrogen by microorganisms. – In: Clark, F. E. and Rosswall, T. (eds), Terrestrial nitrogen cycles. Ecol. Bull. (Stockholm) 33: 179–195.
Paustian, K. 1985. Influence of fungal growth pattern on decomposition and nitrogen mineralization in a model system. – In: Fitter, A. H., Atkinson, D., Read, D. J. and Usher, M. B. (eds), Ecological interactions in soil: Plants, microbes and animals. Brit. Ecol. Soc. Spec. Publ. 4. Blackwell Sci. Publ., Oxford, pp. 159–173.
– and Bonde, T. A. 1987. Interpreting incubation data on nitrogen mineralization from soil organic matter. – INTECOL Bull. 15: 101–112.
– and Schnürer, J. 1987a. Fungal growth response to carbon and nitrogen limitation: I. A theoretical model. – Soil Biol. Biochem. 19: 613–620.
– and Schnürer, J. 1987b. Fungal growth response to carbon and nitrogen limitation: II. Application of a model to laboratory and field data. – Soil Biol. Biochem. 19: 621–629.
–, Andrén, O., Clarholm, M., Hansson, A.-C., Johansson, G., Lagerlöf, J., Lindberg, T., Pettersson, R. and Sohlenius, B. 1990. Carbon and nitrogen budgets of four agroecosystems with annual and perennial crops, with and without N fertilization. – J. Appl. Ecol. In press.
Peat, W. E. 1970. Relationships between photosynthesis and light intensity in the tomato. – Ann. Bot. 34: 319–328.
Perman, O. 1956. Kalken och dess användning i historisk belysning. – Handbok om växtnäring I: 25–33, Tidskriftsaktiebolaget Växtnärings-Nytt, Stockholm (in Swedish).
Perrier, A. 1979. Physical model to simulated energy exchange of plant canopies. – In: Halldin, S. (ed.), Comparison of forest water and energy exchange models. Int. Soc. Ecol. Model., Copenhagen, pp. 101–113.
Persson, H. 1978a. Root dynamics in a young Scots pine stand in central Sweden. – Oikos 30: 508–519.
Persson, J. 1974. Humusbalans i odlad jord (Eng. summary: Humus balance in cultivated soil.) – J. Sci. Agric. Soc. Finland 3: 247–263.
– 1978b. Kulturåtgärdernas inverkan på markorganismerna – markvård eller markförstöring? I. Mikrobaktivitet och humusbalans. – K. Skogs- och Lantbr.-akad. tidskr. 117: 43–48 (in Swedish).
– 1980a. Detaljstudium av den organiska substansens omsättning i ett fastliggande ramförsök – Dept of Soil Sciences, Report 128. Swedish Univ. of Agricultural Sciences, Uppsala (in Swedish).

- 1981. Influence of mineral and organic fertilizers on the humus balance and humus formation. – Coll. Humus-Azote, 7–10 Juillet, 1981 (INRA), Reims, pp. 81–87.
Persson, T. (ed.) 1980b. Structure and function of northern coniferous forests – an ecosystem study. – Ecol. Bull. (Stockholm) 32.
- 1983. Influence of soil animals on nitrogen mineralisation in a northern Scots pine forest. – In: Lebrun, Ph., André, H. M., De Medts, A., Grégorie-Wibo, C. and Wauthy, G. (eds), New trends in soil biology. Dieu-Brichart, Ottignies-Louvain-la-Neuve, pp. 117–126.
- and Lohm, U. 1977. Energetical significance of the Annelids and Arthropods in a Swedish grassland soil. – Ecol. Bull. (Stockholm) 23.
- , Bååth, E., Clarholm, M., Lundkvist, H., Söderström, B. E. and Sohlenius, B. 1980. Trophic structure, biomass dynamics and carbon metabolism of soil organisms in a Scots pine forest. – In: Persson, T. (ed.), Structure and function of northern coniferous forests – an ecosystem study. Ecol. Bull. (Stockholm) 32: 419–462.
Perttu, K. 1981. Radiation cooling and frost risk. – In: Eriksson, G. (ed.), Climatic zones regarding the cultivation of *Picea abies* L. in Sweden. Dept of Forest Genetics, Res. Notes 36. Swedish Univ. of Agricultural Sciences, Umeå.
Petersen, H. 1980. Population dynamics and metabolic characterization of Collembola species in a beech forest ecosystem. – In: Dindal, D. (ed.), Soil biology as related to land use practices. EPA, Washington DC, pp. 806–833.
- and Luxton, M. 1982. A comparative analysis of soil fauna populations and their role in decomposition processes. – In: Petersen, H. (ed.), Quantitative ecology of microfungi and animals in soil and litter. – Oikos 39: 287–388.
Petrusewicz, K. and Macfadyen, A. 1970. Productivity of terrestrial animals – principles and methods. – IBP 13. Blackwell Sci. Publ., Oxford-Edinburgh.
Pettersson, R. 1987. Primary production in arable crops: Above-ground growth dynamics, net production and nitrogen uptake. – Ph. D. Thesis, Dept of Ecology and Environmental Research, Report 31. Swedish Univ. of Agricultural Sciences, Uppsala.
- 1989. Above-ground growth dynamics and net production of spring barley in relation to nitrogen fertilization. – Swed. J. agric. Res. 19: 135–145.
- and Hansson, A.-C. Net primary production of a perennial grass ley (*Festuca pratensis,* Huds.) assessed with different methods and compared with a lucerne ley (*Medicago sativa* L.). – Submitted to J. Appl. Ecol.
- , Hansson, A.-C., Andrén, O. and Steen, E. 1986. Above- and below-ground production and nitrogen uptake in lucerne *(Medicago sativa)*. – Swed. J. agric. Res. 16: 167–177.
Phillipson, J. and Bolton, P. J. 1977. Growth and cocoon production by *Allolobophora rosea* (Oligochaeta: Lumbricidae). – Pedobiologia 17: 70–82.
Piearce, T. G. 1978. Gut contents of some lumbricid earthworms. – Pedobiologia 18: 153–157.
Pimm, S. L. 1982. Food webs. – Chapman and Hall, London.
- and Kitching, R. L. 1988. The determinants of food chain lengths. – Ecology 50: 302–307.
Poth, M. A. and Focht, D. D. 1985. ^{15}N kinetic analysis of N_2O production by *Nitrosomonas europaea:* an examination of nitrifier denitrification. – Appl. Environ. Microbiol. 49: 1134–1141.
Poulsen, E. and Hansen, P. 1962. Undersøgelser over jordens nitratindhold. – Tidsskrift Planteavl 65: 206–234 (in Danish).
Powlson, D. S., Pruden, G., Johnston, A. E. and Jenkinson, D. S. 1986. The nitrogen cycle in the Broadbalk wheat experiment: recovery and losses of ^{15}N-labelled fertilizer applied in spring and inputs of nitrogen from the atmosphere. – J. agric. Sci., Cambridge 107: 591–609.
- , Brookes, P. C. and Christensen, B. T. 1987. Measurement of soil microbial biomass provides an early indication of changes in total soil organic matter due to straw incorporation. – Soil Biol. Biochem. 19: 159–164.
Prestidge, R. A. 1982. Instar duration, adult consumption, oviposition and nitrogen utilization efficiencies of leaf hoppers feeding on different quality food (Auchenorrhyncha: Homoptera). – Ecol. Ent. 7: 91–101.
Purvis, G. and Curry, J. P. 1980. Successional changes in the arthropod fauna of a new ley pasture established on previously cultivated arable land. – J. Appl. Ecol. 17: 309–321.
Pussard, M. and Delay, F. 1985. Dynamique de population d'amibes libres endogées (amoebida, protozoa) I. – Evaluation du degré d'activité en microcosme et observations préliminaires sur la dynamique de population de quelques espéces. – Protistologia 21: 5–15.
- , Alabouvette, C. and Pons, R. 1979. Etude préliminaire d'une amibe mycophage *Thecamoeba granifera* s. sp. *minor* (Thecamoebidae, Amoebida). – Protistologia 15: 139–149.
Reinecke, A. J. and Ljungström, P.-O. 1969. An ecological study of the earthworms from the banks of Mei Mooi River in Potchefstroom, South Africa. – Pedobiologia 9: 106–111.
Rennie, R. J. 1982. Quantifying dinitrogen (N_2) fixation in soybeans by ^{15}N isotope dilution: The question of the non-fixing control plant. – Can. J. Bot. 60: 856–861.
- , Rennie, D. A. and Fried, M. 1978. Concepts of ^{15}N usage in dinitrogen fixation studies. – In: Isotopes in biological dinitrogen fixation. FAO/IAEA, Vienna, pp. 107–133.
Richards, B. N. 1974. Introduction to the soil ecosystem. – Longman, New York.
Rickman, R. W., Klepper, B. and Belford, R. K. 1984. Developmental relationships among roots, leaves and tillers in winter wheat. – In: Day, W. and Atkins, R. K. (eds), Wheat growth and modelling. Plenum Press, New York, pp. 83–98.
Riha, S. J., Campbell, G. S. and Wolfe, J. 1986. A model of competition for ammonia among heterotrophs, nitrifiers and roots. – Soil Sci. Soc. Am. J. 50: 1463–1466.
Risser, P. G., Birney, E. C., Blocka, H. D., May, S. W., Parton, W. J. and Wiens, J. A. 1981. Producers. – In: The true praire ecosystem. US/IBP Synthesis Series 16: 155–187.
Roberts, M. J., Long, S. P., Tieszen, L. L and Beadle, C. L. 1985. Measurements of plant biomass and net primary production. – In: Coombs, J., Hall, D. O., Long, S. P. and Scurlock, J. M. O. (eds), Techniques in bioproductivity and photosynthesis. Pergamon Press, Oxford, pp. 1–18.
Robertson, G. P. 1982. Nitrification in forested ecosystems. – Phil. Trans. R. Soc. London 296: 445–457.
- and Vitousek, P. 1981. Nitrification potentials in primary and secondary succession. – Ecology 62: 376–386.
Robertson, L. A. and Kuenen, J. G. 1984. Aerobic denitrification: a controversy reviewed. – Arch. Microbiol. 139: 351–354.
Rodskjer, N. and Tuvesson, M. 1975. Observations of temperature in winter wheat at Ultuna, Sweden, 1968–72. – Swed. J. agric. Res. 5: 223–234.
Rolston, D. E., Sharply, A. N., Toy, D. W. and Broadbent, F. E. 1982. Field measurement of denitrification. III. Rates during irrigation cycles. – Soil Sci. Soc. Am. J. 46: 289–296.
Rosswall, T. and Clarholm, M. 1974. Characteristics of tundra bacterial populations and a comparison with populations from forest and grassland soils. – In: Holding, A. J., Heal, O. W., MacLean, S. F. and Flanagan, P. W. (eds), Soil organisms and decomposition in tundra. Tundra Biome Steering Committee, Stockholm, pp. 325–340.
- and Heal, O. W. (eds) 1975. Structure and function of tundra ecosystems. – Ecol. Bull. (Stockholm) 20.
- and Granhall, U. 1980. Nitrogen cycling in a subartic ombrotrophic mire. – In: Sonesson, M. (ed.), Ecology of a subarctic mire. Ecol. Bull. (Stockholm) 30: 209–234.
- and Paustian, K. 1984. Cycling of nitrogen in modern agricultural systems. – Pl. Soil 76: 3–21.
- , Bak, F., Baldocchi, D., Cicerone, R. J., Conrad, R.,

Ehhalt, D. H., Firestone, M. K., Galbally, I. E., Galchenko, V. F., Groffman, P., Papen, H., Reeburgh, W. S. - and Sanhueza, E. 1989. What regulates production and consumption of trace gases in ecosystems: biology or physicochemistry? – In: Andreae, M. O. and Schimel, D. S. (eds), Exchange of trace gases between terrestrial ecosystems and the atmosphere. Wiley, Chichester, pp. 73–95.
Rovira, A. D. 1969. Plant root exudates. – Bot. Rev. 35: 35–55.
Rowe, R., Todd, R. and Waide, J. 1977. Microtechnique for most probable number analyses. – Appl. Environ. Microbiol. 33: 675–680.
Rundgren, S. 1975. Vertical distribution of lumbricids in southern Sweden. – Oikos 26: 299–306.
- 1977. Seasonality of emergence in lumbricids in southern Sweden. – Oikos 28: 49–55.
Russell, E. W. 1977. The role of organic matter in soil fertility. – Phil. Trans. Roy. Soc., London B 281: 209–219.
- 1978. Arable agriculture and soil deterioration. – Trans. 11th Int. Congr. Soil Sci. 3: 216–227.
Ryden, J. C. 1983. Denitrification loss from a grassland soil in the field receiving different rates of nitrogen as ammonium nitrate. – J. Soil Sci. 34: 355–365.
- 1985. Denitrification loss from managed grasslands. – In: Golterman, H. L. (ed.), Denitrification in the nitrogen cycle. NATO Conference Series I: Ecology 9: 121–134. Plenum Press, New York.
- and Lund, L. J. 1980. Nature and extent of directly measured denitrification losses from irrigated vegetable crop production unis. – Soil. Sci. Soc. Am. J. 44: 505–511.
-, Lund, L. J., Letey, J. and Focht, D. D. 1979. Direct measurement of denitrification loss from soils. II. Development and application of field method. – Soil Sci. Soc. Am. J. 43: 110–118.
Ryding, S.-O. (ed.) 1988. Jordbrukets inverkan på luft- och vattenmiljön. Kunskapssammanställning om växtnäringsförluster källor, effekter, åtgärder. – Lantbrukarnas Riksförbund Rapport, Stockholm (in Swedish).
Ryl, B. 1977. Enchytraeids on rye and potato fields in Turew. – Ecol. Pol. 25: 519–529.
- 1980. Enchytraeid (Enchytraeidae, Oligochaeta) populations of soils of chosen crop-fields in the vicinity of Turew (Poznan region). – Pol. Ecol. Studies 6: 277–291.
Ryszkowski, L. 1984. Primary production in agroecosystems. – In: Options Mediterraneennes. Int. Centre Adv. Mediterranean Agronomic Studies, Zaragoza, pp. 77–94.
- 1985. Impoverishment of soil fauna due to agriculture. – INTECOL Bull. 12: 7–17.
Saggar, S., Bettany, J. R. and Stewart, J. W. B. 1981. Measurement of microbial sulphur in soil. – Soil Biol. Biochem. 13: 493–498.
Santos, P. F., Phillips, J. and Whitford, W. G. 1981. The role of mites and nematodes in early stages of buried litter decomposition in a desert. – Ecology 62: 664–669.
Sarathchandra, S. U. 1979. A simplified method for estimating ammonium oxidizing bacteria. – Pl. Soil 52: 305–309.
Satchell, J. E. 1967. Lumbricidae. – In: Burges, A. and Raw, F. (eds), Soil biology. Academic Press, London, pp. 259–322.
Sauerbeck, D. R. 1982. Influence of crop rotation, manurial treatment and soil tillage on the organic matter content of German soils. – In: Boels, D., Davies, D. B. and Johnston, A. E. (eds), Soil degradation. Balkema, Rotterdam, pp. 163–178.
- and Johnen, B. G. 1977. Root formation and decomposition during plant growth. – In: Soil organic matter studies, Vol. 1. IAEA, Vienna, pp. 141–148.
-, Johnen, B. G. and Six, R. 1976. Atmung, Abbau und Ausscheidungen von Weizenwurzeln im laufe ihrer Entwicklung. – Landwirtsch. Forsch. Sonderh. 32: 49–58.
Scheu, S. 1987. The role of substrate feeding earthworms (Lumbricidae) for bioturbation in a beechwood soil. – Oecologia (Berl.) 72: 192–196.
Schimel, D., Simkins, S., Rosswall, T., Mosier, A. R. and Parton, W. J. 1989. Scale and the measurement of nitrogen gas fluxes from terrestrial ecosystems. – In: Rosswall, T. H., Woodmansee, R. G. and Risser, P. G. (eds), Scales and global change. Wiley, Chichester, pp. 179–193.
Schmidt, E. L. 1982. Nitrification in soil. – In: Stevenson, F. J. (ed.), Nitrogen in agricultural soils. Agronomy 22: 253–285.
Schnürer, J. 1985. Fungi in arable soil, their role in carbon and nitrogen cycling. – Ph. D. Thesis, Dept of Microbiology, Report 29. Swedish Univ. of Agricultural Sciences, Uppsala.
- and Paustian, K. 1986. Modelling fungal growth in relation to nutrient limitations in soil. – In: Megusàr, F. and Gantar, M. (eds), Perpectives in microbial ecology. Slovene Soc. Microbiol., Ljubljana, pp. 123–130.
- and Rosswall, T. 1987. Mineralization of nitrogen from ^{15}N labelled fungi, soil microbial biomass and roots and its uptake by barley plants. – Pl. Soil 102: 71–78.
-, Clarholm, M. and Rosswall, T. 1985. Microbial biomass and activity in agricultural soils with different carbon contents. – Soil Biol. Biochem. 17: 611–618.
-, Clarholm, M. and Rosswall, T. 1986a. Fungi, bacteria and protozoa in soil from four arable cropping systems. – Biol. Fert. Soils 2: 119–126.
-, Clarholm, M., Boström, S. and Rosswall, T. 1986b. Effects of moisture on soil microorganisms and nematodes – a field experiment. – Microb. Ecol. 12: 217–230.
Schubert, K. R. and Evans, H. J. 1976. Hydrogen evolution: A major factor affecting the efficiency of nitrogen fixation in nodulated symbionts. – Proc. Nat. Acad. Sci. USA 73: 1207–1211.
Seastedt, T. R. 1984. The role of microarthropods in decomposition and mineralization processes. – Ann. Rev. Ent. 29: 25–46.
Setälä, H., Haimi, J. and Huhta, V. 1988. A microcosm study on the respiration and weight loss in birch litter and raw humus as influenced by soil fauna. – Biol. Fertil. Soils 5: 282–287.
Sextone, A. J., Parkin, T. B. and Tiedje, J. M. 1985. Temporal response of soil denitrification rates to rainfall and irrigation. – Soil Sci. Soc. Am. J. 49: 99–103.
Sharma, R. D. 1971. Studies on the plant parasitic nematode *Tylenchorhynchus dubius*. – Meded. Landbouwhogesch., Wageningen 71: 1–154.
Sharpley, A. N., Sycrs, H. K. and Springett, N. A. 1979. Effect of surface casting earthworms on the transport of phosphorus and nitrogen in surface run off from pasture. – Soil Biol. Biochem. 11: 459–462.
Shields, J. A. and Paul, E. A. 1973. Decomposition of ^{14}C-labelled plant material under field conditions. – J. Soil Sci. 53: 297–306.
-, Paul, E. A., Lowe, W. E. and Parkinson, D. 1973. Turnover of microbial tissue in soil under field conditions. – Soil Biol. Biochem. 5: 753–764.
Simmelsgaard, S. E. 1980. Transport af naeringsstoffer til draen og undergrund i relation til vandbalance. – Bilag Statens Planteavlsmøde, Statens Planteavlsforsøg, 8–14, Köpenhamn (in Danish).
Sims, P. L. and Coupland, R. T. 1979. Producers. – In: Coupland, R. T. (ed.), Grassland ecosystems in the world: Analysis of grasslands and their use. Cambridge Univ. Press, Cambridge, pp. 49–72.
Singh, J. S., Lauenroth, W. K., Hunt, H. W. and Swift, D. M. 1984. Bias random errors in estimators of net root production: a simulation approach. – Ecology 65: 1760–1764.
Singh, K. P. and Shekhar, C. 1986. Seasonal pattern of total soil respiration, its fractionation and soil carbon balance in a wheat-maize rotation cropland at Varanasi. – Pedobiologia 29: 305–318.
Slemr, F. and Seiler, W. 1984. Field measurements of NO and

NO$_2$ emissions from fertilized and unfertilized soils. – J. Atmos. Chem. 2: 1–24.
Smith, K. A. 1977. Soil aeration. – Soil Sci. 123: 284–291.
Smith, M. S. and Tiedje, J. M. 1979. The effect of roots on soil denitrification. – Soil Sci. Soc. Am. J. 43: 951–955.
Smith, S. J., Young, L. B. and Miller, G. E. 1977. Evaluation of soil nitrogen mineralization potentials under modified field conditions. – Soil Sci. Soc. Am. J. 41: 74–76.
Söchtig, H. G. and Sauerbeck, D. R. 1982. Soil organic matter properties and turnover of plant residues as influenced by soil type, climate and farming practice. – In: Boels, D., Davies, D. B. and Johnston, A. E. (eds), Soil degradation. Balkema, Rotterdam, pp. 145–161.
Söderlund, R. 1984. Wet deposition of nitrogen compounds. – In: Svensson, B. (ed.), Ecology of Arable Land, Progress Report 1983, Swedish Univ. of Agricultural Sciences, Uppsala, pp. 92–96.
Söderström, B. E. 1977. Vital staining of fungi in pure cultures and in soil with fluorescein diacetate. – Soil Biol. Biochem. 9: 59–63.
Sørensen, L. H. 1982. Mineralization of organically bound nitrogen in soil as influenced by plant growth and fertilization. – Pl. Soil 65: 51–61.
Sohlenius, B. 1979. A carbon budget for nematodes, rotifers and tardigrades in a Swedish coniferous forest soil. – Holarct. Ecol. 2: 30–40.
– and Boström, S. 1984. Colonization, population development and metabolic activity of nematodes in buried barley straw. – Pedobiologia 27: 67–78.
– and Boström, S. 1986. Short-term dynamics of nematode communities in arable soil. – Influence of nitrogen fertilization in barley crops. – Pedobiologia 29: 183–191.
– and Sandor, A. 1987. Vertical distribution of nematodes in arable soil under grass (Festuca pratensis) and barley (Hordeum disticum). – Biol. Fert. Soils 3: 19–25.
–, Boström, S. and Sandor, A. 1987. Long-term dynamics of nematode communities in arable soil under four cropping systems. – J. Appl. Ecol. 24: 131–144.
–, Boström, S. and Sandor, A. 1988. Carbon and nitrogen budgets of nematodes in arable soil. – Biol. Fert. Soils 6: 1–8.
Soil Survey Staff. 1975. Soil taxonomy. A basic system of soil classification for making and interpreting soil surveys. – USDA Handbook No. 436, U.S. Government Printing Office, Washington D.C.
Sokal, R. R. and Rohlf, F. S. 1969. Biometry. – Freeman, San Francisco.
Sollins, P., Grier, C. C., McCorison, F. M., Cromack Jr., K., Fogel, R. and Fredriksen, R. L. 1980. The internal element cycles of an old-growth Douglas-fir ecosystem in western Oregon. – Ecol. Monogr. 50: 261–285.
Sommers, L. E., Gilmour, C. M., Wilding, R. E. and Beck, S. M. 1980. The effect of water potential on decomposition processes in soil. – In: Water potential relations in soil microbiology. SSSA Spec. Publ. 9, pp. 97–117.
Sonesson, M. 1980 (ed.). Ecology of a subarctic mire. – Ecol. Bull. (Stockholm) 30.
Sparling, G. P., Cheshire, M. V. and Mundie, C. M. 1982. Effect of barley plants on the decomposition of ^{14}C-labelled soil organic matter. – J. Soil Sci. 33: 89–100.
Sporrong, U. 1988. Människan som landskapsomvandlare. – Mem. Soc. Fauna Flora Fennica 64: 15–24 (in Swedish).
Sprent, J. I. 1971. Effects of water stress on nitrogen fixation in root nodules. – In: Lie, T. A. and Mulder, E. G. (eds), Biological nitrogen fixation in natural and agricultural habitats. Pl. Soil. Spec. Vol., pp. 225–228.
Stanford, G. and Smith, S. J. 1972. Nitrogen mineralization potential of soil. – Soil Sci. Soc. Am. Proc. 36: 465–472.
– and Epstein, E. 1974. Nitrogen mineralization – water relations in soils. – Soil Sci. Soc. Am. Proc. 38: 103–107.
Statistical Yearbook 1951 and 1981. – Statistiska Centralbyrån, Stockholm (in Swedish).

Steele, K. W., Wilson, A. T. and Saunders, W. M. M. 1980. Nitrification activity in New Zealand grassland soils. – N. Z. J. Agric. Res. 23: 249–256.
Steen, E. 1983. Soil animals in relation to agricultural practices and soil productivity. – Swed. J. agric. Res. 13: 157–165.
– 1984. Variation of root growth in a grass ley studied with a mesh bag technique. – Swed. J. agric. Res. 14: 93–97.
– 1988a. Sammanhållet system för livsmedelsproduktion. – In: Hjorth, P. (ed.) Samhällets ämnesomsättning. Forskningsrådsnämndens rapport 2: 31–33, Stockholm (in Swedish).
– 1988b. Ekologisk miljövård. – Miljö och hälsa 4: 15–18 (in Swedish).
– and Larsson, K. 1986. Carbohydrates in roots and rhizomes of perennial grasses. – New Phytol. 104: 339–346.
– and Lindén, B. 1987. Role of fine roots in the nitrogen economy of sugar beet. – J. Agron. Crop Sci. 158: 1–7.
–, Jansson, P.-E. and Persson, J. 1984. Experimental site of the 'Ecology of Arable Land' project. – Acta Agric. Scand. 34: 153–166.
Stefanson, R. C. 1972a. Soil denitrification in sealed soil-plant systems. I. Effect of plant, soil water content and soil organic matter content. – Pl. Soil 33: 113–127.
– 1972b. Soil denitrification in sealed soil-plant systems. II. Effect of soil water content and applied nitrogen. – Pl. Soil 33: 129–140.
– 1972c. Soil denitrification in sealed soil-plant systems. III. Effect of disturbed and undisturbed soil samples. – Pl. Soil 33: 141–149.
Steineck S. 1988. Flytgödsel till vall. – Dept of Soil Sciences, Report 172. Swedish Univ. of Agricultural Sciences, Uppsala (in Swedish).
Stevenson, F. J. 1982. Origin and distribution of nitrogen in soil. – In: Stevenson, F. J. (ed.), Nitrogen in agricultural soils. Agronomy 22: 1–42.
Stinner, B. R. and Crossley, D. A. 1980. Comparison of mineral element cycling under till and no-till practices: An experimental approach to agroecosystem analysis. – In: Dindal, D. (ed.), Soil biology as related to land use practices. Environment Protection Agency, Washington D.C., pp. 280–287.
–, Crossley Jr., D. A., Odum, E. P. and Todd, R. L. 1984. Nutrient budgets and internal cycling of N, P, K, Ca and Mg in conventional tillage, no-tillage, and old-field ecosystems in the Georgia piedmont. – Ecology 65: 354–369.
Stoskopf, N. C. 1981. Understanding crop production. – Reston, Reston, VA, pp. 91–109.
Stotzky, G., and Norman, A. G. 1964. Factors limiting microbial growth activities in soil. III. Supplementary substrate additions. – Can. J. Microbiol. 10: 143–147.
Stout, J. A., Bawden, A. D. and Coleman, D. C. 1984. Rates and pathways of mineral nitrogen transformation in a soil from pasture. – Soil Biol. Biochem. 6: 127–131.
Stout, J. D. and Heal, O. W. 1967. Protozoa. – In: Burges, A. and Raw, F. (eds), Soil biology. Academic Press, London, pp. 149–196.
Stuckey, I. H. 1941. Seasonal growth of grass roots. – Am. J. Bot. 28: 486–491.
Sundman, V. and Sivelä, S. 1978. A comment on the membrane filter technique for estimation of length of fungal hyphae in soil. – Soil Biol. Biochem. 10: 399–401.
Svensson, B. H. and Rosswall, T. 1980 Energy flow through the subarctic mire at Stordalen. – In: Sonesson, M. (ed.), Ecology of a subarctic mire. Ecol. Bull. (Stockholm) 30: 283–302.
–, Klemedtsson, L. K. and Rosswall, T. 1985. Preliminary field denitrification studies on nitrate-fertilized and nitrogen-fixing crops. – In: Golterman, H. L. (ed.), Denitrification in the nitrogen cycle. NATO Conf. Series: Ecology Vol. 9. Plenum Press, New York, pp. 157–169.
–, Boström, U.-L. and Klemedtsson, L. 1986. Potential for higher rates of denitrification in earthworm casts than in the

surrounding soil. – Biol. Fert. Soils 2: 147–149.
–, Klemedtsson, L., Simkins, S., Paustian, K. and Rosswall, T. 1990. Soil denitrification in three cropping systems with different nitrogen and carbon supply. I. Rate-distribution frequencies, comparison between systems and seasonal N-losses. – Submitted to Biogeochemistry.
Svensson, J. 1979. Storage, retrieval and analyses of continuously recorded eco-system data. – In: Halldin, S. (ed.), Comparison of forest water and energy exchange models. Int. Soc. Ecol. Model., Copenhagen, pp. 27–33.
Svensson, R. and Wigren, M. 1986a. Observations on the decline of some farmland weeds. – Mem. Soc. Fauna Flora Fennica 62: 63–67.
– and Wigren, M. 1986b. A survey of the history, biology and preservation of some retreating synantropic plants. – Acta Univ. Ups. Symb. Bot. Ups. 25: 1–74.
– and Wigren, M. 1986c. A changing flora – a matter of human concern. – Acta Univ. Ups. Symb. Bot. Ups. 27: 241–251.
Swift, M. J., Heal, O. W. and Anderson, J. M. 1979. Decomposition in terrestrial ecosystems. – Blackwell, Oxford.
Syers, J. K., Sharpley, A. N. and Keeney, D. R. 1979. Cycling of nitrogen by surface-casting earthworms in a pasture ecosystem. – Soil Biol. Biochem. 11: 181–185.
Taesler, R. 1972. Byggnadsklimatdata. – Beckman tryckerier AB, Stockholm (in Swedish).
Tanji, K. K. 1982. Modeling of the soil nitrogen cycle. – In: Stevenson, F. J. (ed.), Nitrogen in agricultural soils. Agronomy 22: 721–772.
Terry, R. E. and Tate, R. L. 1980. Denitrification as a pathway for nitrate removal from organic soils. – Soil Sci. 129: 162–166.
Thies, W., Becker, K. W. and Meyer, B. 1978. Bilanz von markiertem Dünger-N ($^{15}NH_4$ und $^{15}NO_3$) in natürlich gelagerten Sandlysimetern sowie zeitlicher Verlauf des dünger- und bodenbürtigen N-austrags in Vergleich Bewuchsbrache. – Landwirtschaftl. Forsch., Sonderheft 2: 55–62.
Tiedje, J. M. 1982. Denitrification. – In: Page, A. L., Miller, R. H. and Keeney, D. R. (eds), Methods of soil analysis. Part 2: Chemical and microbial properties (2nd ed.). Soil Sci. Soc. Am., Madison, pp. 1011–1024.
– 1988. Ecology of denitrification and dissimilatory nitrate reduction to ammonium. – In: Zehnder, A. J. B. (ed.), Biology of anaerobic microorganisms. Wiley, New York, pp. 179–244.
–, Sexstone, A. J. Parkin, T. B., Revsbech, N. P. and Shelton, D. R. 1984. Anaerobic processes in soil. – Pl. Soil 76: 197–212.
Tinker, P. B. 1983. The effect of additions of inorganic fertilizer N on the supply of inorganic N from the soil. – Rothamsted Exp. Stn. Report 1982 1: 262–263.
Tisdall, J. M., and Oades, J. M. 1982. Organic matter and water-stable aggregates in soils. – J. Soil Sci. 33: 141–163.
Torello, W. A., Wehner, D. J. and Turgoen, A. J. 1983. Ammonia volatilization from fertilized turfgrass stands. – Agron. J. 75: 454–456.
Tottman, D. R. and Makpeace, R. J. 1979. An explanation of the decimal code for the growth stages of cereals, with illustrations. – Ann. Appl. Biol. 87: 213–224.
Troughton, A. 1957. The underground organs of herbage grasses. – Commonwealth Bureau of Pastures and Crops. Hurley, Berkshire, Bulletin 44.
Tryselius, O. 1971. Runoff map of Sweden. Average annual runoff for the period 1931–60. – SMHI-Meddelanden, Serie C, Nr. 7. Norrköping.
Turner, G. L. and Gibson, A. H. 1980. Measurement of nitrogen fixation by indirect means. – In: Bergersen, F. J. (ed.), Methods for evaluating biological nitrogen fixation. Wiley, Chichester, pp. 111–138.
Uhlen, G. 1978. Nutrient leaching and surface runoff in field lysimeters on a cultivated soil. I. Runoff measurements, water composition and nutrient balances. – Sci. Rep. Agric. Univ. Norway, Vol. 57 (27): 1–26.
Usher, M. B. 1985. Population and community dynamics in the soil ecosystem. – In: Fitter, A. H., Atkinson, D., Read, D. J. and Usher, M. B. (eds), Ecological interactions in soil: Plants, microbes and animals. Brit. Ecol. Soc. Spec. Publ. 4. Blackwell Sci. Publ., Oxford, pp. 339–354.
Valera, L. and Alexander, M. 1961. Nutrition and physiology of denitrifying bacteria. – Pl. Soil 15: 268–280.
Vallis, I., Harper, L. A., Catchpoole, V. R. and Weier, K. L. 1982. Volatilization of ammonia from urine patches in a subtropical pasture. – Aust. J. Agric. Res. 33: 97–107.
van de Bund, C. F. 1970. Influence of crop and tillage on mites and springtails in arable soil. – Neth. J. agric. Sci. 18: 308–314.
van Veen, J. A. and Paul, E. A. 1981. Organic carbon dynamics in grassland soils. I. Background information and computer simulation. – Can. J. Soil Sci. 61: 185–201.
–, Ladd, J. N. and Frissel, M. J. 1984. Modelling C and N turnover through the microbial biomass in soil. – Pl. Soil 76: 257–274.
Vance, C. P., Heichel, G. H., Barnes, D. K., Bryan, J. W. and Johnsson, L. E. 1979. Nitrogen fixation, nodule development and vegetative regrowth of alfalfa (*Medicago sativa* L.) following harvest. – Pl. Physiol. 64: 1–8.
Vancura, V. and Kunc, F. 1977. The effect of streptomycin and actidione in the rhizosphere and non-rhizosphere soil. – Z. Bakt. Abt. II 132: 472–478.
Vickerman, G. P. 1978. The arthropod fauna of undersown grass and cereal fields. – Sci. Proc. Roy. Dublin Soc. 6A: 273–283.
Visser, S., Whittaker, J. B. and Parkinson, D. 1981. Effect of collembolan grazing on nutrient release and respiration of a leaf litter inhabiting fungus. – Soil Biol. Biochem. 13: 215–218.
Voroney, R. P. 1983. Decomposition of crop residues. – Thesis, Univ. of Saskatchewan, Saskatoon.
– and Paul, E. A. 1984. Determination of k_C and k_N in situ for calibration of the chloroform fumigation-incubation method. – Soil Biol. Biochem. 16: 9–14.
Wallwork, J. A. 1970. Ecology of soil animals. – McGraw Hill, London.
Walter, D. E., Hudgens, R. A. and Freckman, D. W. 1986. Consumption of nematodes by fungivorous mites, *Tyrophagus* spp. (Acarina: Astigmata: Acaridae). – Oecologia (Berl.) 70: 357–361.
–, Hunt, H. W. and Elliott, E. T. 1988. Guilds or functional groups? An analysis of predatory arthropods from a shortgrass steppe soil. – Pedobiologia 31: 247–260.
Walton, P. D. 1983. Production and management of cultivated forages. – Reston, VA, pp. 161–166.
Warembourg, F. R. and Paul, E. A. 1977. Seasonal transfers of assimilated ^{14}C in grassland: Plant production and turnover, soil and plant respiration. – Soil Biol. Biochem. 9: 295–301.
Wasilewska, L. 1967. Analysis of the occurrence of nematodes in alfalfa crops. II. Abundance and quantitative relations between species and ecological groups of species. – Ekol. Pol. Ser. A 15: 347–371.
– 1979. The structure and function of soil nematode communities in natural ecosystems and agrocenoses. – Pol. Ecol. Stud. 5: 97–145.
Wasylik, A. 1980. Occurrence and vertical distribution of soil mites in potato fields. – Pol. Ecol. Stud. 6: 655–663.
Watson, D. J., Thorne, G. N. and French, S. A. W. 1958. Physiological causes of differences in grain yield between varieties of barley. – Ann. Bot. 22: 321–352.
Watson, S. W., Valois, F. W. and Waterbury, J. B. 1981. The family Nitrobacteriaceae. – In: Starr, M. P., Stolp, H., Trüper, H. G., Balows, A. and Schegel, H. G. (eds), The procaryotes. Vol. 1, pp. 1005–1022.
Weaver, J. S. 1926. Root development of field crops. – McGraw Hill, New York.
Wessén, B. and Berg, B. 1986. Long-term decomposition of

barley straw: chemical changes and ingrowth of fungal mycelium. – Soil Biol. Biochem. 18: 53–59.
Whitehead, A. G. and Fraser, J. E. 1972. Injury to field beans (*Vicia faba* L.) by *Tylenchorhynchus dubius*. – Pl. Pathol. 21: 112–113.
Wiegert, R. G. and Evans, F. C. 1964. Primary production and the disappearance of dead vegetation on an old field in southeastern Michigan. – Ecology 45: 49–63.
Wiklert, P. 1960. Studier av rotutvecklingen hos några nyttoväxter med särskild hänsyn till markstrukturen. – Grundförbättring 3: 113–148 (in Swedish).
Williams, E. J., Parrish, D. D., Buhr, M. P. and Fehsenfeld, F. C. 1988. Measurements of soil NO_x emissions in central Pennsylvania. – J. Geophys. Res. 93: 9539–9546.
Witty, J. F. 1983. Estimating N_2-fixation in the field using ^{15}N-labelled fertilizer: Some problems and solutions. – Soil Biol. Biochem. 15: 631–639.
Wivstad, M., Mårtensson, A. M. and Ljunggren, H. D. 1987. Field measurement of symbiotic nitrogen fixation in an established lucerne ley using ^{15}N and an acetylene reduction method. – Pl. Soil. 97: 93–104.
Woldendorp, J. W. 1963. The influence of living plants on denitrification. – Meded. Landbouwhogesch., Wageningen 63: 1–100.
– 1981. Nutrients in the rhizosphere. – In: Agricultural yield potentials in continental climates. Proc. 16th Coll. Int. Potash Inst., Warsaw, pp. 99–125.
Woodmansee, R. G., Dodd, J. L., Bowman, R. A. and Clark, F. E. 1978. Nitrogen budget of a shortgrass prairie ecosystem. – Oecologia (Berl.) 34: 363–376.
Yearbook of Agricultural Statistics 1987. – Statistiska Centralbyrån, Stockholm (in Swedish).
Yeates, G. W. 1979. Soil nematodes in terrestrial ecosystems. – J. Nematol. 11: 213–229.
– 1982. Variation of pasture nematode populations over thirty-six months in a summer dry silt loam. – Pedobiologia 24: 329–346.
– 1984. Variation of pasture nematode populations over thirty-six months in a summer moist silt loam. – Pedobiologia 27: 207–219.
Yoshinari, T. and Knowles, R. 1976. Acetylene inhibition of nitrous oxide reduction by denitrifying bacteria. – Biochem. Biophys. Res. Comm. 69: 705–710.
Zadoks, J. C., Chang, T. and Konzak, C. F. 1974. A decimal code for the growth stages of cereals. – Weed Res. 14: 415–421.
Zafirou, O. C. and McFarland, M. 1981. Nitric oxide from nitrite photolysis in the central equatorial Pacific. – J. Geophys. Res. 86: 3173–3182.

Appendix 1. Management calendar

1978
May: Fertilization with 100 kg ha^{-1} N, 30 kg ha^{-1} P, and 30 kg ha^{-1} K. Sowing of barley.
Sep.: Harvest.
Oct.: Ploughing.

1979
May: Fallow (ploughing and harrowing).
8 Aug.: General soil survey sampling.
10 Aug.: Treatment with Roundup (4 l ha^{-1} glyphosate 36%) in blocks A–E.
10–25 Sep.: Installation of rubber lysimeters in block A.
5 Oct.: Fertilization with 90 kg ha^{-1} P and 168 kg ha^{-1} K.

1980
12–14 May: Pipe draining.
19–22 May: Harrowing and N-fertilization with calcium nitrate in blocks A–E:
B0: 0 kg ha^{-1} N — Barley undersown with lucerne (LL): 30 kg ha^{-1} N.
B120: 120 kg ha^{-1} N — Barley undersown with meadow fescue (GL): 60 kg ha^{-1} N.
23 May: Sowing of blocks A–E with barley (*Hordeum distichum* L., cv. Gunilla, 200 kg ha^{-1}).
27 May: Sowing of blocks A–E with lucerne (*Medicago sativa* L, cv. Vertus) 25 kg ha^{-1} and meadow fescue (*Festuca pratensis*, Huds., cv. Svalöfs sena) 20 kg ha^{-1}. The lucerne seed was inoculated with a mixture of *Rhizobium meliloti* strains.
29–30 May: Installation of measurement station in block F.
25 Jun.: Treatment with Roundup (4 l ha^{-1} glyphosate 36%) in blocks F–G.
1 Jul.: Treatment with Croneton (Etiofencarb 50%) against aphids in barley blocks A–E (1 l ha^{-1}).
15–16 Jul.: Ploughing and harrowing of blocks F–G.
18 Aug.: Harrowing of blocks F–G.
19 Aug.: Sowing of meadow fescue and lucerne in blocks F–G with the same cultivar and rates as in blocks A–E.
5 Sep.: Harvest of barley in blocks A–E.
9 Sep.: Removal of straw.
23–24 Oct.: Ploughing of barley in blocks A–E.

1981
18–19 May: Disc harrowing of blocks A–E, F$_1$, F$_2$, G$_3$ and G$_4$.
N-fertilization in blocks A–G:
B0: 0 kg ha^{-1} N   LL: 0 kg ha^{-1} N
B120: 120 kg ha^{-1} N   GL: 120 kg ha^{-1} N.
PK-fertilization in blocks F–G with 90 kg ha^{-1} P and 168 kg ha^{-1} K.
20–21 May: Sowing of barley in blocks A–G, cv. Tellus (200 kg ha^{-1}).
Resowing of the leys in blocks A–E:
LL: cv. Sverre (20 kg ha^{-1}).
GL: cv. Mimer (25 kg ha^{-1}).
29 Jun.: Cutting of high weeds in leys in blocks A–E.
7 Jul.: N-fertilization of grass leys in all blocks (80 kg ha^{-1} N).
13 Aug.: Harvest of leys in all blocks.
5 Sep.: Harvest of barley in all blocks.
6 Sep.: Removal of straw in barley blocks.
15–16 Sep.: Trimming of stubble to 8 cm height in all barley plots.
29 Sep.: Stubble harrowing in all barley plots.
14 Oct.: Ploughing of barley plots in all blocks.
14–18 Dec.: Installation of cement lysimeters (L, Fig. 2.14).
21 Dec.: Clearing of open ditches surrounding the field.

1982
25 Jan.: Clearing of open ditches, continued.
10–14 May: Installation of field soil-core lysimeters (M, Fig. 2.14).
17 May: Harrowing and fertilization in blocks A–G:
B0: 0 kg ha^{-1} N, 28 kg ha^{-1} P and 52 kg ha^{-1} K.

	B120: 120 kg ha^{-1} N, 28 kg ha^{-1} P and 52 kg ha^{-1} K.
	LL: 0 kg ha^{-1} N, 40 kg ha^{-1} P and 128 kg ha^{-1} K.
	GL: 120 kg ha^{-1} N, 40 kg ha^{-1} P and 128 kg ha^{-1} K.
19 May:	Harrowing, sowing and packing of barley cv. Gunilla (200 kg ha^{-1}).
24 May:	Fertilization with ^{15}N enriched calcium nitrate in B120 and GL plots above field rubber lysimeters A3 and A4 (120 kg ha^{-1} N, 1.402% a.e. (B120), 0.876% a.e. (GL)).
16 Jun.:	Harvest of leys.
18 Jun.:	Fertilization of the grass leys (80 kg ha^{-1} N). Fertilization with ^{15}N enriched calcium nitrate in GL above the field rubber lysimeter A3 (80 kg ha^{-1} N, 1.011% a.e.).
2 Jul.:	Treatment with Certol Trippel (Dichloroprop + MCPA + Dicamba, 35:15:9 against weeds in barley (3 kg ha^{-1}).
9 Jul.:	Treatment with Roundup (glyphosate 36%) in lucerne in blocks C–F and G (6.3 l ha^{-1}).
27 Jul.:	Ploughing and packing of the lucerne plots C4, D4, E4, F4 and G1.
12 Aug.:	Harrowing of C4, D4, E4, F4 and G1.
18 Aug.:	Harvest of leys in blocks F and G.
23 Aug.:	Harvest of leys in blocks A–E.
24 Aug.:	Harrowing and sowing of red clover in plots C4, D4, E4, F4 and G1 (20 kg ha^{-1}).
5 Sep.:	Harvest of barley and packing of plots sown with red clover.
6 Sep.:	Removal of straw.
17 Sep.:	Stubble harrowing in barley plots.
5 Oct.:	Ploughing.
18 Oct.:	Harrowing.
	1983
19 May:	Harrowing.
22–24 May:	Harrowing, fertilization, sowing, and packing of barley cv. Gunilla (200 kg ha^{-1}):
	B0: 0 kg ha^{-1} N, 14 kg ha^{-1} P, and 26 kg ha^{-1} K.
	B120: 120 kg ha^{-1} N, 28 kg ha^{-1} P, and 52 kg ha^{-1} K.
	LL: 0 kg ha^{-1} N, 56 kg ha^{-1} P, and 104 kg ha^{-1} K.
	GL: 120 kg ha^{-1} N, 56 kg ha^{-1} P, and 104 kg ha^{-1} K.
21 Jun.:	Harvest of leys.
4 Jul.:	Fertilization of grass leys (80 kg ha^{-1} N).
6 Jul.:	Treatment with MCPA in barley (1.5 l ha^{-1}).
25 Aug.:	Harvest of leys.
25 Sep.:	Harvest of barley.
5 Oct.:	Stubble harrowing in barley plots.
24 Oct.:	Ploughing.
27 Oct.:	Harrowing.
	1984
15 May:	Ploughing clover plots in C4, D4, E4, F4, G1 and the field rubber lysimeter.
16 May:	Fertilization of B120: 120 kg ha^{-1} N, GL: 120 kg ha^{-1} N.
17 May:	Harrowing, sowing of all ploughed areas and packing of barley: cv. Gunilla (200 kg ha^{-1}).
18 Jun.:	Harvest of leys.
19 Jun.:	Treatment with MCPA 750 (2.0 l ha^{-1}).
2 Jul.:	Fertilization of grass leys (80 kg ha^{-1} N).
9 Aug.:	Harvest of leys.
10 Aug.:	Harvest, cutting and ploughing of blocks F and G.
11 Aug.:	Cutting, rotary cultivation and ploughing of ½ GL in block A–E and ½ LL in blocks A and B. Clod-crushing in blocks A–E.
6 Sep.:	Harvest of barley.
19 Sep.:	Stubble harrowing in barley plots.
15 Oct.:	Ploughing.
	1985
7 May:	Digging of cement lysimeters.
21–22 May:	Harrowing, fertilization, sowing and packing of barley: cv. Gunilla (200 kg ha^{-1}). B120: 120 kg ha^{-1} N, GL: 120 kg ha^{-1} N.
28 Jun.:	Harvest of leys.
9 Jul.:	Treatment with Pirimor (Pirimicarb 50%), 0.5 kg ha^{-1}.
22 Aug.:	Harvest of leys.
17 Sep.:	Harvest of barley.
20–21 Sep.:	Removal of straw.
11–13 Nov.:	Ploughing.

Appendix 2. Scientific and technical/administrative personnel

Scientific

Abiotic processes
Alvenäs, G. Dept of Soil Sciences, Swedish University of Agricultural Sciences, Box 7014, S-75007 Uppsala.
Jansson, P.-E. Dept of Soil Sciences, Swedish University of Agricultural Sciences, Box 7014, S-75007 Uppsala.
Johnsson, H. Dept of Soil Sciences, Swedish University of Agricultural Sciences, Box 7072, S-75007 Uppsala.
Thoms-Järpe, C. Swedish Meteorological and Hydrological Inst., S-10675 Norrköping.

Ammonia transport
Ferm, M. Swedish Water and Air Pollution Sciences Inst., Box 5207, S-40224 Gothenburg.

Decomposition
Andrén, O. Dept of Ecology and Environmental Research, Swedish Univ. of Agr. Sci., Box 7072, S-75007 Uppsala.
Berg, B. Dept of Ecology and Environmental Research, Swedish Univ. of Agr. Sci., Box 7072, S-75007 Uppsala.
Persson, J. Dept of Soil Sciences, Swedish Univ. of Agr. Sci., Box 7014, S-75007 Uppsala.

Denitrification
Klemedtsson, L. Dept of Microbiology, Swedish Univ. of Agr. Sci., Box 7025, S-75007 Uppsala.
Rosswall, T. Dept of Water and Environmental Research, Univ. of Linköping, S-58183 Linköping.
Simkins, S. Dept of Plant and Soil Science, Univ. of Massachusetts, Stockbridge, Amherst, MA 01003, USA.
Svensson, B. Dept of Microbiology, Swedish Univ. of Agr. Sci., Box 7025, S-75007 Uppsala.

Dry deposition/emission
Granat, L. Dept of Meteorology, Univ. of Stockholm, S-10691 Stockholm.
Johansson, C. Dept of Meteorology, Univ. of Stockholm, S-10691 Stockholm.

Ecosystem budgets
Paustian, K. Dept of Ecology and Environmental Research, Swedish Univ. of Agr. Sci., Box 7072, S-75007 Uppsala.

Fauna
Andrén, O. Dept of Ecology and Environmental Research, Swedish Univ. of Agr. Sci., Box 7072, S-75007 Uppsala.
Boström, S. Dept of Zoology, Univ. of Stockholm, S-11386 Stockholm.
Boström, U. Dept of Ecology and Environmental Research, Swedish Univ. of Agr. Sci., Box 7072, S-75007 Uppsala.
Gustafsson A.-K. Dept of Animal Nutrition and Management, Swedish Univ. of Agr. Sci., Box 7024, S-75007 Uppsala.
Lagerlöf, J. Dept of Ecology and Environmental Research, Swedish Univ. of Agr. Sci., Box 7072, S-75007 Uppsala.
Lofs-Holmin, A. Dept of Ecology and Environmental Re-

search, Swedish Univ. of Agr. Sci., Box 7072, S-75007 Uppsala.
Palmborg, C. Dept of Ecology and Environmental Research, Swedish Univ. of Agr. Sci., Box 7072, S-75007 Uppsala.
Sohlenius, B. Dept of Zoology, Univ. of Stockholm, S-10691 Stockholm.

Field site management
Johansson, R. Dept of Ecology and Environmental Research, Swedish Univ. of Agr. Sci., Box 7072, S-75007 Uppsala.

Leaching
Bergström, L. Dept of Soil Sciences, Swedish Univ. of Agr. Sci., Box 7072, S-75007 Uppsala.
Brink, N. Dept of Soil Sciences, Swedish Univ. of Agr. Sci., Box 7072, S-75007 Uppsala.
Gustafson, A. Dept of Soil Sciences, Swedish Univ. of Agr. Sci., Box 7072, S-75007 Uppsala.

Microorganisms
Clarholm, M. Dept of Ecology and Environmental Research, Swedish Univ. of Agr. Sci., Box 7072, S-75007 Uppsala.
Norén, B. (deceased) Dept of Microbiology, Swedish Univ. of Agr. Sci., Box 7025, S-75007 Uppsala.
Rosswall, T. Dept of Water and Environmental Research, Univ. of Linköping, S-58183 Linköping.
Schnürer, J. Dept of Microbiology, Swedish Univ. of Agr. Sci., Box 7025, S-75007 Uppsala.
Wessén, B. Pegasus Lab AB, Kungsgatan 113, S-75321 Uppsala.

Modelling of biotic processes
Paustian, K. Dept of Ecology and Environmental Research, Swedish Univ. of Agr. Sci., Box 7072, S-75007 Uppsala.

Nitrification
Berg, P. Inst. for Future Studies, Hagagatan 23 A, S-11347 Stockholm.
Rosswall, T. Dept of Water and Environmental Research, Univ. of Linköping, S-58183 Linköping.

Nitrogen fixation
Ljunggren, H. Dept of Microbiology, Swedish Univ. of Agr. Sci., Box 7025, S-75007 Uppsala.
Mårtensson, A. Dept of Microbiology, Swedish Univ. of Agr. Sci., Box 7025, S-75007 Uppsala.
Wivstad, M. Dept of Crop Production Science, Swedish Univ. of Agr. Sci., Box 7043, S-75007 Uppsala.

Nitrogen mineralization
Bonde, T. Dept of Water and Environmental Research, Univ. of Linköping, S-58183 Linköping.
Börjesson, I. Dept of Ecology and Environmental Research, Swedish Univ. of Agr. Sci., Box 7072, S-75007 Uppsala.
Clarholm, M. Dept of Ecology and Environmental Research, Swedish Univ. of Agr. Sci., Box 7072, S-75007 Uppsala.
Lindberg, T. Dept of Microbiology, Swedish Univ. of Agr. Sci., Box 7025, S-75007 Uppsala.

Primary production
Andrén, O. Dept of Ecology and Environmental Research, Swedish Univ. of Agr. Sci., Box 7072, S-75007 Uppsala.
Hansson, A.-C. Dept of Ecology and Environmental Research, Swedish Univ. of Agr. Sci., Box 7072, S-75007 Uppsala.
Johansson, G. Dept of Soil Sciences, Swedish Univ. of Agr. Sci., Box 7014, S-75007 Uppsala.
Pettersson, R. Dept of Ecology and Environmental Research, Swedish Univ. of Agr. Sci., Box 7072, S-75007 Uppsala.
Steen, E. Dept of Ecology and Environmental Research, Swedish Univ. of Agr. Sci., Box 7072, S-75007 Uppsala.

Statistical consultant
Ekbohm, G. Dept of Statistics, Data Processing and Agricultural Extension, Swedish Univ. of Agr. Sci., Box 7013, S-75007 Uppsala.

Wet deposition
Söderlund, R. Dept of Meteorology, Univ. of Stockholm, S-10691 Stockholm.

Technical/administrative
Agerberg, A. Dept of Ecology and Environmental Research, Swedish Univ. of Agr. Sci., Box 7072, S-75007 Uppsala.
Al-Windi, I. Dept of Ecology and Environmental Research, Swedish Univ. of Agr. Sci., Box 7072, S-75007 Uppsala.
Bäcklin, L. Dept of Meteorology, Univ. of Stockholm, S-10691 Stockholm.
Brolin, M.-B. Dept of Microbiology, Swedish Univ. of Agr. Sci., Box 7025, S-75007 Uppsala.
Carlsson, B. Dept of Ecology and Environmental Research, Swedish Univ. of Agr. Sci., Box 7072, S-75007 Uppsala.
Eriksson, G. Dept of Microbiology, Swedish Univ. of Agr. Sci., Box 7025, S-75007 Uppsala.
Fontana, S. Dept of Microbiology, Swedish Univ. of Agr. Sci., Box 7025, S-75007 Uppsala.
Geidnert, S. Dept of Microbiology, Swedish Univ. of Agr. Sci., Box 7025, S-75007 Uppsala.
Hansson, M. Dept of Ecology and Environmental Research, Swedish Univ. of Agr. Sci., Box 7072, S-75007 Uppsala.
Holmberg, Y. Kjettslinge, S-74060 Örbyhus.
Jansson, Å. Dept of Microbiology, Swedish Univ. of Agr. Sci., Box 7025, S-75007 Uppsala.
Johansson, B. Dept of Microbiology, Swedish Univ. of Agr. Sci., Box 7025, S-75007 Uppsala.
Juremalm, I. Dept of Soil Sciences, Swedish Univ. of Agr. Sci., Box 7014, S-75007 Uppsala.
Lundquist, A. Dept of Microbiology, Swedish Univ. of Agr. Sci., Box 7025, S-75007 Uppsala.
Niklasson, R. Dept of Soil Sciences, Swedish Univ. of Agr. Sci., Box 7072, S-75007 Uppsala.
Ohlsson, I. Dept of Microbiology, Swedish Univ. of Agr. Sci., Box 7025, S-75007 Uppsala.
Sköld, A. Dept of Microbiology, Swedish Univ. of Agr. Sci., Box 7025, S-75007 Uppsala.
Staaf, I. Dept of Ecology and Environmental Research, Swedish Univ. of Agr. Sci., Box 7072, S-75007 Uppsala.
Söderqvist, M. Dept of Forest Products, Swedish Univ. of Agr. Sci., Box 7008, S-75007 Uppsala.
Tahvanainen, K. Dept of Microbiology, Swedish Univ. of Agr. Sci., Box 7025, S-75007 Uppsala.
Wiklund, H. Dept of Microbiology, Swedish Univ. of Agr. Sci., Box 7025, S-75007 Uppsala.
Zhang, Liquan Dept of Ecology and Environmental Research, Swedish Univ. of Agr. Sci., Box 7072, S-75007 Uppsala.
Ziverts, I. Dept of Microbiology, Swedish Univ. of Agr. Sci., Box 7025, S-75007 Uppsala.
Åberg, K. Dept of Microbiology, Swedish Univ. of Agr. Sci., Box 7025, S-75007 Uppsala.
Öberg, L. Kjettslinge, S-74060 Örbyhus.

Appendix 3. Project publications

1980
Andrén, O. and Lagerlöf, J. 1980. The abundance of soil animals (Microarthropoda, Enchytraeidae, Nematoda) in a crop rotation dominated by ley and in a rotation with varied crops. – In: Dindal, D. (ed.), Soil Biology Related to Land Use Practices. Environment Protection Agency, Washington D.C., pp. 274–279.
Rosswall, T. 1980. Ecological systems of arable land. – Paper presented at the 'Scientific Forum' of the Conference on

Security and Cooperation in Europe, Hamburg, FRG. 18 February – 3 March, 1980. Nat/Foo 15 (22). Conference Secretariat, Hamburg, 12 pp + 2 appendices.

1981

Clarholm, M. 1981. Protozoan grazing of bacteria in soil. – Impact and importance. – Microb. Ecol. 7: 343–350.
Clark, F. E. and Rosswall. T. 1981 (eds), Terrestrial Nitrogen Cycles. Processes, Ecosystem Strategies and Management Impacts. – Ecol. Bull. (Stockholm) 33.
Klemedtsson, L. 1981. Analysis of N_2O production from soil systems. – In: Summary of papers presented at the 2nd Nordic Head-Space Symposium, Kongsberg, Norway 12–14 October 1980. Sentralinstitutt for Industriell Forskning, Oslo, pp. G1–G7 (In Swedish).
Rosswall, T. 1981 (ed.), Ecology of Arable Land. The Role of Organisms in Nitrogen Cycling. – Progress Report 1980. Swedish Univ. of Agr. Sci., Uppsala.
Steen, E. 1981. Ecology of Arable Land – A New Research Project. – Research Information Centre, Swedish Univ. of Agr. Sci., Uppsala. Report 8 :1–8 (In Swedish).

1982

Berg, P., Klemedtsson, L. K. and Rosswall, T. 1982. Inhibitory effect of low partial pressures of acetylene on nitrification. – Soil Biol. Biochem. 14: 301–303.
Boström, S. 1982. A Study of the Nematode Fauna under Barley and Grass Ley in an Experimental Field in Central Sweden. – M. Sc. Thesis, Dept of Zoology, Univ. of Stockholm, Stockholm.
Hansson, A.-C. 1982. Primary production and nitrogen contents above and below-ground in barley in four different cropping systems. – Dept of Ecology and Environmental Research, Report 22. Swedish Univ. of Agr. Sci., Uppsala.
Rosswall, T. 1982. Microbial regulation of the biogeochemical nitrogen cycle. – Pl. Soil 67: 15–34.
– 1982. Nitrogen cycling in terrestrial ecosystems. – Informatore Botanico Italiano 14: 99–109.
– 1982. The role of microorganisms and animals in the regulation of the biogeochemical nitrogen cycle in agricultural soil. – K. Skogs- och Lantbr.-akad. tidskr. Suppl 14: 26–31 (In Swedish, English summary).
– 1982 (ed.), Ecology of Arable Land. The Role of Organisms in Nitrogen Cycling. – Progress Report 1981. Swedish Univ. of Agr. Sci., Uppsala.
Schnürer, J. and Rosswall, T. 1982. Fluorescein diacetate hydrolysis as a measure of total microbial activity in soil and litter. – Appl. Envir. Microbiol. 43: 1256–1261.
Torstensson, L. 1982. Decomposition of glyphosate in agricultural soils. – In: Weeds and Weed Control. 23rd Swedish Weed Conference, Uppsala 27–29 January 1982. Vol. 2, Reports, pp. 385–392 (In English) and pp. 396–403 (In Swedish).

1983

Andrén, O. and Lagerlöf, J. 1983. Soil fauna (microarthropods, enchytraeids, nematodes) in Swedish agricultural cropping systems. – Acta Agr. Scand. 33: 33–52.
– and Lagerlöf, J. 1983. Succession of soil microarthropods in decomposing barley straw. – In: Lebrun, Ph., André, H. M., DeMedts, A., Grégoire-Wibo, C. and Wauthy, G. (eds), New Trends in Soil Biology. Dieu-Brichart, Ottignies – Louvain-la-Neuve, pp. 644–646.
Clarholm, M. 1983. Dynamics of Soil Bacteria in Relation to Plants, Protozoa and Inorganic Nitrogen. – Ph. D. Thesis, Dept of Microbiology, Report 17. Swedish Univ. of Agr. Sci., Uppsala.
– 1983. Microflora in the root zone. – In: The root and the environment of the root II. Research Information Centre, Swedish Univ. of Agr. Sci., Uppsala. Report 47, pp. 71–76 (In Swedish).
Ferm, M. 1983. Ammonia volatilization from arable land – An evaluation of the chamber technique. – In: Observation and Measurement of Atmospheric Contaminants. WMO Spec. Environ. Rep. 16. WHO, Geneva, pp. 145–172.
Hansson, A.-C. 1983. Input of organic matter and nitrogen to an arable soil through root production. – In: Böhm, W., Kutschera, L. and Lichtenegger, E. (eds.), Root Ecology and its Practical Application, Proc. Int. Symp. Bundesanstalt für Alpenländische Landwirtschaft, Gumpenstein, Irdning, pp. 757–760.
Lofs-Holmin, A. 1983. Earthworm population dynamics in different agricultural rotations. – In: Satchell, J. E. (ed.), Earthworm Ecology from Darwin to Vermiculture. Chapman and Hall, London, pp. 151–160.
– 1983. Earthworms in Arable Land. Methods for Studying Agricultural Impacts, Especially Pesticides. – Ph. D. Thesis, Dept of Ecology and Environmental Research, Swedish Univ. of Agr. Sci., Uppsala.
– 1983. Influence of agricultural practicies on earthworms. – Acta Agr. Scand. 33: 225–234.
Magnusson, C. and Lagerlöf, J. 1983. Fauna in soil and rhizosphere. – In: The Root and The Environment of The Root II. Research Information Centre, Swedish Univ. of Agr. Sci., Uppsala. Report 47, pp. 77–92 (In Swedish).
Parton, W. J., Persson, J. and Anderson, D. W. 1983. Simulation of organic matter changes in Swedish soils. – In: Lauenroth, W. K., Skogerboe, G. V. and Flug, M. (eds), Analysis of Ecological Systems: State-of-the Art in Ecological Modelling. Elsevier, Amsterdam, pp. 511–516.
Persson, J. and Rosswall, T. 1983. Opportunities for research in agricultural ecosystems – Sweden. – In: Lowrance, R. R., Todd, R. L., Asmussen, L. E. and Leonard, R. A. (eds), Nutrient Cycling in Agricultural Ecosystems. Univ. of Georgia Coll. Agr. Exp. Stn. Spec. Publ. 23: 61–71.
Rosswall, T. 1983. Global balances of nitrogen and phosphorus. – K. Skogs- och Lantbr.-akad. Tidskr. 122: 287–292 (In Swedish, English summary).
– 1983. The nitrogen cycle. – In: Bolin, B. and Cook, R. B. (eds), The Major Biogeochemical Cycles and Their Interactions. SCOPE Report 21. Wiley, Chichester, pp. 46–50.
– 1983 (ed.), Ecology of Arable Land. The Role of Organisms in Nitrogen Cycling. – Progress Report 1982. Swedish Univ. of Agr. Sci., Uppsala.
Steen, E. 1983. Soil animals in relation to agricultural practices and soil productivity. – Swed. J. agr. Res. 13: 157–165.
– 1983. The net stocking method for studying quantitative and qualitative variation with time of grass roots. – In: Böhm, W., Kutschera, L. and Lichtenegger, E. (eds), Root Ecology and its Practical Application. Proc. Int. Symp. Bundesanstalt für Alpenländische Landwirtschaft, Gumpenstein, Irdning, pp. 63–74.
Svensson, B. H. 1983. Denitrification. – In: Eriksson, T. (ed.), Nitrogen Prognosis – Present State and Directions for the Future. – K. Skogs- och Lantbr.-akad. Rap. 1983 (6): 60–62 (In Swedish).
Wessén, B. 1983. Decomposition of Some Forest Leaf Litters and Barley Straw – Some Rate Determining Factors. – Ph. D. Thesis, Dept of Microbiology, Report 19. Swedish Univ. of Agr. Sci., Uppsala.

1984

Andrén, O. 1984. REF – A Computerized System For Handling Scientific References (Under VAX/VMS Operating System). – Dept of Ecology and Environmental Research, Report 18. Swedish Univ. of Agr. Sci., Uppsala (In Swedish).
– 1984. Soil Mesofauna of Arable Land and Its Significance For Decomposition of Organic Matter. – Ph. D. Thesis, Dept of Ecology and Environmental Research, Report 16. Swedish Univ. of Agr. Sci., Uppsala.
Clarholm, M. 1984. Heterotrophic, free-living protozoa: Neglected microorganisms with an important task in regulating bacterial populations. – In: Klug, M. J. and Reddy,

C. A. (eds), Current Perspectives in Microbial Ecology. Am. Soc. Microbiol., Washington D.C., pp. 321–326.

Hansson, A.-C. and Steen, E. 1984. Methods of calculating root production and nitrogen uptake in an annual crop. – Swed. J. agr. Res. 14: 191–200.

Johansson, C. and Galbally, I. E. 1984. Production of nitric oxide in loam under aerobic and anaerobic conditions. – Appl. Env. Microbiol. 47: 1284–1289.

– and Granat, L. 1984. Emission of nitric oxide from arable land. – Tellus 36: 25–37.

Larsson, K. and Steen, E. 1984. Nitrogen and carbohydrates in grass roots sampled with a mesh bag technique. – Swed. J. agr. Res. 14: 159–164.

Mårtensson, A. 1984. Biological nitrogen fixation in lucerne during the establishment year. – Fakta mark-växter 25. Swedish Univ. of Agr. Sci., Uppsala (In Swedish).

– and Ljunggren, H. D. 1984. A comparison between the acetylene reduction method, the isotope dilution method and the total nitrogen difference method for measuring nitrogen fixation in lucerne (*Medicago sativa* L.) – Pl. Soil 81: 177–184.

– and Ljunggren, H. D. 1984. Nitrogen fixation in an establishing alfalfa (*Medicago sativa* L.) ley in Sweden, estimated by three different methods. – Appl. Env. Microbiol. 48: 702–707.

Rosswall, T. and Paustian, K. 1984. Cycling of nitrogen in modern agricultural systems. – Pl. Soil 76: 3–21.

– and Steen, E. 1984. Ecology of arable land. – The role of organisms in nitrogen cycling. – Fakta mark-växter 5. Swedish Univ. of Agr. Sci., Uppsala (In Swedish).

Sohlenius, B. and Boström, S. 1984. Colonization, population development and metabolic activity of nematodes in buried barley straw. – Pedobiologia 27: 67–78.

Steen, E. 1984. Variation of root growth in a grass ley studied with a mesh bag technique. – Swed. J. agr. Res. 14: 93–97.

– , Jansson, P.-E. and Persson, J. 1984. Experimental site of the 'Ecology of Arable Land' project. – Acta Agr. Scand. 34: 153–166.

Svensson, B. 1984. Ecology of Arable Land. The Role of Organisms in Nitrogen Cycling. – Progress Report 1983. Swedish Univ. of Agr. Sci., Uppsala.

1985

Andrén, O. 1985. Microcomputer-controlled extractor for soil micro-arthropods. – Soil Biol. Biochem. 17: 27–30.

– and Schnürer, J. 1985. Barley straw decomposition with varied levels of microbial grazing by *Folsomia fimetaria* (L.) (Collembola, Isotomidae). – Oecologia (Berl.) 68: 57–62.

Berg, P. and Rosswall, T. 1985. Ammonium oxidizer numbers, potential and actual oxidation rates in two Swedish arable soils. – Biol. Fert. Soils 1: 131–140.

Bergström, L. 1985. Ecology of arable land – Nitrogen losses. – Research Information Centre, Swedish Univ. of Agr. Sci., Uppsala. Report 63: 8, pp. 1–13 (In Swedish).

Carter, A., Lagerlöf, J. and Steen, E. 1985. Effects of major disturbances in different agricultural cropping systems on soil macroarthropods. – Acta Agr. Scand. 35: 67–77.

Clarholm, M. 1985. Interactions of bacteria, protozoa and plants leading to mineralization of soil nitrogen. – Soil Biol. Biochem. 17: 181–187.

– 1985. The role of microorganisms in the turnover of soil organic matter. – Sveriges Utsädesförenings Tidskrift 95: 241–247 (In Swedish).

– 1985. Possible roles for roots, bacteria, protozoa and fungi in supplying nitrogen to the plants. – In: Fitter, A. H., Atkinson, D., Read, D. J. and Usher, M. B. (eds), Ecological Interactions in Soil: Plants, Microbes and Animals. British Ecol. Soc. Spec. Publ. 4. Blackwell, Oxford, pp. 355–365.

Lagerlöf, J. 1985. Fauna in the agricultural landscape and its conservation. – Fauna Flora 80: 149–158 (In Swedish, English summary).

– 1985. The importance of the soil fauna in arable land. – Research Information Centre, Swedish Univ. of Agr. Sci., Uppsala. Report 63: 9, pp. 1–9 (In Swedish).

– and Andrén, O. 1985. Succession and activity of microarthropods and enchytraeids during barley straw decomposition. – Pedobiologia 28: 343–357.

Mårtensson, A. 1985. Ecology of *Rhizobium*. Methods for Strain Identification and Quantification of Nitrogen Fixation. – Ph. D. Thesis, Dept of Microbiology, Report 30. Swedish Univ. of Agr. Sci., Uppsala.

Paustian, K. 1985. Influence of fungal growth pattern on decomposition and nitrogen mineralization in a model system. – In: Fitter, A. H., Atkinson, D., Read, D. J. and Usher, M. B. (eds), Ecological Interactions in Soil: Plants, Microbes, and Animals. British Ecol. Soc. Spec. Publ. 4. Blackwell, Oxford, pp. 159–174.

Rosswall, T. 1985. Interaction and integration – the role of microbiology in ecological research. – In: Cooley, J. (ed.), Trends in Ecological Research for the 1980's. NATO Conference Series: Ecology Vol. 6. Plenum Press, New York, pp. 19–34.

– 1985. Carrying capacity of agricultural and forestry ecosystems. – Lantbruksforskarna om framtiden. LUI:s skriftserie 2: 93–116 (In Swedish).

– 1985. Ecology of arable land – a multidisciplinary research project. – Research Information Centre, Swedish Univ. of Agr. Sci., Uppsala. Report 63: 6, pp. 1–8 (In Swedish).

– 1985. Ecology of arable land – The nitrogen cycle – Research Information Centre, Swedish Univ. of Agr. Sci., Uppsala. Report 63: 10, pp. 1–9 (In Swedish).

Schnürer, J. 1985. Fungi in Arable Soil, Their Role in Carbon and Nitrogen Cycling. – Ph. D. Thesis, Dept of Microbiology, Report 29. Swedish Univ. of Agr. Sci., Uppsala.

– , Clarholm, M. and Rosswall, T. 1985. Microbial biomass and activity in agricultural soils with different carbon contents. – Soil Biol. Biochem. 17: 611–618.

Steen, E. 1985. Role of plant cover in the biology and fertility of Swedish arable soils. – INTECOL Bull. 12: 48–56.

– 1985. Root and rhizome dynamics in a perennial grass crop during an annual growth cycle. – Swed. J. agr. Res. 15: 25–30.

– 1985. Ecology of arable land – Primary production. – Research Information Centre, Swedish Univ. of Agr. Sci., Uppsala. Report 63: 7, pp. 1–6 (In Swedish).

Svensson, B. H., Klemedtsson, L. and Rosswall, T. 1985. Preliminary field denitrification studies of nitrate-fertilized and nitrogen-fixing crops. – In: Golterman, H. L. (ed.), Denitrification in the Nitrogen Cycle. NATO Conference Series: Ecology Vol. 9. Plenum Press, New York, pp. 157–169.

1986

Alvenäs, G., Johnsson, H. and Jansson, P.-E. 1986. Meteorological Conditions and Soil Climate of Four Cropping Systems. Measurements and Simulations From the Project "Ecology of Arable Land". – Dept of Ecology and Environmental Research, Report 24. Swedish Univ. of Agr. Sci., Uppsala.

Berg, P. 1986. Nitrifier Populations and Nitrification Rates in Agricultural Soil. – Ph. D. Thesis, Dept of Microbiology, Report 34. Swedish Univ. of Agr. Sci., Uppsala.

Bergström, L. 1986. Distribution and temporal changes of mineral nitrogen in soils supporting annual and perennial crops. – Swed. J. agr. Res. 16: 105–112.

Boström, S. and Sohlenius, B. 1986. Short-term dynamics of nematode communities in arable soil. Influence of a perennial and an annual cropping system. – Pedobiologia 29: 345–357.

Boström, U. and Lofs-Holmin, A. 1986. Growth of earthworms (*Allolobophora caliginosa*) fed shoots and roots of barley, meadow fescue and lucerne. Studies in relation to particle size, protein, crude fiber content and toxicity. –

Pedobiologia 29: 1–12.
Curry, J. P. 1986. Above-ground arthropod fauna of four Swedish cropping systems and its role in carbon and nitrogen cycling. – J. Appl. Ecol. 23: 853–870.
Ferm, M. 1986. Concentration Measurements and Equilibrium Studies of Ammonium, Nitrate and Sulphur Species in Air and Precipitation. – Ph. D. Thesis, Dept of Inorganic Chemistry, Univ. of Göteborg, Sweden.
Hansson, A.-C. and Andrén, O. 1986. Below-ground plant production in a perennial grass ley *(Festuca pratensis)* assessed with different methods. – J. Appl. Ecol. 23: 657–666.
Jansson, P.-E. and Thoms-Järpe, C. 1986. Simulated and measured soil water dynamics of unfertilized and fertilized barley. – Acta Agr. Scand. 36: 162–172.
Klemedtsson, L. 1986. Denitrification in Arable Soil with Special Emphasis on the Influence of Plant Roots. – Ph. D. Thesis, Dept of Microbiology, Report 32. Swedish Univ. of Agr. Sci., Uppsala.
– , Simkins, S. and Svensson, B. H. 1986. Tandem thermal-conductivity and electron-capture detectors and non-linear calibration curves in quantitative N_2O analysis. – J. Chrom. 361: 107–116.
Pettersson, R., Hansson, A.-C., Andrén, O. and Steen, E. 1986. Above- and below-ground production and nitrogen uptake in lucerne *(Medicago sativa)* – Swed. J. agr. Res. 16: 167–177.
Schnürer, J. and Paustian, K. 1986. Modelling fungal growth in relaton to nutrient limitations in soil. – In: Megusár, F. and Gantar, M. (eds), Perspectives in Microbial Ecology. Slovene Soc. Microbiol., Ljubljana, pp. 123–130.
– , Clarholm, M., Boström, S. and Rosswall, T. 1986. Effects of moisture on soil microorganisms and nematodes – a field experiment. – Microb. Ecol. 12: 217–230.
– , Clarholm, M. and Rosswall, T. 1986. Fungi, bacteria and protozoa in soil from four arable cropping systems. – Biol. Fert. Soils 2: 119–126.
Sohlenius, B. and Boström, S. 1986. Short-term dynamics of nematode communities in arable soil. – Influence of nitrogen fertilization in barley crops. – Pedobiologia 29: 183–191.
Svensson, B. H., Boström, U.-L. and Klemedtsson, L. 1986. Potential for higher rates of denitrification in earthworm casts than in the surrounding soil. – Biol. Fert. Soils 2: 147–149.
Wessén, B. and Berg, B. 1986. Long-term decomposition of barley straw: chemical changes and ingrowth of fungal mycelium. – Soil Biol. Biochem. 18: 53–59.

1987
Andrén, O. 1987. Decomposition of shoot and root litter of barley, lucerne and meadow fescue under field conditions. – Swed. J. agr. Res. 17: 113–122.
– and Paustian, K. 1987. Barley straw decomposition in the field: a comparison of models. – Ecology 68: 1190–1200.
– , Hansson, A.-C. and Pettersson, R. 1987. Contributions to soil organic matter from four arable crops. – INTECOL Bull. 15: 41–47.
Berg, P. and Rosswall, T. 1987. Seasonal variations in abundance and activity of nitrifiers in four arable cropping systems. – Microb. Ecol. 13: 75–87.
Bergström, L. 1987. Leaching of 15-N labeled nitrate fertilizer applied to barley and a grass ley. – Acta Agr. Scand. 37: 199–206.
– 1987. Nitrate leaching and drainage from annual and perennial crops in tile-drained plots and lysimeters. – J. Environ. Qual. 16: 11–18.
– 1987. Transport and Transformation of Nitrogen in an Arable Soil. – Ph. D. Thesis, Ekohydrologi 23. Div. of Water Management, Swedish Univ. of Agr. Sci., Uppsala.
– , Jansson, P.-E., Johnsson, H. and Paustian, K. 1987. A model for simulation of nitrogen dynamics in soil and nitrate leaching. – Fakta mark-växter 4. Swedish Univ. of Agr. Sci., Uppsala (Swedish and English versions).
Bonde, T. and Rosswall, T. 1987. Seasonal variation of potentially mineralizable nitrogen in four cropping systems. – Soil Sci. Soc. Amer. J. 51: 1508–1514.
Boström, U. 1987. Growth of earthworms *(Allolobophora caliginosa)* in soil mixed with either barley, lucerne or meadow fescue at various stages of decomposition. – Pedobiologia 30: 311–321.
Hansson, A.-C. 1987. Roots of Arable Crops: Production, Growth Dynamics and Nitrogen Content. – Ph. D. Thesis, Department of Ecology and Environmental Research, Report 28. Swedish Univ. of Agr. Sci., Uppsala.
– and Andrén, O. 1987. Root dynamics in barley, lucerne and meadow fescue investigated with a mini-rhizotron technique. – Pl. Soil 103: 33–38.
– Pettersson, R. and Paustian, K. 1987. Shoot and root production and nitrogen uptake in barley, with and without nitrogen fertilization. – J. Agron. Crop Sci. 158: 163–171.
Johnsson, H., Bergström, L., Jansson, P.-E. and Paustian, K. 1987. Simulated nitrogen dynamics and losses in a layered agricultural soil. – Agr. Ecosys. Environ. 18: 333–356.
Klemedtsson, L., Berg, P., Clarholm, M., Schnürer, J. and Rosswall, T. 1987. Microbial nitrogen transformations in the root environment of barley. – Soil Biol. Biochem. 19: 551–558.
– , Svensson, B. H. and Rosswall, T. 1987. Dinitrogen and nitrous oxide production by denitrification and nitrification in soil with and without roots. – Pl. Soil 99: 303–319.
Lagerlöf, J. 1987. Ecology of Soil Fauna in Arable Land. Dynamics and Activity of Microarthropods and Enchytraeides in Four Cropping Systems. – Ph. D. Thesis, Dept of Ecology and Environmental Research, Report 29. Swedish Univ. of Agr. Sci., Uppsala.
– and Lofs-Holmin, A. 1987. Relationships between earthworms and soil mesofauna during decomposition of crop residues. – In: Striganova, B. R. (ed.), Soil Fauna and Soil Fertility. Proc. 9th Int. Coll. Soil Zoology. Moscow "Nauka", pp. 377–381.
Paustian, K. 1987. Theoretical Analyses of C and N Cycling in Soil. – Ph. D. Thesis, Dept of Ecology and Environmental Research, Report 30. Swedish Univ. of Agr. Sci., Uppsala.
– and Bonde, T. A. 1987. Interpreting incubation data on nitrogen mineralization from soil organic matter. – INTECOL Bull. 15: 101–112.
– and Schnürer, J. 1987. Fungal growth response to carbon and nitrogen limitation: I. A theoretical model. – Soil Biol. Biochem. 19: 613–620.
– and Schnürer, J. 1987. Fungal growth response to carbon and nitrogen limitation: II. Application of a model to laboratory and field data. – Soil Biol. Biochem. 19: 621–629.
Pettersson, R. 1987. Primary Production in Arable Crops: Above-ground Growth Dynamics, Net Production and Nitrogen Uptake. – Ph. D. Thesis, Dept of Ecology and Environmental Research, Report 31. Swedish Univ. of Agr. Sci., Uppsala.
Schnürer, J. 1987. The influence of agriculture on the soil microorganisms. – K. Skogs- och Lantbr.-akad. tidskr. Suppl. 19: 29–35 (In Swedish).
– and Rosswall, T. 1987. Mineralization of nitrogen from ^{15}N labelled fungi, soil microbial biomass and roots and its uptake by barley plants. – Pl. Soil 102: 71–78.
Sohlenius, B. and Sandor, A. 1987. Vertical distribution of nematodes in arable soil under grass *(Festuca pratensis)* and barley *(Hordeum distichum)*. – Biol. Fert. Soils 3: 19–25.
– , Boström, S. and Sandor, A. 1987. Long-term dynamics of nematode communities in arable soil under four cropping systems. – J. Appl. Ecol. 24: 131–144.
Wivstad, M., Mårtensson, A. M. and Ljunggren, H. D. 1987. Field measurement of symbiotic nitrogen fixation in an established lucerne ley using ^{15}N and an acetylene reduction method. – Pl. Soil 97: 93–104.

1988

Andrén, O. 1988. Ecology of arable land – an integrated project. – Ecol. Bull. (Copenhagen) 39: 131–133.

–, Paustian, K. and Rosswall, T. 1988. Soil biotic interactions in the functioning of agroecosystems. – Agr. Ecosys. Environ. 24: 57–68.

Bergström, L. and Johnsson, H. 1988. Simulated nitrogen dynamics and nitrate leaching in a perennial grass ley. – Pl. Soil 105: 273–281.

Bonde, T. A. and Lindberg, T. 1988. Nitrogen mineralization kinetics in soil during long-term aerobic laboratory incubations: a case study. – J. Env. Qual. 17: 414–417.

–, Schnürer, J. and Rosswall, T. 1988. Microbial biomass as a fraction of potentially mineralizable nitrogen in soils from long-term field experiment. – Soil Biol. Biochem. 20: 447–452.

Boström, S. 1988. Morphological and systematic studies of the family Cephalobidae (Nematoda: Rhabditida). – Ph. D. Thesis, Dept of Zoology, Univ. of Stockholm, Sweden.

Boström, U. 1988. Ecology of Earthworms in Arable Land. Population Dynamics and Activity in Four Cropping Systems. – Ph. D. Thesis, Dept of Ecology and Environmental Research, Report 34. Swedish Univ. of Agr. Sci., Uppsala.

– 1988. Growth and cocoon production by the earthworm *Aporrectodea caliginosa* in soil mixed with various plant materials. – Pedobiologia 32: 77–80.

Clarholm, M., Gustafson, A. and Fleischer, S. 1988. Does agriculture kill fish? – Possible ways to decrease nitrogen leaching from land to water. – Ecol. Bull. (Copenhagen) 39: 139–140.

Hansson, A.-C. 1988. Fate of ^{15}N labelled fertilizer in a perennial meadow fescue ley. – Ecol. Bull. (Copenhagen) 39: 146–148.

Johansson, C. 1988. Exchange of Nitrogen Oxides between the Atmosphere and Terrestrial Surfaces. – Ph. D. Thesis, Department of Meteorology, Univ. of Stockholm, Sweden.

Klemedtsson, L., Svensson, B. H. and Rosswall, T. 1988. A method of selective inhibition to distinguish between nitrification and denitrification as sources of nitrous oxide in soil. – Biol. Fert. Soils 6: 112–119.

–, Svensson, B. H. and Rosswall, T. 1988. Relationships between soil moisture content and nitrous oxide production during nitrification and denitrification. – Biol. Fert. Soils 6: 106–111.

Lagerlöf, J. and Andrén, O. 1988. Abundance and activity of soil mites (Acari) in four cropping systems. – Pedobiologia 32: 129–145.

Risser, P. G., Rosswall, T. and Woodmansee, R. G. 1988. Spatial and temporal variability in biospheric and geospheric processes: A summary. – In: Rosswall, T., Woodmansee, R. G. and Risser, P. G. (eds), Scales and Global Change. SCOPE Vol. 35. Wiley, Chichester, pp. 1–10.

Robertson, K., Schnürer, J., Clarholm, M., Bonde, T. and Rosswall, T. 1988. Microbial biomass in relation to C and N mineralization during laboratory incubations. – Soil Biol. Biochem. 20: 281–286.

Schimel, D. S., Simkins, S., Rosswall, T., Mosier, A. R. and Parton, W. J. 1988. Scale and the measurement of nitrogen-gas fluxes from terrestrial ecosystems. – In: Rosswall, T., Woodmansee, R. G. and Risser, P. G. (eds), Scales and Global Change. SCOPE Vol. 35. Wiley, Chichester, pp. 179–193.

Sohlenius, B., Boström, S. and Sandor, A. 1988. Carbon and nitrogen budgets of nematodes in arable soil. – Biol. Fert. Soils 6: 1–8.

1989

Clarholm, M. and Bergström, L. (eds), 1989. Ecology of Arable Land – Perspectives and Challenges. – Developments in Plant and Soil Sciences 39. Kluwer Academic Publishers, Dordrecht.

Galbally, I. E. and Johansson, C. 1989. A model relating laboratory measurements of rates of nitric oxide production and field measurements of nitric oxide emission from soils. – J. Geophys. Res.

Hansson, A.-C. and Pettersson, R. 1989. Uptake and above- and below-ground allocation of soil mineral-N and fertilizer–^{15}N in a perennial grass ley (*Festuca pratensis* Huds.). – J. Appl. Ecol. 26: 259–271.

Lagerlöf, J. and Scheller, U. 1989. Abundance and activity of Pauropoda and Symphyla (Myriapoda) in four cropping systems. – Pedobiologia 33: 315–321.

–, Andrén, O. and Paustian, K. 1989. Dynamics and contribution to carbon flows of Enchytraeidae (Oligochaeta) under four cropping systems. – J. Appl. Ecol. 26: 183–199.

Lindberg, T., Bonde, T. A., Bergström, L., Pettersson, R., Rosswall, T. and Schnürer, J. 1989. Distribution of ^{15}N in the soil-plant system during a four-year field lysimeter study with barley (*Hordeum distichum* L.) and perennial meadow fescue (*Festuca pratensis* Huds.). – Pl. Soil 119: 23–37.

Lundin, L.-C. 1989. Water and heat flows in frozen soils. Basic theory and operational modeling. – Acta Univ. Ups. 186. Uppsala Univ., Uppsala.

Pettersson, R. 1989. Above-ground growth dynamics and net production of spring barley in relation to nitrogen fertilization. – Swed. J. agr. Res. 19: 135–145.

Rosswall, T., Bak, F., Baldocchi, D., Cicerone, R. J., Conrad, R., Ehhalt, D. H., Firestone, M. K., Galbally, I. E., Galchenko, V. F., Groffman, P., Papen, H., Reeburgh, W. S. and Sanhueza, E. 1989. What regulates production and consumption of trace gases in ecosystems: biology or physiochemistry. – In: Andreae, M. O. and Schimmel, D. S. (eds), Exchange of trace gases between terrestrial ecosystems and the atmosphere. Wiley, Chichester, pp. 73–95.

Sohlenius, B. and Sandor, A. 1989. Ploughing of a perennial grass ley-effects on the nematode fauna. – Pedobiologia 33: 199–210.

In press

Berg, P. and Rosswall, T. 1989. Abiotic factors regulating nitrification in an arable soil. – Biol. Fert. Soils.

Clarholm, M. Effects of plant-bacterial-amoebal interactions on plant uptake of nitrogen under field conditions. – Biol. Fertil. Soils.

Hansson, A.-C., Andrén, O. and Steen, E. Root production of four arable crops in Sweden and its effect on the abundance of soil organisms. – In: Atkinson, D. (ed.), Plant Root Systems: Their Effect on Ecosystem Composition and Structure. Blackwell, Oxford.

Lagerlöf, J. and Andrén, O. Abundance and activity of Collembola, Protura and Diplura (Insecta, Apterygota) in four cropping systems. – Pedobiologia.

Öborn, I. and Johnsson, H. Mollic Gleysol at Kjettslinge. – Dept Forest Soils and Div of Soil Science and Ecochemistry, Soil Description serie, No 2, Swedish Univ. of Agricultural Sciences, Uppsala.

Paustian, K., Andrén, O., Clarholm, M., Hansson, A.-C., Johansson, G., Lagerlöf, J., Lindberg, T., Pettersson, R. and Sohlenius, B. 1990. Carbon and nitrogen budgets of four agroecosystems with annual and perennial crops, with and without N fertilization. – J. Appl. Ecol.

Submitted

Klemedtsson, L., Simkins, S., Svensson, B. H. and Rosswall, T. 1990. Soil denitrification in three cropping systems characterized by differences in carbon and nitrogen supply. II. Water and NO_3^- effects on the denitrification process. Biogeochemistry.

Pettersson, R. and Hansson, A.-C. Net primary production of a perennial grass ley (*Festuca pratensis* Huds.) assessed with different methods and compared with a lucerne ley (*Medicago sativa* L.). – J. Appl. Ecol.

Svensson, B. H., Klemedtsson, L., Simkins, S., Paustian, K. and Rosswall, T. 1990. Soil denitrification in three cropping systems characterized by differences in nitrogen and carbon supply. I. Rate-distribution frequencies, comparison between systems and seasonal N-losses. – Biogeochemistry.